Springer Theses

Recognizing Outstanding Ph.D. Research

Aims and Scope

The series "Springer Theses" brings together a selection of the very best Ph.D. theses from around the world and across the physical sciences. Nominated and endorsed by two recognized specialists, each published volume has been selected for its scientific excellence and the high impact of its contents for the pertinent field of research. For greater accessibility to non-specialists, the published versions include an extended introduction, as well as a foreword by the student's supervisor explaining the special relevance of the work for the field. As a whole, the series will provide a valuable resource both for newcomers to the research fields described, and for other scientists seeking detailed background information on special questions. Finally, it provides an accredited documentation of the valuable contributions made by today's younger generation of scientists.

Theses are accepted into the series by invited nomination only and must fulfill all of the following criteria

- They must be written in good English.
- The topic should fall within the confines of Chemistry, Physics, Earth Sciences, Engineering and related interdisciplinary fields such as Materials, Nanoscience, Chemical Engineering, Complex Systems and Biophysics.
- The work reported in the thesis must represent a significant scientific advance.
- If the thesis includes previously published material, permission to reproduce this must be gained from the respective copyright holder.
- They must have been examined and passed during the 12 months prior to nomination.
- Each thesis should include a foreword by the supervisor outlining the significance of its content.
- The theses should have a clearly defined structure including an introduction accessible to scientists not expert in that particular field.

More information about this series at http://www.springer.com/series/8790

Ye Wang

Advances in State Estimation, Diagnosis and Control of Complex Systems

Doctoral Thesis accepted by
Universitat Politècnica de Catalunya,
Barcelona, Spain

Springer

Author
Dr. Ye Wang
Department of Automatic Control
Institut de Robòtica i Informàtica
Industrial, CSIC-UPC
Universitat Politècnica de Catalunya
Barcelona, Spain

College of Automation
Harbin Engineering University
Harbin, P. R. China

Department of Electrical
and Electronic Engineering
The University of Melbourne
Melbourne, Australia

Supervisors
Prof. Gabriela Cembrano Gennari
Department of Automatic Control
Institut de Robòtica i Informàtica
Industrial, CSIC-UPC
Universitat Politècnica de Catalunya
Barcelona, Spain

Prof. Vicenç Puig Cayuela
Department of Automatic Control
Institut de Robòtica i Informàtica
Industrial, CSIC-UPC
Universitat Politècnica de Catalunya
Barcelona, Spain

ISSN 2190-5053 ISSN 2190-5061 (electronic)
Springer Theses
ISBN 978-3-030-52439-5 ISBN 978-3-030-52440-1 (eBook)
https://doi.org/10.1007/978-3-030-52440-1

This Springer imprint is published by the registered company Springer Nature Switzerland AG
The registered company address is: Gewerbestrasse 11, 6330 Cham, Switzerland

To
my family,
for their love, encouragement, and patience.

Supervisors' Foreword

It is a great pleasure to introduce Dr. Ye Wang thesis work, accepted for publication within Springer Theses after being selected a by Spanish Committee of Automatic Control in 2019 Edition of the Best Ph.D. thesis in Control Engineering. Dr. Wang joined the Advanced Control Group of Universitat Politècnica de Catalunya (UPC) at the Institute of Robotics and Industrial Informatics (IRI) as a master student to do the Master thesis in 2014. A bit later, he got a four-year Ph.D. scholarship from the Spanish Ministry of Science and Education. Finally, he received the UPC Ph.D. degree (Excellent Cum Laude) with the International Mention in Automatic Control, Robotics and Computer Vision in October, 2018.

Dr. Wang's thesis includes a significant amount of original scientific theoretical and practical contributions on estimation, diagnosis and control of complex systems, especially in the mathematical form of descriptor systems. The research is fully motivated by real applications, such as Water Distribution Networks (WDNs) and power systems, which require a control system to provide a proper management able to take into account their specific features and operating limits in presence of uncertainties related to their operation and failures from component malfunctions. Such a control system is expected to provide an optimal operation to obtain efficient and reliable performance. As one of the main contributions of this thesis, it can be highlighted the work devoted to build a set-based framework for uncertain descriptor systems derived from a variety of real systems. This framework, including two types of approaches, has been applied for state estimation and fault diagnosis of such systems. The other part of this thesis is devoted to establish control strategies for operational management of WDNs and power systems taking into account behaviors of those systems. In particular, the thesis develops

- robust state estimation approaches based on set theory for descriptor systems;
- distributed set-membership approach for complex systems;
- fault diagnosis strategies based on set theory for descriptor systems;
- Economic Model Predictive Control (EMPC) for periodic operation with applications to realistic complex systems;
- fault-tolerant strategies for descriptor systems.

A significant part of Dr. Wang's thesis has been published in high-quality top journals and international well-recognized conferences, some of them are fruit of joint collaborations with the research group of Prof. Sorin Olaru from the CentraleSupélec in France and Prof. Teodoro Alamo and David Muñoz de la Peña from the University of Seville in Spain. These fruitful scientific outputs demonstrate the amount of ideas behind the thesis and its quality and appealing within the scientific community. The thesis has also been improved by the valuable comments of the external reviewers Prof. Florin Stoican and Prof. Cristina Stoica Maniu during the Ph.D. thesis evaluation phase.

Barcelona, Spain
March 2020

Prof. Vicenç Puig
Prof. Gabriela Cembrano

Foreword from CEA, the Automatic Control Spanish Association

It is a great pleasure for us to introduce here the work of Ye Wang, awarded with the best Ph.D. Thesis in Control Engineering during the 2017–2019 edition of the award call organized by the control engineering group of Comité Español de Automática (CEA), the Spanish Committee of Automatic Control.

This yearly award is aimed at recognizing outstanding Ph.D. research carried out within the Control Engineering field. At least one of the supervisors must be partner of the Spanish Committee of Automatic Control and member of the Control Engineering group. The jury is composed by three well-known doctors in the control engineering field; two of them are partners of the CEA, while the third one is a foreign professor.

This edition counted submissions from Ph.D. students examined between September 2017 and July 2019. Notably, the scientific production of the candidates consisted of a total of 27 publications in international indexed journals, where 19 of them are ranked in first quartile journals. This shows the high scientific quality of the submitted Ph.Ds. All the thesis presented:

- have been under the international doctorate mention (the candidate must have at least one international research stay out of Spain, write part of the thesis in English, and have foreign members in the Ph.D. jury).
- have been mostly oriented to academic research than to industrial applications.
- were presented at international conferences. Among them, IFAC and IEEE are considered as preferred conferences.

As a result, Ye Wang's Ph.D. thesis was selected as the best one among an excellent group of candidates. His Ph.D. thesis deserves the label "the best of the best", Springer Theses' main motto. Not only his nomination as winner of the Spanish control engineering context has been our great pleasure, but also to recommend his thesis for publication within the Springer Theses collection. On behalf of CEA we wish Ye Wang to continue his outstanding scientific career, keeping his genuine enthusiasm. We also hope that his work may be of inspiration for other students working in the control engineering field.

Madrid, Spain

Prof. Dr. Carlos Balaguer
President of CEA

Dr. Jose Luis Guzman
Dr. Manuel Gil Ortega
Coordinators of the Control Engineering group

Parts of this thesis have been published in the following articles:

Journal Papers

1. Y. Wang, S. Olaru, G. Valmorbida, V. Puig and G. Cembrano. Set-invariance characterizations of discrete-time descriptor systems with application to active mode detection. *Automatica*, 107: 255–263, 2019
2. Y. Wang, Z. Wang, V. Puig, and G. Cembrano. Zonotopic set-membership state estimation for discrete-time descriptor LPV systems. *IEEE Transactions on Automatic Control*, 64(5): 2092–2099, 2019
3. Y. Wang, V. Puig, F. Xu and G. Cembrano. Robust fault detection and isolation based on zonotopic unknown input observer for discrete-time descriptor systems. *Journal of the Franklin Institute*, 356(10): 5293–5314, 2019
4. Y. Wang, D. Muñoz de la Peña, V. Puig and G. Cembrano. Robust economic model predictive control based on a periodicity constraint. *International Journal of Robust and Nonlinear Control*, 29(11): 3296–3310, 2019
5. Y. Wang, V. Puig, and G. Cembrano. Set-membership approach and Kalman observer based on zonotopes for discrete-time descriptor systems. *Automatica*, 93:435–443, 2018
6. Y. Wang, V. Puig and G. Cembrano. Robust fault estimation based on zonotopic Kalman filter for discrete-time descriptor systems. *International Journal of Robust and Nonlinear Control*, 28(16): 5071–5086, 2018
7. Y. Wang, J. R. Salvador, D. Muñoz de la Peña, V. Puig and G. Cembrano. Economic model predictive control based on a periodicity constraint. *Journal of Process Control*, 68: 226–239, 2018
8. Y. Wang, T. Alamo, V. Puig and G. Cembrano. Economic model predictive control with nonlinear constraint relaxation for the management of water distribution networks. *Energies*, 11(4): 991, 2018
9. Y. Wang, V. Puig and G. Cembrano. Non-linear economic model predictive control of water distribution networks. *Journal of Process Control*, 56: 23–34, 2017

Conference Papers

1. Y. Wang, D. Rotondo, V. Puig, G. Cembrano and Y. Zhao. Fault-tolerant control of discrete-time descriptor systems using virtual actuator. *International Conference on Control and Fault-tolerant Systems (Systol)*, 2019, Casablanca, Morocco, pp. 147–152
2. Y. Wang, T. Alamo, V. Puig and G. Cembrano, A distributed set-membership approach based on zonotopes of interconnected systems, *57th IEEE Conference on Decision and Control (IEEE-CDC)*, 2018, Miami, USA, pp. 668–673
3. Y. Wang, D. Muñoz de la Peña, V. Puig and G. Cembrano, Robust periodic economic predictive control based on probabilistic set invariance for descriptor systems, *6th IFAC Conference on Nonlinear Model Predictive Control*, 2018, Madison, USA, pp. 436–441

4. Y. Wang, V. Puig, F. Xu and G. Cembrano. Zonotopic unknown input observer of discrete-time descriptor systems for state estimation and robust fault detection, *10th IFAC Symposium on Fault Detection, Supervision and Safety for Technical Processes (IFAC-SAFEPROCESS)*, 2018, Warsaw, Poland, pp. 307–313
5. Y. Wang, D. Muñoz de la Peña, V. Puig and G. Cembrano, A novel formulation of economic model predictive control for periodic operations, *European Control Conference (ECC)*, 2018, Limassol, Cyprus, pp. 1015–1020
6. Y. Wang, S. Olaru, G. Valmorbida, V. Puig and G. Cembrano, Robust invariant sets and active mode detection for discrete-time uncertain descriptor systems, *56th IEEE Conference on Decision and Control (IEEE-CDC)*, 2017, Melbourne, Australia, pp. 5648–5653
7. Y. Wang, M. Zhou, V. Puig, G. Cembrano and Z. Wang. Zonotopic fault detection observer with H_performance, *36th Chinese Control Conference (CCC)*, 2017, Dalian, China, pp. 7230–7235
8. Y. Wang, Z. Wang, V. Puig and G. Cembrano. Zonotopic fault estimation filter design for discrete-time descriptor systems, *20th IFAC World Congress*, 2017, Toulouse, France, pp. 5055–5060
9. Y. Wang, T. Alamo, V. Puig and G. Cembrano. Periodic economic model predictive control with nonlinear-constraint relaxation for water distribution networks, *IEEE Conference on Control Application (IEEE-CCA)* in *IEEE Multi-Conference on Systems and Control (IEEE-MSC)*, 2016, Buenos Aires, Argentina, pp. 1167–1172

Acknowledgments

It is time to summarize my doctoral journey over past three years. While doing research, I have met and interacted with many smart people all over the world from whom I have learned a large number of professional skills and tried to think how to improve myself. Hereby I would like to express my sincere gratitude to those that supported and helped me during this journey.

First, I want to thank my Ph.D. advisors and dear friends Prof. Vicenç Puig and Dr. Gabriela Cembrano for their continuing supports during this journey. They have given me opportunity to start this challenging and unforgettable period, from which I have learned not only how to enjoy doing research by acquiring knowledge and happiness but also being a good man. I believe everything encountered in this period will benefit for the rest of my life. Especially, I would like to thank Prof. Vicenç Puig for spending a lot of time with me to solve many academic problems as well as share happiness.

Second, I want to thank Prof. David Muñoz de la Peña and Prof. Teodoro Alamo from Universidad de Sevilla, in Spain, for the collaborations carried out within several Spanish projects. During this long-term collaborations, with Prof. David Muñoz de la Peña, I have received productive ideas regarding EMPC and RMPC. And with Prof. Teodoro Alamo, I have learned a lot regarding set-membership approaches and more importantly, how to write magic mathematical results for potential applications. Special thanks go to Prof. Daniel Limón for valuable discussions on interesting results related to MPC stability.

Third, I want to thank Prof. Sorin Olaru from CentraleSupélec, Université Paris Saclay, in France, for fruitful collaboration to contribute set-theoretic results for descriptor systems from which I have learned also how to be a good researcher and stay in professional. I also thank Dr. Giorgio Valmorbida also from CentraleSupélec for pressurizing me to carefully prepare any publication at a top level and sufficiently produce high-quality results.

Fourth, I want to thank Prof. Florin Stoican from University Politehnica of Bucharest, in Romania and Prof. Cristina Nicoleta Maniu from CentraleSupélec, Université Paris Saclay, in France for having reviewed this thesis with providing valuable comments. Especially, Prof. Stoican did a fantastic review with very detailed comments for each chapter of this thesis. Besides, I would like to thank Prof. Ramon Costa-Castelló from Universitat Politècnica de Catalunya, in Spain for joining my Ph.D. thesis committee.

Fifth, I want to thank many colleagues and friends I met at Institut de Robòtica i Informàtica Industrial, CSIC-UPC, with whom I founded not only a professional relationship with discussions of interesting research ideas but also friendship to share happiness in life and relieve pressure encountered from difficulties. Special thanks go to Masoud Pourasghar, Fatemeh Karimi Pour (Asi), José Luis Sampietro, Pau Segovia, Unnikrishnan Raveendran Nair, Wicak Ananduta, Jenny Lorena Díaz. Moreover, I would like to thank Dr. Joaquim Blesa and Dr. Damiano Rotondo in the control group for useful discussions and nice collaborations.

Sixth, I want to thank the research projects that funded my research. I sincerely thank the Spanish State Research Agency (AEI) and the European Regional Development Fund (ERFD) to provide the FPI grant (ref. BES-2014-068319) to fund my doctoral study, and through the projects ECOCIS (ref. DPI2013-48243-C2-1-R), HARCRICS (ref. DPI2014-58104-R), DEOCS (ref. DPI2016-76493-C3-3-R) and SCAV (ref. DPI2017-88403-R). This work is also partially funded by the European Commission through the EFFINET project (ref. FP7-ICT-2012-318556), and by AGAUR of Generalitat de Catalunya through the Advanced Control Systems (SAC) group grant (ref. 2017-SGR-482).

Last but not least, I owe my gratitude to my parents, Hongjia Wang and Aihua Zhang, for encouraging me to pursue my dream to be a scientist. I also express my deepest thankfulness to my lovely wife, Hongyu Qian, for her unlimited love and accompany despite the distance to make this journal not being in loneliness. I want to thank all the members of my family for long-term supporting me. In particular, I would like to express my greatest gratitude to my paternal grandmother, Li Zhang, for raising me up and teaching me being an upright man and having perseverance and courage to overcome any difficulties. Moreover, this dissertation is dedicated to my maternal grandmother, Lanzhen Wu, who passed away in the first year of my Ph.D. study. Due to the distance and tight study schedule, I could not attend her funeral but she is and will be always living in my heart.

Barcelona, Spain
Summer 2018

Ye Wang

Contents

Notation

$\succ (\prec)$	Positive (negative) definite
$\succeq (\preceq)$	Positive (negative) semi-definite
$\mathbb{R}$	Set of real numbers
$\mathbb{R}_+$	Set of non-negative real numbers, defined as $\mathbb{R}_+ \triangleq \mathbb{R} \setminus (-\infty, 0]$
$\mathbb{N}$	Set of natural numbers
$\mathbb{C}$	Set of complex numbers
$\mathbb{Z}$	Set of integer numbers
$\mathbb{Z}_+$	Set of non-negative integer numbers
$\mathbb{Z}_{[a,b]}$	Set $\mathbb{Z}$ in an interval between a and b
$\mathbb{S}^n$	Set of symmetric matrices of dimension n
$\mathbb{S}^n_{\succ 0}$	Set of positive definite symmetric matrices of dimension n
$\mathbb{S}^n_{\succeq 0}$	Set of positive semi-definite symmetric matrices of dimension n
$\mathbb{S}_{\prec 0}$	Set of negative definite symmetric matrices of appropriate dimension
$\mathbb{S}_{\preceq 0}$	Set of negative semi-definite symmetric matrices of appropriate dimension
$\mathbf{B}^n$	Hypercube defined as $\mathbf{B}^n := [-1, 1]^n$
$\mathcal{X} \oplus \mathcal{Y}$	Minkowski sum of two sets $\mathcal{X}$ and $\mathcal{Y}$
$\mathcal{X} \ominus \mathcal{Y}$	Pontryagin difference of two sets $\mathcal{X}$ and $\mathcal{Y}$
$\mathcal{X} \times \mathcal{Y}$	Cartesian product of two sets $\mathcal{X}$ and $\mathcal{Y}$
$\mathcal{X} \subset (\subseteq) \mathcal{Y}$	Set $\mathcal{X}$ is a (strict) subset of set $\mathcal{Y}$
$d_H(\mathcal{X}, \mathcal{Y})$	Hausdorff distance of two sets $\mathcal{X}$ and $\mathcal{Y}$
$\text{Card}(\mathcal{X})$	Cardinality of a set $\mathcal{X}$, that is, the number of elements of $\mathcal{X}$
$X \otimes Y$	Kronecker product of two matrices X and Y
I_n	Identity matrix of dimension n
$x^\top (X^\top)$	Transpose of a vector x (a matrix X)
X^{-1}	Inverse matrix of X
$X^\dagger$	Pseudo-inverse matrix of X
$\text{rank}(X)$	Rank of a matrix X

$\mathrm{tr}(X)$	Trace of a matrix X
$\det(X(z))$	Determinant of a matrix X on variable z
$\deg(\det(X(z)))$	Degree of the determinant of a matrix X on variable z
$\mathbf{vec}(X)$	Vectorization of a matrix X
$\mathbf{He}(X)$	$\mathbf{He}(X) = X + X^\top$
$\underline{\sigma}(X)$	Least/minimum singular value of a matrix X
$\lambda(X)$	Set of eigenvalues of a matrix X, that is, $\lambda(X) := \{z : \det(zI - X) = 0\}$
$\lambda(X, Y)$	Set of generalized eigenvalues, that is, $\lambda(X, Y) := \{z : \det(zX - Y) = 0\}$
$\|X\|_F$	Frobenius norm of a matrix X is defined by $\|X\|_F := \sqrt{\mathrm{tr}(X^\top X)}$
$\|X\|_{F,W}$	Weighted Frobenius norm $\|X\|_{F,W} := \sqrt{\mathrm{tr}(X^\top W X)}$ with $W \in \mathbb{S}_{\succ 0}$
$\underset{j\in N}{\mathrm{cat}}\{X_j\}$	$\underset{j\in N}{\mathrm{cat}}\{X_j\} = [X_{j_1}, \ldots, X_{j_N}]$ for a set of matrices X_j with $j \in N$
$\|x\|_2$	2-norm of a vector x is defined by $\|x\|_2 = \sqrt{x^\mathrm{T} x}$
$\|x\|_{2,W}$	Weighted 2-norm $\|x\|_{2,W} = \sqrt{x^\top W x}$ with $W \in \mathbb{S}_{\succ 0}$
$\|z\|_\infty$	$\mathcal{L}_\infty$ norm (peak norm) of a signal z is defined by $\|z\|_\infty = \sup_k \|z(k)\|$
$\mathrm{mod}(a, b)$	Modulo operator of two scalars a and b
$\mathrm{diag}(\cdot)$	Operator that builds a diagonal matrix with the elements of its argument
$\star$	A term induced by (Hermitian) symmetry in a block matrix

Acronyms

EMPC	Economic Model Predictive Control
FD	Fault Detection
FDI	Fault Detection and Isolation
FE	Fault Estimation
FI	Fault Isolation
GAMS	General Algebraic Modeling System
KKT	Karush–Kuhn–Tucker
KPI	Key Performance Indicator
LMI	Linear Matrix Inequality
LPV	Linear Parameter Varying
LTI	Linear Time Invariant
LTV	Linear Time Varying
MAE	Mean Absolute Error
MPC	Model Predictive Control
mRPI	minimal Robust Positively Invariant
MSE	Mean Square Error
NEMPC	Nonlinear Economic Model Predictive Control
REMPC	Robust Economic Model Predictive Control
RI	Robust Invariant
RMPC	Robust Model Predictive Control
RMS	Root Mean Squared
RNI	Robust Negatively Invariant
RPI	Robust Positively Invariant
SG	Smart Grid
SMAPE	Symmetric Mean Absolute Percentage Error
UIO	Unknown Input Observer
VA	Virtual Actuator
VDA	Virtual Delayed Actuator
VS	Virtual Sensor
WDN	Water Distribution Network

List of Figures

List of Tables

Chapter 1
Introduction

1.1 Motivation

In modern societies, reliable and sustainable operation of certain infrastructures plays a fundamental role in the quality of individual life, economic development and security of nations. Large-scale critical infrastructure systems, especially those located in urban areas, such as water distribution networks (WDNs) and smart grids (SGs), are a subject of increasing concern. Therefore, it is of vital importance to develop management systems that guarantee a reliable and sustainable operation of these infrastructures. On the other hand, for the management of these infrastructures, it is also significant that their operation must use efficiently the resources that they can deliver, e.g., water and electricity, and also be efficient from an economic point of view and guarantee future supply.

The critical nature of these complex systems implies the need for a management able to take into account their specific features and operating limits in presence of uncertainties related to their operation and failures from component malfunctions. Thus, it is of paramount importance to have a control system for the management that, from sensor measurements and available predictions of external influential variables based on a priori knowledge, produces a suitable way to operate the complex system in an efficient, reliable and sustainable manner.

For designing a model-based control system, an appropriate mathematical model is required to represent the most relevant system dynamics. In terms of aforementioned complex systems, system dynamics are usually described by differential/difference equations while static relations also appearing in complex systems based on their topologies lead to the use of algebraic equations. In the literature, the class of systems including not only differential/difference equations but also algebraic equations is called *descriptor*, *singular*, or *differential/difference-algebraic* systems [1, 2].

Together with sensor measurements, the need for state predictions in a control implementation requires a suitable state estimation approach also being able to atten-

Y. Wang, *Advances in State Estimation, Diagnosis and Control of Complex Systems*, Springer Theses, https://doi.org/10.1007/978-3-030-52440-1_1

uate effects from uncertainties in order to achieve robustness. Such a state estimation approach is also useful for implementing fault diagnosis of component malfunctions inside systems. A plausible solution to address a robust state estimation is to use set theory under the assumption that unknown uncertainties are bounded in a deterministic set with a predefined geometrical structure. Based on an iterative procedure, the effects of these bounded uncertainties can be propagated at each time step and explicitly characterized in an updated set. From analysis of these effects, fault diagnosis may also be achieved by means of these set tools. As a result, worst-case scenarios for state and fault predictions can also be used in the controller design.

In many applications, control objectives of complex systems are mainly different from traditional tracking or regulating problems. Thus, the challenge for the optimal controller design is how to obtain an optimal economic cost and meanwhile guarantee closed-loop stability and convergence to a certain steady trajectory. Looking into real systems, the system behavior is usually not only constrained by some limits but also affected by potential periodic behavior. For instance in WDNs, water demands and economic cost of consumed energy follow periodic patterns which may lead to an optimal periodic operation. For this purpose, periodicity can be used in the design of the controller, where the closed-loop convergence can be guaranteed. Furthermore, in terms of the critical nature of complex systems, fault-tolerant capability deserves to be included in the controller design. After having a suitable fault diagnosis block in the control system, a fault hiding strategy can be employed for the system reconfiguration.

The research in this thesis is motivated by real application of WDNs, under the scope of the Spanish project: EConomic Operation of Critical Infrastructure Systems (ECOCIS), which is also in line with the objectives of European research policy developed through the framework program of Horizon 2020, and the Spanish research plan 2013. This doctoral thesis is devoted to investigating economic-oriented optimal control and fault-tolerant control (FTC) strategies for the management of complex systems. To this end, several approaches on robust state estimation and fault diagnosis based on set theory taking into account descriptor models are investigated.

1.2 State of the Art

1.2.1 Descriptor Systems

Due to mass, volume or energy conservation laws, the differential/difference equations describing a dynamical system can be coupled with a set of algebraic equations. As aforementioned, descriptor models may be used for representing this class of systems. Instances of such systems can be found in water systems [3–5], chemical processes [6], electrical circuits [7], aircraft [8], biological systems [9] and economic models [1]. From a theoretical point of view, descriptor systems satisfying a well-posed property, for which a solution exists and is unique, are called *regular* [1].

Regularity, however, does not imply *causality* and some models of interest in economy may be non-causal, see e.g. the Leontief model [1, 10, 11]. In terms of a control system, *stability* [12] is an important property for the analysis of boundedness and convergence of the closed-loop trajectory. In particular, in terms of descriptor systems, *admissibility* guarantees the properties of regularity, causality and stability [1]. For monitoring purposes and for developing control strategies, state estimation is usually required. Some research works on state estimation for discrete-time descriptor systems have been carried out (see as e.g. [13, 14], where system states can be estimated using different versions of Kalman filtering).

1.2.2 Economic Model Predictive Control

Economic model predictive control (EMPC) has attracted an increasing attention during the past decade [15–17]. Unlike conventional model predictive control (MPC) formulations [18, 19], the main control objective of EMPC is to optimize an economic performance index without regulating the system to a given trajectory. Economic cost functions are not necessarily quadratic or positive-definite with respect to a given trajectory as tracking MPC. EMPC has been applied to a variety of industrial applications as a real-time control strategy, see, e.g. drinking-water networks [5, 20–23], wastewater treatment processes [24], SGs [25, 26] and chemical processes [27, 28].

Recently, the closed-loop stability and convergence of EMPC has been widely investigated. Since the cost function in EMPC may not be a quadratic function, the conventional MPC stability analysis as in [19] cannot be directly applied to EMPC. In [15, 29, 30], stability analysis of EMPC has been established under the strong duality and the dissipativity assumptions. Terminal cost and constraint around the optimal steady state are used. In [31], a review is presented for discussing the role of constraints in EMPC, where the convergence of EMPC can be enforced by adding terminal constraints. Besides, EMPC without terminal constraints is studied in [32, 33]. Based on the turnpike and controllability properties, closed-loop convergence is proved. In [34], EMPC with extended prediction horizon is designed based on an auxiliary controller. An additional term with the auxiliary control law is included in the cost function in order to guarantee closed-loop convergence.

From an application point of view, systems may also be affected by disturbances, which implies that a proper robust MPC (RMPC) strategy should be addressed for such systems, for instance [25]. Tube-based techniques have been proposed to guarantee robust constraint satisfaction in the presence of uncertainties for conventional MPC and other applications, as e.g. in distributed approaches [35]. An RMPC was proposed to track periodic trajectories online in [36], where a local control law is used to refine the constraints in order to guarantee recursive feasibility in closed-loop. In recent years, several developments on adjusting the robustness of EMPC have been studied in [37–39], where the strong terminal constraint and cost are used to enforce the periodicity.

1.2.3 Fault Diagnosis

As introduced in [40, 41], fault diagnosis basically consists of the following three essential tasks:

- *Fault Detection* (FD): detection of the occurrence of faults in malfunctioned components that lead to undesired or intolerable behavior of the whole system;
- *Fault Isolation* (FI): localization of different occurred faults;
- *Fault Estimation* (FE): determination of the magnitude of occurred faults.

A fault detection and isolation (FDI) module usually includes robust performance for the system affected by uncertainties. Robust FDI aims at minimizing the sensitivity to uncertainties (such as modeling errors, process disturbances, measurement noise as so on) while maximizing fault sensitivity to achieve great FDI performance. For this aim, different approaches have been studied. One category relies on the use of robust control techniques, for instance using $\mathcal{H}_\infty$ and $\mathcal{H}_-$ norms (see e.g. [41–44]). Among these references, a generated adaptive threshold is usually computed for the decision making of the FDI alarm. An over-approximation of the decision-making threshold may lead to wrong FDI results. Alternatively, another category of FDI strategies is built under a set-based framework, such as [45–47]. System uncertainties are considered as unknown but bounded in predefined sets (intervals, zonotopes and polytopes) and the resulting uncertain states and generated residuals are propagated also in bounded sets [48]. Due to the simple computation load, zonotopes are usually chosen as the geometrical sets for bounding uncertain states or residuals [49, 50]. Under this framework, robustness and fault sensitivity of the FDI strategy can be achieved by checking the consistency between the system model and the measurement information. Unknown input observer (UIO) is a well-known tool for designing a robust FDI strategy that can be achieved by generating residuals with decoupled unknown inputs [51]. The design of UIO has been well-discussed for a variety of systems with different structures (see e.g. [51–54]). In the design of UIO for implementing an FDI strategy, robustness and fault sensitivity should also be taken into account. As discussed in [55], it has shown the potential of linking UIO with a set-based framework, where unknown inputs are divided into two groups: the one can be decoupled using UIO transformation matrices; the other cannot be decoupled but bounded using invariant sets. Besides, an extension to robust FDI based on set-based UIO has been studied in [56].

FE has been studied by a large amount of approaches during the past decades, see e.g. [40, 41, 57–60]. A suitable FE with robust performance against system uncertainties is necessary for implementing an active fault-tolerant control system [61–63]. By means of alternative robust control techniques, robust FE has been implemented in a variety of systems as e.g. [54, 64, 65], where the effects of uncertainties are bounded and therefore FE results can be obtained with the minimum estimation error. In the literature, several FE approaches for different types of descriptor systems have been investigated. In [62], a Lyapunov-based robust FE approach is developed for Lipschitz non-linear descriptor systems. Robust FE approaches for linear descriptor systems can be found in [53, 64, 66]. Besides, FE approaches have also studied for linear parameter-varying (LPV) systems [67–70] and switched descriptor systems [71]. From these existing approaches, it can be seen that the obtained estimation results only include punctual values. In terms of set-based approaches, with considering system uncertainties bounded in a predefined set, the uncertain variables are propagated by operating these sets. Regarding the possible application to robust FE, as a benefit from using a set-based approach, the obtained estimation results can be characterized in a deterministic set that includes not only punctual values but also worst-case bounds. The robustness against uncertainties can be achieved by shrinking the size of these sets.

1.2.4 Fault-Tolerant Control

An increasing number of research works in the control field focus on satisfying reliability, safety and fault tolerance of critical complex systems. In many situations, the consequences of a minor fault in a control system can be catastrophic. According to [40], FTC techniques can be divided into two types: *passive* and *active*. Passive FTC technique, also known as robust approach, aims to find a control law able to cope with the occurred faults considering them as system perturbations. Compared with active FTC technique, neither FD, FI and FE modules nor reconfiguration/accommodation are required for passive FTC. A literature review including a comparison of different approaches according to different criteria is addressed in [72]. In this reference, several active FTC techniques that can be found in the literature are considered including linear quadratic, pseudo-inverse method, intelligent control, gain-scheduling approach, model following, adaptive control, multiple model, integrated diagnostic and control, eigenstructure assignment, feedback linearization/dynamic inversion, MPC, quantitative feedback theory and variable structure control/sliding model control.

Faults may appear in actuators, sensors and other system components. Typically, the active FTC scheme can be divided into four parts:

- a reconfigurable controller;
- a fault diagnosis scheme;
- a controller reconfiguration mechanism;
- a command governor.

The inclusion of both the fault diagnosis scheme and the controller reconfiguration within a general control system is the main difference between active FTC and passive FTC. Hence, some key issues of active FTC have to be considered:

- a controller that can be reconfigured, for instance, an MPC strategy provides an alternative and flexible framework and it is quite easy to be reconfigured;
- a fault diagnosis scheme with high sensitivity to faults and robustness against model uncertainties, variations of the operating conditions and external disturbances;

- a reconfiguration mechanism that allows recovering the fault-free system performance as much as possible within admissible performance degradation.

In recent years, the *fault-hiding* paradigm has been proposed as an active FTC strategy to obtain fault tolerance [73]. In this paradigm, the faulty plant is reconfigured by inserting a reconfiguration block, named *virtual actuator* (VA) in the case of actuator faults and *virtual sensor* (VS) when sensor faults occurred. VA and VS aim at hiding the faults from controller and sensor failures, so that it approximately recovers the same plant as before faults occurred. This active FTC strategy has been extended successfully to many classes of systems, e.g. LPV systems [74], hybrid systems [75], Takagi-Sugeno systems [76], piecewise affine systems [77] and uncertain systems [78].

1.2.5 Set-Based Approaches

Research on set-based state estimation has been quite active for the last decades, e.g. [49, 79–85] among others. In the literature, set-based state estimation approaches can be classified according to whether they follow a set-membership or an interval observer-based paradigm. A set-membership approach relies on over-bounding the uncertain estimated states considering unknown-but-bounded uncertainties [86]. An interval observer-based approach bounds the set of estimated states by means of an observer structure in which the gain is designed assuming that uncertainties are modeled in a deterministic way (as e.g. using intervals for bounding them [87]) or in a stochastic way (as e.g. using the Kalman filtering [88, 89]). From the application point of view, the set-based approaches are very popular in the fault diagnosis framework, e.g. [47, 90].

Zonotopes are a special class of geometrical sets. The symmetry properties of zonotopes help to reduce the computational load of using them in an iterative way. Worst-case state estimation for dynamical systems using zonotopes is investigated in [82]. A state bounding observer based on zonotopes is introduced in [91]. The zonotopic observer in combination with Kalman filtering is addressed in [80, 92]. Moreover, a set-membership approach based on zonotopes is proposed for dynamical systems in [49, 79].

On the other hand, set invariance theory has played an essential role in automatic control with a variety of applications to control systems, which is widely used for guaranteeing the stability and achieving desired performance [93–95]. For systems affected by disturbances, different techniques in set invariance theory are used for the computation of invariant sets. These techniques have been applied to linear dynamical systems [96, 97], LPV systems [98, 99], switched systems [98, 100, 101], and nonlinear systems [102–105]. In particular, ultimate boundedness methods are used to compute invariant sets with relative low complexity [101, 106]. In this context, an iterative strategy is proposed in [107], which leads to approximations of minimal

RI sets for linear systems and its generalization to discrete-time descriptor systems is the motivation of the present work.

Furthermore, set-invariance characterizations are instrumental for control strategies, such as reference governor design [108], FDI [45, 47], FTC [107, 109–111] and RMPC [35, 112]. A remarkable application of RI sets is on mode detection of systems subject to multiple modes of operation. Indeed, since different operating modes lead to different RI sets, the distance between these sets can be used for monitoring and mode detection. Due to the fact that the RI sets of different modes may overlap, an additive input signal can be conveniently designed to separate a parametrization of the RI sets, represented by tubes of trajectories [113]. In this case, the set-based mode detection mechanism is called *active*. In the literature, this mechanism is also called active fault diagnosis, which can be found in [114, 115]. A set of additive inputs are designed to guarantee fault diagnosis outputs that are only consistent with one faulty scenario. These additive inputs can be obtained from the solution to a mixed-integer quadratic program or using a multi-parametric approach, see e.g. [108, 116].

1.3 Thesis Objectives

According to the motivation of this thesis and state of the art, specific thesis objectives are summarized as follows:

(i) *Develop robust state estimation approaches based on set theory for descriptor systems;*
(ii) *Improve the limitation of set-membership approach for complex systems;*
(iii) *Investigate fault diagnosis strategies based on set theory for descriptor systems;*
(iv) *Contribute to EMPC strategies for periodic operation with applications to realistic complex systems;*
(v) *Include fault-tolerant capability in the controller design for descriptor systems.*

1.4 Thesis Outline

The contents of this thesis are organized into 3 parts, as shown in Fig. 1.1. Part I deals with the contributions on the objectives (i)–(ii). Part II refers to the contributions on the objective (iii). Finally, Part III summarizes the contributions towards the objectives (iv)–(v). The road map of this thesis is shown in Fig. 1.2, which gives the general scheme and illustrates the connections among chapters. Specifically, the contents of Chaps. 2–10 are summarized as follows:

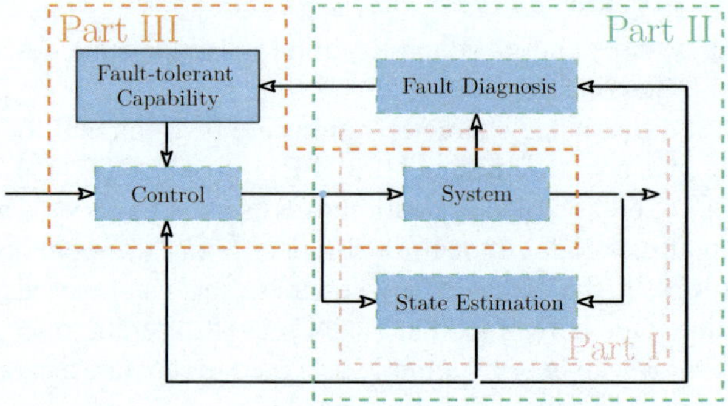

Fig. 1.1 General scheme of thesis

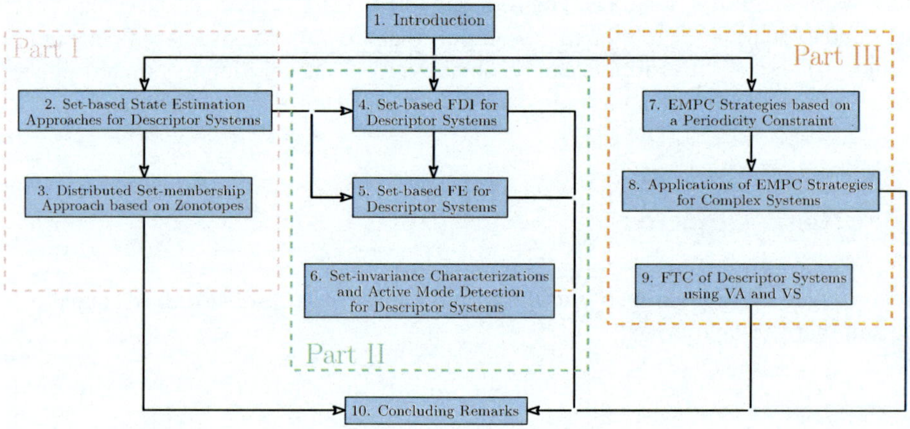

Fig. 1.2 Road map of thesis chapters

Chapter 2: Set-Based State Estimation Approaches for Descriptor Systems

This chapter proposes a general set-based framework for discrete-time descriptor systems with application to robust state estimation. Specifically, set-membership approach based on zonotopes and zonotopic Kalman observer are extended to descriptor systems subject to unknown-but-bounded uncertainties as well as unknown inputs. The relationship between these two approaches is discussed. As another extension, the zonotopic set-membership approach is also investigated for discrete-time LPV descriptor systems with a new zonotope minimization criterion.

Chapter 3: Distributed Set-Membership Approach Based on Zonotopes

This chapter presents a distributed approach to overcome the weakness of the set-membership approach for potential applications to large-scale systems. Instead of bounding uncertain system states in a single zonotope, a set of distributed zonotopes is defined to only bound uncertain states in each agent. Each distributed zonotope is only corrected by the measurement information of each agent. Besides, considering the coupled states, each distributed zonotope is able to send its information to all its neighbors.

Chapter 4: Set-Based Fault Detection and Isolation for Descriptor Systems

This chapter applies the set-based approach to FDI for discrete-time descriptor systems under the set-based framework in Chap. 2. In addition to achieving robustness against uncertainties, the design of the FD gain also takes into account sensitivity to faults. Two different criteria of fault sensitivity are investigated for the design of the FD observer gain. Besides, a bank of zonotopic UIOs is designed for FI.

Chapter 5: Set-Based Fault Estimation for Descriptor Systems

This chapter applies the set-based approach to FE for discrete-time descriptor systems under the set-based framework in Chap. 2. The fault detectability indices and matrix are used for the identification of the occurred actuator faults. Under the framework of zonotopic Kalman filter, the optimal filter gain for FE is computed. Moreover, boundedness of zonotopic FE is proved to guarantee that the estimation results do not diverge.

Chapter 6: Set-Invariance Characterizations and Active Mode Detection for Descriptor Systems

This chapter systematically presents a general set-invariance framework for discrete-time descriptor systems considering both causal and non-causal parts. In addition to RPI sets, an RNI set for non-causal descriptor systems is defined. The computation of these sets is based on ultimate bounds. Moreover, an active mode detection mechanism is proposed for discrete-time descriptor systems based on set invariance theory. Active detection inputs are designed using optimization methods.

Chapter 7: Economic Model Predictive Control Strategies Based on a Periodicity Constraint

This chapter addresses a novel formulation of EMPC for periodic operation in both nominal and robust cases. With the convex analysis, the closed-loop properties including recursive feasibility, robust constraint satisfaction as well as convergence are discussed. Moreover, an optimality certificate is also provided to check if the periodic steady trajectory is optimal.

Chapter 8: Applications of Economic Model Predictive Control Strategies for Complex Systems

This chapter collects three application results of EMPC strategies for complex systems, such as WDNs and SGs. These complex systems can be modeled by difference-algebraic equations in a descriptor form. First, a two-layer control strategy for real-time implementation with a realistic simulator is proposed, which includes a nonlinear EMPC (NEMPC) in the upper layer and a pumping scheduling approach in the lower layer. Second, an iterative approach of nonlinear constraint relaxation is proposed for dealing with nonlinear algebraic equation in WDN and therefore an integration to EMPC is implemented. Finally, a robust EMPC (REMPC) for periodic operation is applied to a micro SG.

Chapter 9: Fault-Tolerant Control of Discrete-Time Descriptor Systems Using Virtual Actuator and Virtual Sensor

This chapter designs an FTC controller for discrete-time descriptor systems. First, for the use of state feedback, an observer-based delayed state-feedback controller is proposed for this class of systems taking into account the algebraic loop appeared in the implementation that prevents using a standard state feedback. Improved admissibility conditions are proposed for discrete-time descriptor system with state delay. Then, system reconfiguration of discrete-time descriptor systems subject to actuator and sensor faults is based on VA and VS.

1.5 Background

In this section, some necessary definitions, mathematical tools and properties are introduced, which will be used in this thesis.

1.5.1 Properties of Discrete-Time Descriptor Systems

Consider the discrete-time linear time-invariant (LTI) descriptor system with additive disturbances

$$Ex(k+1) = Ax(k) + B_w w(k), \tag{1.1}$$

where $x \in \mathbb{R}^n$ and $w \in \mathbb{R}^q$ denote the state and disturbance vectors, respectively, $k \in \mathbb{N}$. $A \in \mathbb{R}^{n \times n}$, $B_w \in \mathbb{R}^{n \times q}$ and $E \in \mathbb{R}^{n \times n}$ with $\text{rank}(E) = r \leq n$.

The definitions and lemmas below are related to the trajectories and solutions of a descriptor system (1.1).

Definition 1.1 (*Regularity*) The descriptor system (1.1) is said to be regular if it has a unique solution defined as an application $x(k) : \mathbb{N} \to \mathbb{R}^n$, $\forall k \in \mathbb{N}$ which satisfies (1.1) for any disturbance realization $w(k) : \mathbb{N} \to \mathbb{R}^q$ and a compatible initial state $x(0)$.

From the above definition, if the system (1.1) is regular, then it has a unique solution for the disturbance-free case ($w \equiv 0$). The matrix pair (E, A) is also called to be regular.

Definition 1.2 (*Causality*) The regular descriptor system (1.1) is said to be causal if $x(k)$, $\forall k \in \mathbb{N}$ is determined completely by the initial condition $x(0)$ and $w(j)$, for $j = 0, \ldots, k$. Otherwise, it is said to be non-causal.

Definition 1.3 (*Asymptotic stability*) The regular descriptor system (1.1) is said to be asymptotically stable for the disturbances-free case ($w \equiv 0$) if $\lim_{k \to \infty} x(k) = 0$.

Definition 1.4 (*Admissibility*) The descriptor system (1.1) for the disturbances-free case ($w \equiv 0$) is said to be admissible if it is regular, causal and asymptotically stable.

Lemma 1.1 ([1]) *For the matrix pair (E, A) of the descriptor system* (1.1), *the following properties hold:*

- *(Regularity) the pair (E, A) is regular if $\exists z \in \mathbb{C}$, $\det(zE - A)$ is not identically zero;*
- *(Causality) the pair (E, A) is causal if $\exists z \in \mathbb{C}$, $\deg(\det(zE - A)) = \text{rank}(E)$;*
- *(Asymptotic stability) the pair (E, A) is asymptotically stable if $|\nu| < 1$, $\forall \nu \in \lambda(E, A)$.*

In the following, admissibility is not part of the assumption, i.e. the study concerns both causal and non-causal descriptor systems.

Assumption 1.1 The descriptor system (1.1) (the matrix pair (E, A)) is regular and asymptotically stable in the disturbance-free case ($w \equiv 0$).

The following suitable transformations are established, which decompose the descriptor system (1.1) in subsystems.

Definition 1.5 (*Equivalence of descriptor systems*) Consider two descriptor systems respectively defined by the triplets (E, A, B_w) and $(\tilde{E}, \tilde{A}, \tilde{B}_w)$. If there exists a pair of non-singular matrices $Q \in \mathbb{R}^{n \times n}$ and $P \in \mathbb{R}^{n \times n}$ satisfying

$$QEP = \tilde{E}, \ QAP = \tilde{A}, \ QB_w = \tilde{B}_w, \tag{1.2}$$

then these two systems are called restricted equivalent under the transformation (Q, P).

For the descriptor system (1.1), two standard restricted equivalent forms are presented [2, Chap. 2].

1.5.1.1 Dynamics Decomposition Form

Consider the descriptor system (1.1) with $\text{rank}(E) = r$. There always exists a transformation (Q, P) yielding

$$QEP = \begin{bmatrix} I_r & 0 \\ 0 & 0 \end{bmatrix}, \ QAP = \begin{bmatrix} A_1 & A_2 \\ A_3 & A_4 \end{bmatrix}, \ QB_w = \begin{bmatrix} B_{w1} \\ B_{w2} \end{bmatrix}, \tag{1.3}$$

with $A_1 \in \mathbb{R}^{r \times r}$, $A_2 \in \mathbb{R}^{r \times (n-r)}$, $A_3 \in \mathbb{R}^{(n-r) \times r}$, $A_4 \in \mathbb{R}^{(n-r) \times (n-r)}$, $B_{w1} \in \mathbb{R}^{r \times q}$ and $B_{w2} \in \mathbb{R}^{(n-r) \times q}$.

Lemma 1.2 (Dynamics decomposition form [2]) *The descriptor system* (1.1) *is causal if and only if there exists a transformation* (Q, P) *yielding* (1.3) *with a non-singular block matrix* A_4.

Based on the above lemma, an equivalent causal descriptor system in a standard dynamical form is presented in the following.

Lemma 1.3 (Equivalent causal descriptor system) *The causal descriptor system* (1.1) *with* $\text{rank}(E) = r$ *can be transformed into the following form*

$$\tilde{x}(k+1) = \tilde{A}\tilde{x}(k) + \tilde{B}_w \tilde{w}(k), \tag{1.4}$$

where

$$\tilde{A} = \begin{bmatrix} A_1 - A_2 A_4^{-1} A_3 & 0 \\ -A_4^{-1} A_3 \left(A_1 - A_2 A_4^{-1} A_3\right) & 0 \end{bmatrix}, \tag{1.5a}$$

$$\tilde{B}_w = \begin{bmatrix} B_{w1} - A_2 A_4^{-1} B_{w2} & 0 \\ -A_4^{-1} A_3 \left(B_{w1} - A_2 A_4^{-1} B_{w2}\right) & -A_4^{-1} B_{w2} \end{bmatrix}. \tag{1.5b}$$

and A_1, A_2, A_3, A_4, B_{w1}, B_{w2} *are defined in* (1.3) *and*

$$\tilde{x}(k) = \begin{bmatrix} \tilde{x}_1(k) \\ \tilde{x}_2(k) \end{bmatrix} = P^{-1}x(k), \; \tilde{w}(k) = \begin{bmatrix} w(k) \\ w(k+1) \end{bmatrix}, \tag{1.6}$$

with $\tilde{x}_1(k) \in \mathbb{R}^r$, $\tilde{x}_2(k) \in \mathbb{R}^{(n-r)}$.

Proof The proof can be found in [117, Lemma 3].

1.5.1.2 Kronecker Canonical Form

The regular descriptor system (1.1) also allows the transformation in the so-called Kronecker canonical form according to the following lemma.

Lemma 1.4 (Kronecker canonical form [1]) *The descriptor system* (1.1) *is regular if and only if there exists a transformation* $(\bar{Q}, \bar{P})$ *yielding*

$$\bar{Q}E\bar{P} = \begin{bmatrix} I_p & 0 \\ 0 & \bar{N} \end{bmatrix}, \; \bar{Q}A\bar{P} = \begin{bmatrix} \bar{A} & 0 \\ 0 & I \end{bmatrix}, \; \bar{Q}B_w = \begin{bmatrix} \bar{B}_{w1} \\ \bar{B}_{w2} \end{bmatrix}, \tag{1.7}$$

with $\bar{A} \in \mathbb{R}^{p\times p}$, $\bar{B}_{w1} \in \mathbb{R}^{p\times q}$, $\bar{B}_{w2} \in \mathbb{R}^{(n-p)\times q}$. *Moreover,* $\bar{N} \in \mathbb{R}^{(n-p)\times(n-p)}$ *is a nilpotent matrix (that is there exists a scalar* $s > 0$ *such that* $\bar{N}^s = 0$ *and* $\bar{N}^{s-1} \neq 0$, $s \leq n-p$*) and* $p \leq r = \text{rank}(E)$.

Computationally efficient and numerically stable methods exist to obtain these transformations as reported in [57, 118].

Lemma 1.5 (Causality [1]) *The descriptor system* (1.1) *transformed in the Kronecker canonical form* (1.7) *is causal if and only if* $\bar{N} = 0$.

1.5.2 Zonotopes

Definition 1.6 (*Zonotope*) The r-order zonotope $\mathcal{Z} \subset \mathbb{R}^n$ in n-dimensional space is defined with its center $p \in \mathbb{R}^n$ and a segment matrix $H \in \mathbb{R}^{n\times r}$ as

$$\mathcal{Z} = \langle p, H\rangle = \{p + Hz, \|z\|_\infty \leq 1\}. \tag{1.8}$$

Definition 1.7 (*Interval hull*) Given a zonotope $\mathcal{Z} = \langle p, H\rangle \subset \mathbb{R}^n$, the interval hull $rs(H) \in \mathbb{R}^{n\times n}$ is defined as an aligned box such that the inclusion property holds: $\langle p, H\rangle \subset \langle p, rs(H)\rangle$, where $rs(H)$ is a diagonal matrix with diagonal elements of $rs(H)_{i,i} = \sum_{j=1}^{r} \left|H_{i,j}\right|, i = 1, \ldots, n$.

Define $\mathbf{B}^r = [-1, +1]^r \subset \mathbb{R}^r$ as a r-order hypercube. Using the Minkowski sum, the zonotope $\mathcal{Z}$ in (1.8) can also be defined by $\mathcal{Z} = p \oplus H\mathbf{B}^r$. Besides, the following properties hold:

$$\langle p_1, H_1\rangle \oplus \langle p_2, H_2\rangle = \langle p_1 + p_2, [H_1 \quad H_2]\rangle, \tag{1.9a}$$

$$L\langle p, H\rangle = \langle Lp, LH\rangle, \tag{1.9b}$$

$$\langle p, H\rangle \subseteq \langle p, rs(H)\rangle, \tag{1.9c}$$

where L is a matrix of appropriate dimension.

Definition 1.8 (F_W-*radius*) Given a zonotope $\mathcal{Z} = \langle p, H\rangle \subset \mathbb{R}^n$ and a weighting matrix $W \in \mathbb{S}^{n_x}$, the F_W-radius of $\mathcal{Z}$ is defined using the weighted Frobenius norm of H as

$$\ell_{F,W} = \|\langle p, H\rangle\|_{F,W} = \|H\|_{F,W}. \tag{1.10}$$

Definition 1.9 (W-*radius*) Given a zonotope $\mathcal{Z} = \langle p, H\rangle \subset \mathbb{R}^n$ and a weighting matrix $W \in \mathbb{S}^{n_x}$, the W-radius of $\mathcal{Z}$ is defined by

$$\ell_W = \max_{z\in\mathcal{Z}} \|z - p\|_W^2 = \max_{b\in\mathbf{B}^r} \|Hb\|_W^2. \tag{1.11}$$

Definition 1.10 (*Radius*) Given a zonotope $\mathcal{Z} = \langle p, H\rangle$, the radius is defined by

$$\ell = \max_{z\in\mathcal{Z}} \|z - p\|^2 = \max_{b\in\mathbf{B}^r} \|Hb\|^2. \tag{1.12}$$

In order to reduce the order of a zonotope $\mathcal{Z} = \langle p, H\rangle \subset \mathbb{R}^n$, the weighted reduction operator $\downarrow_{q,W}(H)$ proposed in [80] is used, where q specifies the maximum number of columns of H and $W \in \mathbb{S}^n_{>0}$ is a weighting matrix. The inclusion property $\langle p, H\rangle \subset \langle p, \downarrow_{q,W}(H)\rangle$ also holds. The procedure for implementing the operator $\downarrow_{q,W}(H)$ is summarized as follows:

- Sort the column of segment matrix H in decreasing order: $\downarrow_W(H) = [h_1, h_2, \ldots, h_r]$, $\|h_j\|^2_W \geq \|h_{j+1}\|^2_W$, where $\|h_j\|_W$ is the weighted 2-norm of h_j;
- Take the first q-column of $\downarrow_W(H)$ and enclose a set $H_<$ generated by remaining columns $(r - q \geq n)$ into an aligned box (interval hull) as follows:

$$\begin{aligned}
&\textbf{If } r \leq q, \textbf{ then } \downarrow_{q,W}(H) = \downarrow_W(H),\\
&\textbf{Else } \downarrow_{q,W}(H) = [H_>, rs(H_<)] \in \mathbb{R}^{n\times(q+n)},\\
&\quad H_> = \left[h_1, \ldots, h_q\right], \ H_< = \left[h_{q+1}, \ldots, h_r\right].
\end{aligned}$$

1.5.3 Set Invariance Theory

The set-based notions are introduced for discrete-time descriptor systems. For a regular and stable descriptor system (1.1), consider that the additive disturbances are unknown but bounded in a known set

$$w(k) \in \mathcal{W} = \left\{w \in \mathbb{R}^q : |w| \leq \overline{w}\right\}, \ \forall k \in \mathbb{N}, \tag{1.13}$$

with a given $\overline{w} \in \mathbb{R}^q$.

As a consequence of the boundedness of the disturbances and the stability of the dynamics, the system trajectories eventually converge to a bounded region of the state space [94] for the forward trajectories. Given an initial state $x(0)$ and the unique solution to (1.1) (note that the discrete-time domain of the solution may include negative values for backward propagations), the following definitions are introduced in terms of the set-based analysis.

Definition 1.11 (*RI set*) A set $\Omega \in \mathbb{R}^n$ is said to be robust invariant (RI) with respect to the system (1.1) if $x(0) \in \Omega$ implies $x(k) \in \Omega$, $\forall w(k) \in \mathcal{W}$ and $\forall k \in \mathbb{Z}$.

Definition 1.12 (*RPI set*) A set $\Omega \in \mathbb{R}^n$ is said to be robust positively invariant (RPI) with respect to the system (1.1) if $x(0) \in \Omega$ implies $x(k) \in \Omega$, $\forall w(k) \in \mathcal{W}$ and $\forall k \in \mathbb{N}$.

Definition 1.13 (*mRPI set*) An RPI set $\Omega_\infty \in \mathbb{R}^n$ is said to be minimal RPI (mRPI) with respect to the system (1.1) if it is contained in every closed RPI set.

Definition 1.14 (*L-step RNI set*) A set $\Omega \in \mathbb{R}^n$ is L-step robust negatively invariant (RNI) with respect to the system (1.1) if $x(L) \in \Omega$ implies $x(L + k) \in \Omega$, $\forall w(k) \in \mathcal{W}$ and $\forall k \in \mathbb{Z}_{[-L,0]}$.

For dynamical LTI systems (i.e. the system (1.1) with $E = I_n$), the mRPI sets are characterized as a limit set of a sequence of sets and lacks finite determinedness. A number of strategies to approximate the mRPI sets have been proposed [96, 106, 107]. The iterative strategy proposed in [107] yields a polytopic approximation of the mRPI set and will be extended here for the class of descriptor systems.

1.5.4 Linear Algebra

Let X, A, B and C be matrices of appropriate dimensions. The following matrix calculus regarding the matrix trace holds:

$$\frac{\partial}{\partial X}\mathrm{tr}\left(AX^\top B\right) = A^\top B^\top, \tag{1.14a}$$

$$\frac{\partial}{\partial X}\mathrm{tr}\left(AXBX^\top C\right) = BX^\top CA + B^\top X^\top A^\top C^\top. \tag{1.14b}$$

For two matrices X and Y, it holds

$$\mathrm{tr}\left(X^\top Y\right) = \mathbf{vec}(X)^\top \mathbf{vec}(Y) = \mathbf{vec}(Y)^\top \mathbf{vec}(X).$$

The Kronecker product of X and Y is denoted by $X \otimes Y$. Consider matrices A, B and X, the following properties hold:

$$\mathbf{vec}\,(AXB) = \left(B^\top \otimes A\right)\mathbf{vec}\,(X),$$
$$\mathbf{vec}\,(AB) = (I \otimes A)\,\mathbf{vec}\,(B).$$

1.5.5 $\mathcal{H}_-$ Index and Generalized KYP Lemma

To formulate the fault sensitivity, the $\mathcal{H}_-$ index is used. The definition of the $\mathcal{H}_-$ index and generalized KYP lemma are introduced in the following.

Definition 1.15 ($\mathcal{H}_-$ *index of discrete-time systems* [41]) Given a transfer function $G_{yu}(z)$ of discrete-time systems as $G_{yu}(z) = \mathcal{C}(zI - \mathcal{A})^{-1}\mathcal{B} + \mathcal{D}$ between signals y_k and u_k with $z = e^{j\theta}$ and $\forall k \in \mathbb{N}$, the $\mathcal{H}_-$ index of $G_{yu}(z)$ is defined by

$$\left\|G_{yu}(z)\right\|_- := \inf_{u\neq 0}\frac{\|y\|_2}{\|u\|_2} = \inf_{\theta}\underline{\sigma}\left(G_{yu}(e^{j\theta})\right). \tag{1.15}$$

By this definition, the $\mathcal{H}_-$ index between signals y and u, $k \in \mathbb{N}$ can also be presented by $\left\|G_{yu}(z)\right\|_- \geq \beta$ with $\beta > 0$, that is

Table 1.1 Selections of matrices Ξ in different frequency domains

	LF	MF	HF
Θ	$\lvert\theta\rvert \leq \theta_l$	$\theta_1 \leq \theta \leq \theta_2$	$\lvert\theta\rvert \geq \theta_h$
Ξ	$\begin{bmatrix} -P & Q \\ Q & P-2\cos(\theta_l)Q \end{bmatrix}$	$\begin{bmatrix} -P & e^{j\theta_c}Q \\ e^{-j\theta_c}Q & P-2\cos(\theta_w)Q \end{bmatrix}$	$\begin{bmatrix} -P & -Q \\ -Q & P+2\cos(\theta_h)Q \end{bmatrix}$

LF: low-frequency domain, MF: middle-frequency domain, HF: high-frequency domain

$$\sum_{k=0}^{\infty} y(k)^\top y(k) \geq \beta^2 \sum_{k=0}^{\infty} u(k)^\top u(k). \tag{1.16}$$

Lemma 1.6 (Generalized KYP lemma for discrete-time systems [119]) *Given a transfer function $G(z)$ of discrete-time systems as $G(z) = \mathcal{C}(zI - \mathcal{A})^{-1}\mathcal{B} + \mathcal{D}$ with $z = e^{j\theta}$, a symmetric matrix Π of appropriate dimension. The following statements are equivalent:*

(1) For a finite-frequency domain $\forall\theta \in \Theta$, the following condition holds:

$$\begin{bmatrix} G(e^{j\theta}) \\ I \end{bmatrix}^\top \Pi \begin{bmatrix} G(e^{j\theta}) \\ I \end{bmatrix} \prec 0, \ \forall\theta \in \Theta. \tag{1.17}$$

(2) There exist Hermitian matrices P and Q such that $Q \succ 0$ and

$$\begin{bmatrix} \mathcal{A} & \mathcal{B} \\ I & 0 \end{bmatrix}^\top \Xi \begin{bmatrix} \mathcal{A} & \mathcal{B} \\ I & 0 \end{bmatrix} + \begin{bmatrix} \mathcal{C} & \mathcal{D} \\ 0 & I \end{bmatrix}^\top \Pi \begin{bmatrix} \mathcal{C} & \mathcal{D} \\ 0 & I \end{bmatrix} \prec 0, \tag{1.18}$$

where the selections of Ξ are presented in Table 1.1, and $\theta_c = \frac{\theta_1+\theta_2}{2}$, $\theta_w = \frac{\theta_2-\theta_1}{2}$.

References

1. Dai L (1989) Singular control systems. Springer, Berlin Heidelberg, Germany
2. Duan G (2010) Analysis and design of descriptor linear systems. Springer, New York, USA
3. Araujo J, Barros P, Dorea C (2012) Design of observers with error limitation in discrete-time descriptor systems: a case study of a hydraulic tank system. IEEE Trans Control Syst Technol 20(4):1041–1047
4. Puig V, Ocampo-Martinez C, Pérez R, Cembrano G, Quevedo J, Escobet T (2017) Real-time monitoring and operational control of drinking-water systems. Springer
5. Wang Y, Puig V, Cembrano G (2017) Non-linear economic model predictive control of water distribution networks. J Process Control 56:23–34
6. Biegler L, Campbell S, Mehrmann V (2012) Control and optimization with differential-algebraic constraints. Society for Industrial and Applied Mathematics, Philadelphia, USA
7. Riaza R (2008) Differential-algebraic systems: analytical aspects and circuit applications. World Scientific Publishing Company, New York, USA
8. Stevens B, Lewis FL, Johnson E (2016) Aircraft control and simulation: dynamics, controls design, and autonomous systems. Wiley-Blackwell, New York, USA

9. Zhang Q, Liu C, Zhang X (2012) Complexity, analysis and control of singular biological systems. Springer, London, UK
10. Luenberger D, Arbel A (1977) Singular dynamic Leontief systems. Econometrica 45(4):991–995
11. Zhang L, Lam J, Zhang Q (1999) Lyapunov and Riccati equations of discrete-time descriptor systems. IEEE Trans Autom Control 44(11):2134–2139
12. Halanay A, Rasvan V (2000) Stability and stable oscillations in discrete time systems. CRC Press
13. Hsieh C (2013) State estimation for descriptor systems via the unknown input filtering method. Automatica 49(5):1281–1286
14. Ishihara J, Terra M, Bianco A (2010) Recursive linear estimation for general discrete-time descriptor systems. Automatica 46(4):761–766
15. Angeli D, Amrit R, Rawlings J (2012) On average performance and stability of economic model predictive control. IEEE Trans Autom Control 57(7):1615–1626
16. Ellis M, Liu J, Christofides P (2017) Economic model predictive control: theory, formulations and chemical process applications. Springer
17. Mayne D (2014) Model predictive control: recent developments and future promise. Automatica 50(12):2967–2986
18. Maciejowski J (2002) Predictive control with constraints. Prentice-Hall
19. Rawlings J, Mayne D (2009) Model predictive control: theory and design. Wis. Nob Hill Pub. cop, Madison
20. Cembrano G, Quevedo J, Salamero M, Puig V, Figueras J, Martí J (2004) Optimal control of urban drainage systems. A case study. Control Eng Pract 12(1):1–9
21. Cembrano G, Wells G, Quevedo J, Perez R, Argelaguet R (2000) Optimal control of a water distribution network in a supervisory control system. Control Eng Pract 8(10):1177–1188
22. Ocampo-Martinez C, Puig V, Cembrano G, Quevedo J (2013) Application of MPC strategies to the management of complex networks of the urban water cycle. IEEE Control Syst 33(1):15–41
23. Pereira M, Muñoz de la Peña D, Limon D, Alvarado I, Alamo T (2016) Application to a drinking water network of robust periodic MPC. Control Eng Pract 57:50–60
24. Zeng J, Liu J (2015) Economic model predictive control of wastewater treatment processes. Ind Eng Chem Res 54(21):5710–5721
25. Pereira M, Limon D, Muñoz de la Peña D, Limon D (2017) Robust economic model predictive control of a community micro-grid. Renew Energy 100:3–17
26. Pereira M, Limon D, Muñoz de la Peña D, Valverde L, Alamo T (2015) Periodic economic control of a nonisolated microgrid. IEEE Trans Ind Electron 62(8):5247–5255
27. Liu S, Zhang J, Liu J (2015) Economic MPC with terminal cost and application to an oilsand primary separation vessel. Chem Eng Sci 136:27–37
28. Santander O, Elkamel A, Budman H (2016) Economic model predictive control of chemical processes with parameter uncertainty. Comput Chem Eng 95:10–20
29. Amrit R, Rawlings J, Angeli D (2011) Economic optimization using model predictive control with a terminal cost. Annu Rev Control 35(2):178–186
30. Müller M, Angeli D, Allgöwer F (2014) On the performance of economic model predictive control with self-tuning terminal cost. J Process Control 24(8):1179–1186
31. Ellis M, Durand H, Christofides P (2016) Elucidation of the role of constraints in economic model predictive control. Annu Rev Control 41:208–217
32. Grüne L (2013) Economic receding horizon control without terminal constraints. Automatica 49(3):725–734
33. Grüne L, Stieler M (2014) Asymptotic stability and transient optimality of economic MPC without terminal conditions. J Process Control 24(8):1187–1196
34. Liu S, Liu J (2016) Economic model predictive control with extended horizon. Automatica 73:180–192
35. Mayne D, Seron M, Raković S (2005) Robust model predictive control of constrained linear systems with bounded disturbances. Automatica 41(2):219–224

36. Pereira M, Muñoz de la Peña D, Limon D, Alvarado I, Alamo T (2017) Robust model predictive controller for tracking changing periodic signals. IEEE Trans Autom Control 62(10):5343–5350
37. Bayer F, Müller M, Allgöwer F (2018) On optimal system operation in robust economic MPC. Automatica 88:98–106
38. Broomhead T, Manzie C, Shekhar R, Hield P (2015) Robust periodic economic MPC for linear systems. Automatica 60:30–37
39. Huang R, Biegler L, Harinath E (2012) Robust stability of economically oriented infinite horizon NMPC that include cyclic processes. J Process Control 22(1):51–59
40. Blanke M, Kinnaert M, Lunze J, Staroswiecki M (2016) Diagnosis and fault-tolerant control. Springer, Berlin Heidelberg
41. Ding S (2013) Model-based fault diagnosis techniques. Springer, London, UK
42. Chadli M, Abdo A, Ding S (2013) Fault detection filter design for discrete-time Takagi-Sugeno fuzzy system. Automatica 49(7):1996–2005
43. Ichalal D, Marx B, Ragot J, Maquin D (2014) Fault detection, isolation and estimation for Takagi-Sugeno nonlinear systems. J Frankl Inst 351(7):3651–3676
44. Lee KS, Park TG (2015) Robust fault detection observer design under fault sensitivity constraints. J Frankl Inst 352(5):1791–1810
45. Blanchini F, Casagrande D, Giordano G, Miani S, Olaru S, Reppa V (2017) Active fault isolation: a duality-based approach via convex programming. SIAM J Control Optim 55(3):1619–1640
46. Raka S, Combastel C (2013) Fault detection based on robust adaptive thresholds: a dynamic interval approach. Annu Rev Control 37(1):119–128
47. Xu F, Puig V, Ocampo-Martinez C, Stoican F, Olaru S (2014) Actuator-fault detection and isolation based on set-theoretic approaches. J Process Control 24(6):947–956
48. Puig V, Quevedo J, Escobet T, Nejjari F, De Las Heras S (2008) Passive robust fault detection of dynamic processes using interval models. IEEE Trans Control Syst Technol 16(5):1083–1089
49. Alamo T, Bravo J, Camacho E (2005) Guaranteed state estimation by zonotopes. Automatica 41(6):1035–1043
50. Scott J, Raimondo D, Marseglia G, Braatz R (2016) Constrained zonotopes: a new tool for set-based estimation and fault detection. Automatica 69:126–136
51. Chen J, Patton R (1999) Robust model-based fault diagnosis for dynamic systems. Springer, New York, USA
52. Hou M, Muller PC (1992) Design of observers for linear systems with unknown inputs. IEEE Trans Autom Control 37(6):871–875
53. Koenig D (2005) Unknown input proportional multiple-integral observer design for linear descriptor systems: application to state and fault estimation. IEEE Trans Autom Control 50(2):212–217
54. Rotondo D, Witczak M, Puig V, Nejjari F, Pazera M (2016) Robust unknown input observer for state and fault estimation in discrete-time Takagi-Sugeno systems. Int J Syst Sci 47(14):3409–3424
55. Xu F, Tan J, Wang X, Puig V, Liang B, Yuan B (2016) A novel design of unknown input observers using set-theoretic methods for robust fault detection. In: American control conference (ACC), Boston, USA, pp 5957–5961
56. Xu F, Tan J, Wang X, Puig V, Liang B, Yuan B (2018) Mixed active/passive robust fault detection and isolation using set-theoretic unknown input observers. IEEE Trans Autom Sci Eng 15(2):863–871
57. Varga A (2017) Solving fault diagnosis problems. Springer
58. Wan Y, Keviczky T (2019) Real-time fault-tolerant moving horizon air data estimation for the reconfigure benchmark. IEEE Trans Control Syst Technol 27(3):997–1011
59. Wan Y, Keviczky T (2019) Real-time nonlinear moving horizon observer with pre-estimation for aircraft sensor fault detection and estimation. Int J Robust Nonlinear Control 29(16):5394–5411

60. Wan Y, Keviczky T, Verhaegen M (2018) Fault estimation filter design with guaranteed stability using Markov parameters. IEEE Trans Autom Control 63(4):1132–1139
61. Gao Z (2015) Fault estimation and fault-tolerant control for discrete-time dynamic systems. IEEE Trans Ind Electron 62(6):3874–3884
62. Gao Z, Ding S (2007) Actuator fault robust estimation and fault-tolerant control for a class of nonlinear descriptor systems. Automatica 43(5):912–920
63. Lan J, Patton R (2016) A new strategy for integration of fault estimation within fault-tolerant control. Automatica 69:48–59
64. Wang Z, Rodrigues M, Theilliol D, Shen Y (2015) Fault estimation filter design for discrete-time descriptor systems. IET Control Theory Appl 9(10):1587–1594
65. Zhang K, Jiang B, Shi P, Xu J (2015) Fault estimation observer design for discrete-time systems in finite-frequency domain. Int J Robust Nonlinear Control 25(9):1379–1398
66. Wang Z, Shi P, Lim C (2017) Robust fault estimation observer in the finite frequency domain for descriptor systems. Int J Control 1590–1599
67. Lopez-Estrada F, Ponsart J, Astorga-Zaragoza C, Camas-Anzueto J, Theilliol D (2015) Robust sensor fault estimation for descriptor-LPV systems with unmeasurable gain scheduling functions: application to an anaerobic bioreactor. Int J Appl Math Comput Sci 25(2):233–244
68. Rodrigues M, Hamdi H, Theilliol D, Mechmeche C, BenHadj Braiek N (2015) Actuator fault estimation based adaptive polytopic observer for a class of LPV descriptor systems. Int J Robust Nonlinear Control 25(5):673–688
69. Shi F, Patton R (2015) Fault estimation and active fault tolerant control for linear parameter varying descriptor systems. Int J Robust Nonlinear Control 25(5):689–706
70. Wang Z, Rodrigues M, Theilliol D, Shen Y (2015) Actuator fault estimation observer design for discrete-time linear parameter-varying descriptor systems. Int J Adapt Control Signal Process 29(2):242–258
71. Koenig D, Marx B, Varrier S (2016) Filtering and fault estimation of descriptor switched systems. Automatica 63:116–121
72. Zhang Y, Jiang J (2008) Bibliographical review on reconfigurable fault-tolerant control systems. Annu Rev Control 32(2):229–252
73. Lunze J, Steffen T (2006) Control reconfiguration after actuator failures using disturbance decoupling methods. IEEE Trans Autom Control 51(10):1590–1601
74. Rotondo D, Nejjari F, Puig V (2014) A virtual actuator and sensor approach for fault tolerant control of LPV systems. J Process Control 24(3):203–222
75. Rotondo D, Puig V, Nejjari F, Romera J (2015) A fault-hiding approach for the switching quasi-LPV fault-tolerant control of a four-wheeled omnidirectional mobile robot. IEEE Trans Ind Electron 62(6):3932–3944
76. Rotondo D, Nejjari F, Puig V (2016) Fault tolerant control of a proton exchange membrane fuel cell using Takagi-Sugeno virtual actuators. J Process Control 45:12–29
77. Richter J, Heemels W, van de Wouw N, Lunze J (2011) Reconfigurable control of piecewise affine systems with actuator and sensor faults: stability and tracking. Automatica 47(4):678–691
78. Rotondo D, Cristofaro A, Johansen T (2018) Fault tolerant control of uncertain dynamical systems using interval virtual actuators. Int J Robust Nonlinear Control 28(2):611–624
79. Alamo T, Bravo J, Redondo M, Camacho E (2008) A set-membership state estimation algorithm based on DC programming. Automatica 44(1):216–224
80. Combastel C (2015) Zonotopes and Kalman observers: gain optimality under distinct uncertainty paradigms and robust convergence. Automatica 55:265–273
81. Jaulin L, Kieffer M, Didrit O, Walter E (2001) Applied interval analysis, with examples in parameter and state estimation, robust control and robotics. Springer
82. Puig V, Cuguero P, Quevedo J (2001) Worst-case state estimation and simulation of uncertain discrete-time systems using zonotopes. In: European control conference (ECC), pp 1691–1697
83. Raïssi T, Ramdani N, Candau Y (2004) Set membership state and parameter estimation for systems described by nonlinear differential equations. Automatica 40(10):1771–1777

84. Thabet R, Raïssi T, Combastel C, Efimov D, Zolghadri A (2014) An effective method to interval observer design for time-varying systems. Automatica 50(10):2677–2684
85. Wan Y, Puig V, Ocampo-Martinez C, Wang Y, Braatz R (2018) Probability-guaranteed set-membership state estimation for polynomially uncertain linear time-invariant systems. In: 57th IEEE conference on decision and control (IEEE-CDC), Miami, USA, pp 2291–2296
86. Schweppe F (1968) Recursive state estimation: unknown but bounded errors and system inputs. IEEE Trans Autom Control 13(1):22–28
87. Efimov D, Perruquetti W, Raïssi T, Zolghadri A (2013) Interval observers for time-varying discrete-time systems. IEEE Trans Autom Control 58(12):3218–3224
88. Kalman R (1960) A new approach to linear filtering and prediction problems. ASME J Basic Eng 82(1):35–45
89. Kalman R, Bucy R (1961) New results in linear filtering and prediction theory. ASME J Basic Eng 83(1):95–108
90. Puig V (2010) Fault diagnosis and fault tolerant control using set-membership approaches: application to real case studies. Int J Appl Math Comput Sci 20(4):619–635
91. Combastel C (2003) A state bounding observer based on zonotopes. In: European control conference (ECC), Cambridge, UK, pp 2589–2594
92. Combastel C (2015) Merging Kalman filtering and zonotopic state bounding for robust fault detection under noisy environment. In: 9th IFAC symposium on fault detection. Supervision and safety for technical processes (SAFEPROCESS). France, Paris, pp 289–295
93. Blanchini F (1999) Set invariance in control. Automatica 35(11):1747–1767
94. Kolmanovsky I, Gilbert E (1998) Theory and computation of disturbance invariant sets for discrete-time linear systems. Math Probl Eng 4(4):317–367
95. Limon D, Alamo T, Camacho E (2002) Stability analysis of systems with bounded additive uncertainties based on invariant sets: stability and feasibility of MPC. In: American control conference (ACC), Alaska, USA, pp 364–369
96. Raković S, Kerrigan E, Kouramas K, Mayne D (2005) Invariant approximations of the minimal robust positively invariant set. IEEE Trans Autom Control 50(3):406–410
97. Stoican F, Oară C, Hovd M (2015) RPI approximations of the mRPI set characterizing linear dynamics with zonotopic disturbances. In: Developments in model-based optimization and control: distributed control and industrial applications. Springer, pp 361–377
98. Seron M, De Doná J (2015) On robust stability and set invariance of switched linear parameter varying systems. Int J Control 88(12):2588–2597
99. Seron M, De Doná J (2016) On invariant sets and closed-loop boundedness of Lure-type nonlinear systems by LPV-embedding. Int J Robust Nonlinear Control 26(5):1092–1111
100. Blanchini F, Casagrande D, Miani S (2010) Modal and transition dwell time computation in switching systems: a set-theoretic approach. Automatica 46(9):1477–1482
101. Heidari R, Braslavsky J, Seron M, Haimovich H (2016) Ultimate bound minimisation by state feedback in discrete-time switched linear systems under arbitrary switching. Nonlinear Anal Hybrid Syst 21:84–102
102. Alamo T, Cepeda A, Fiacchini M, Camacho E (2009) Convex invariant sets for discrete-time Lur'e systems. Automatica 45(4):1066–1071
103. Bravo J, Limon D, Alamo T, Camacho E (2005) On the computation of invariant sets for constrained nonlinear systems: an interval arithmetic approach. Automatica 41(9):1583–1589
104. Fiacchini M, Alamo T, Camacho E (2010) On the computation of convex robust control invariant sets for nonlinear systems. Automatica 46(8):1334–1338
105. Fiacchini M, Alamo T, Camacho E (2012) Invariant sets computation for convex difference inclusions systems. Syst Control Lett 61(8):819–826
106. Kofman E, Haimovich H, Seron M (2007) A systematic method to obtain ultimate bounds for perturbed systems. Int J Control 80(2):167–178
107. Olaru S, De Doná J, Seron M, Stoican F (2010) Positive invariant sets for fault tolerant multisensor control schemes. Int J Control 83(12):2622–2640
108. Stoican F, Olaru S, Seron M, De Doná J (2012) Reference governor design for tracking problems with fault detection guarantees. J Process Control 22(5):829–836

109. Seron M, De Doná J, Olaru S (2012) Fault tolerant control allowing sensor healthy-to-faulty and faulty-to-healthy transitions. IEEE Trans Autom Control 57(7):1657–1669
110. Stoican F, Olaru S (2013) Set-theoretic fault-tolerant control in multisensor systems. Wiley-ISTE
111. Xu F, Olaru S, Puig V, Ocampo-Martinez C, Niculescu S (2017) Sensor-fault tolerance using robust MPC with set-based state estimation and active fault isolation. Int J Robust Nonlinear Control 27(8):1260–1283
112. Mayne D, Raković S, Findeisen R, Allgöwer F (2006) Robust output feedback model predictive control of constrained linear systems. Automatica 42(7):1217–1222
113. Kofman E, Seron M, Haimovich H (2008) Control design with guaranteed ultimate bound for perturbed systems. Automatica 44(7):1815–1821
114. Raimondo D, Marseglia G, Braatz R, Scott J (2016) Closed-loop input design for guaranteed fault diagnosis using set-valued observers. Automatica 74:107–117
115. Scott J, Findeisen R, Braatz R, Raimondo D (2014) Input design for guaranteed fault diagnosis using zonotopes. Automatica 50(6):1580–1589
116. Marseglia G, Raimondo D (2017) Active fault diagnosis: a multi-parametric approach. Automatica 79:223–230
117. Wang Y, Olaru S, Valmorbida G, Puig V, Cembrano G (2017) Robust invariant sets and active mode detection for discrete-time uncertain descriptor systems. In: 56th IEEE conference on decision and control (IEEE-CDC), Melbourne, Australia, pp 5648–5653
118. Gerdin M (2004) Computation of a canonical form for linear differential-algebraic equations. Technical report, Linköping University
119. Iwasaki T, Hara S (2005) Generalized KYP lemma: unified frequency domain inequalities with design applications. IEEE Trans Autom Control 50(1):41–59

Part I
State Estimation

Chapter 2
Set-Based State Estimation Approaches for Descriptor Systems

This chapter proposes a general set-based framework for robust state estimation of discrete-time descriptor systems, which builds a bridge to fault diagnosis and control design problems. Specifically, a set-membership state estimator and a zonotopic Kalman observer are investigated. The considered LTI descriptor systems are affected by three types of system uncertainties: unknown inputs and unknown-but-bounded system disturbances and measurement noise. One limitation for the use of zonotopic approaches in real applications is that some system disturbances are unknown and it may not be possible to bound them in a predefined zonotope as a priori. To overcome this problem, two classes of unknown system disturbances are considered: (i) bounded disturbances in a zonotope; (ii) unbounded disturbances, which are considered to be unknown inputs and can be decoupled in the observer design. As shown in Fig. 2.1, two set-based approaches with different criteria are studied and therefore the relationship between both approaches is also established. In particular, it is proved that the zonotopic observer in the current estimation type is equivalent to the set-membership approach. Besides, the set-membership approach is extended for discrete-time LPV descriptor systems, where a new zonotope minimization criterion based on the $\mathcal{L}_\infty$ norm is defined.

2.1 Set-Membership Approach and Zonotopic Kalman Observer for Discrete-Time Descriptor Systems

Consider the discrete-time descriptor linear system as

$$Ex(k+1) = Ax(k) + Bu(k) + Dw(k) + D_d d(k), \tag{2.1a}$$
$$y(k) = Cx(k) + Fv(k), \tag{2.1b}$$

Y. Wang, *Advances in State Estimation, Diagnosis and Control of Complex Systems*, Springer Theses, https://doi.org/10.1007/978-3-030-52440-1_2

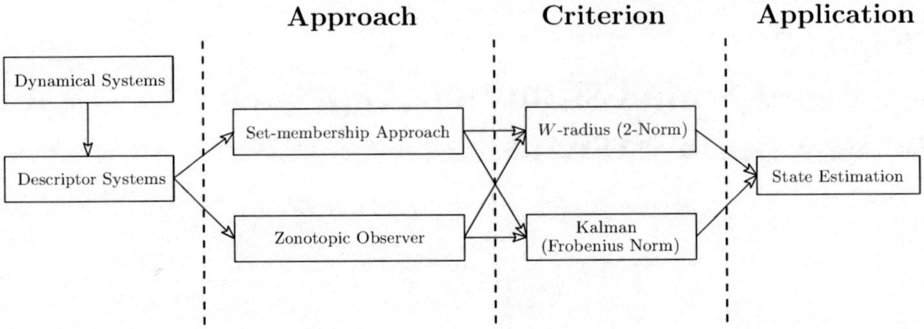

Fig. 2.1 Set-based state estimation scheme

where $x \in \mathbb{R}^{n_x}$ denotes the vector of system states, $u \in \mathbb{R}^{n_u}$ denotes the vector of known inputs, $d \in \mathbb{R}^{n_d}$ denotes the vector of unknown inputs, $y \in \mathbb{R}^{n_y}$ denotes the vector of measurement outputs, $E \in \mathbb{R}^{n_x \times n_x}$, $A \in \mathbb{R}^{n_x \times n_x}$, $B \in \mathbb{R}^{n_x \times n_u}$, $C \in \mathbb{R}^{n_y \times n_x}$, $D \in \mathbb{R}^{n_x \times n_w}$, $D_d \in \mathbb{R}^{n_x \times n_d}$ and $F \in \mathbb{R}^{n_y \times n_v}$. For the descriptor system (2.1), E may be a singular matrix and $\text{rank}(E) \leq n_x$.

Assumption 2.1 The initial state $x(0)$ is assumed to be in the inclusion zonotope $\mathcal{X}(0) = \langle p(0), H(0) \rangle$, where $p(0) \in \mathbb{R}^{n_x}$ and $H(0) \in \mathbb{R}^{n_x \times n_x}$ are the center and generator matrix of this zonotope.

Assumption 2.2 The system disturbance vector $w(k) \in \mathbb{R}^{n_w}$ and the measurement noise vector $v(k) \in \mathbb{R}^{n_v}$ are assumed to be unknown but bounded by zonotopes $w(k) \in \mathcal{W} = \langle 0, I_{n_w} \rangle$, $v(k) \in \mathcal{V} = \langle 0, I_{n_v} \rangle$, $\forall k \in \mathbb{N}$.

Assumption 2.3 For the descriptor system (2.1), the unknown input $d(k)$, $\forall k \in \mathbb{N}$ can be decoupled, and matrices E, C and D_d satisfy the following rank condition [1].

$$\text{rank} \begin{bmatrix} I_{n_x} \otimes \begin{bmatrix} E & D_d \\ C & 0 \end{bmatrix}^{\top} \\ \textbf{vec}\left(\begin{bmatrix} I_{n_x} \\ 0 \end{bmatrix} \right) \end{bmatrix} = n_x \cdot \text{rank} \begin{bmatrix} E & D_d \\ C & 0 \end{bmatrix}. \tag{2.2}$$

Thus, there exists a non-empty set of solutions of matrices T and N satisfying

$$TE + NC = I_{n_x}, \tag{2.3a}$$
$$TD_d = 0. \tag{2.3b}$$

In this chapter, we investigate state estimation approaches based on zonotopes for descriptor system (2.1). We propose two set-based approaches to use zonotope bounding uncertain states with unknown-but-bounded disturbances and noise as well as unbounded disturbances (as unknown inputs).

2.1.1 Set-Membership Approach Based on Zonotopes for Discrete-Time Descriptor Systems

We now propose a set-membership state estimation approach based on zonotopes for discrete-time descriptor system (2.1). This approach uses the structure of the parameterized intersection zonotope for implementing the measurement consistency test including unknown inputs. Some preliminary definitions are introduced as follows.

Definition 2.1 (*Uncertain state set*) Given the descriptor system (2.1) with $x(0) \in \langle p(0), H(0)\rangle$, $w(k) \in \mathcal{W}$, $\forall k \in \mathbb{N}$, the uncertain state set $\bar{\mathcal{X}}(k)$ is defined by

$$\bar{\mathcal{X}}(k) = \left\{x \in \mathbb{R}^{n_x} \mid Ex \in A\bar{\mathcal{X}}(k-1) \oplus Bu(k-1) \oplus D_d d(k-1) \oplus D\mathcal{W}\right\}.$$

Definition 2.2 (*Measurement state set*) Given the descriptor system (2.1), a measurement output vector $y(k)$ and $v(k) \in \mathcal{V}$, $\forall k \in \mathbb{N}$, the measurement state set $\mathcal{P}(k)$ is defined by

$$\mathcal{P}(k) = \left\{x \in \mathbb{R}^{n_x} \mid Cx - y(k) = F\alpha, \forall \alpha \in \mathbf{B}^{n_v}\right\}.$$

Definition 2.3 (*Exact uncertain state set*) Given the descriptor system (2.1), a measurement output vector $y(k)$, $w(k) \in \mathcal{W}$ and $v(k) \in \mathcal{V}$, $\forall k \in \mathbb{N}$, the exact uncertain state set $\mathcal{X}(k)$ is defined by $\mathcal{X}(k) = \bar{\mathcal{X}}(k) \cap \mathcal{P}(k)$.

Since $d(k)$ is an unknown input vector, it is impossible to directly characterize the uncertain state set from Definition 2.1. Meanwhile, the goal is to approximate the exact uncertain state set $\mathcal{X}(k)$ by an outer approximation for the descriptor system (2.1) through implementing a measurement consistency test. In general, the proposed set-membership approach includes three steps: (i) *prediction step*; (ii) *measurement step*; (iii) *correction step*.

More specifically, assuming $x(k) \in \mathcal{X}(k) \subseteq \hat{\mathcal{X}}(k) = \langle \hat{p}(k), \hat{H}(k)\rangle$ at time step $k \in \mathbb{N}$ that also satisfies $x(0) \in \mathcal{X}(0) = \langle p(0), H(0)\rangle$ when $k = 0$, these three steps are implemented as follows: (i) compute the predicted uncertain state set $\bar{\mathcal{X}}(k+1)$; (ii) compute the measurement state set $\mathcal{P}(k+1)$ with a measurement output vector $y(k+1)$; (iii) find an intersection zonotope $\hat{\mathcal{X}}(k+1)$ satisfying $\{\bar{\mathcal{X}}(k+1) \cap \mathcal{P}(k+1)\} \subseteq \hat{\mathcal{X}}(k+1)$, where $\hat{\mathcal{X}}(k+1)$ is a parameterized intersection zonotope with respect to a correction matrix $\Lambda \in \mathbb{R}^{n_x \times n_y}$. The structure of this intersection zonotope is defined as follows.

Theorem 2.1 (Intersection zonotope for descriptor systems) *Given the descriptor system (2.1), a measurement output vector $y(k+1)$, $x(0) \in \mathcal{X}(0)$, $w(k) \in \mathcal{W}$, $v(k) \in \mathcal{V}$, $\forall k \in \mathbb{N}$, $x(k) \in \langle \hat{p}(k), \hat{H}(k)\rangle \subseteq \langle \hat{p}(k), \bar{H}(k)\rangle$ with $\bar{H}(k) = \downarrow_{q,W} (\hat{H}(k))$, $T \in \mathbb{R}^{n_x \times n_x}$ and $N \in \mathbb{R}^{n_x \times n_y}$ satisfying (2.3). Then, for any correction matrix $\Lambda \in \mathbb{R}^{n_x \times n_y}$, $x(k+1) \in \left\{\bar{\mathcal{X}}(k+1) \cap \mathcal{P}(k+1)\right\} \subseteq \hat{\mathcal{X}}(k+1) = \langle \hat{p}(k+1), \hat{H}(k+1)\rangle$, where*

$$\hat{p}(k+1) = (I - \Lambda C)\,TA\hat{p}(k) + (I - \Lambda C)\,TBu(k) \\ + (N + \Lambda - \Lambda CN)\,y(k+1), \tag{2.4a}$$

$$\hat{H}(k+1) = \left[(I - \Lambda C)\,TA\bar{H}(k),\ (I - \Lambda C)\,TD,\ (I - \Lambda C)\,NF,\ \Lambda F\right]. \tag{2.4b}$$

Proof For any $x(k+1) \in \left\{\hat{\mathcal{X}}(k+1) \cap \mathcal{P}(k+1)\right\}$, we know that $x(k+1) \in \hat{\mathcal{X}}(k+1)$ and $x(k+1) \in \mathcal{P}(k+1)$. Considering the descriptor system (2.1a) with the inclusion $x(k) \in \langle \hat{p}(k), \hat{H}(k)\rangle \subseteq \langle \hat{p}(k), \downarrow_{q,W} (\hat{H}(k))\rangle$, there exists a vector $s_1 \in \mathbf{B}^{q+n_w}$ such that

$$Ex(k+1) = A\hat{p}(k) + Bu(k) + D_d d(k) + \left[A\bar{H}(k),\ D\right] s_1.$$

Besides, from $x(k+1) \in \mathcal{P}(k+1)$, there exists a vector $\alpha \in \mathbf{B}^{n_v}$ such that

$$Cx(k+1) - y(k+1) = F\alpha. \tag{2.5}$$

Consider the rank condition (2.2) is satisfied. With a pair of matrices T and N satisfying (2.3), (2.1) and (2.5) can be combined leading to

$$(TE + NC)x(k+1) = TA\hat{p}(k) + TBu(k) + TD_d d(k) + Ny(k+1) \\ + \left[TA\bar{H}(k),\ TD\right] s_1 + NF\alpha.$$

Set $R(k) = \left[TA\bar{H}(k),\ TD,\ NF\right]$ and $\beta = \left[s_1^{\mathrm{T}},\ \alpha^{\mathrm{T}}\right]^{\mathrm{T}}$. According to (2.3), the above equation can be simplified to be

$$x(k+1) = TA\hat{p}(k) + TBu(k) + Ny(k+1) + R(k)\beta. \tag{2.6}$$

Therefore, with $\Lambda \in \mathbb{R}^{n_x \times n_y}$ and a correction term $\Lambda CR(k)\beta$, we add and substitute $CR(k)\beta$ in (2.6) to obtain

$$x(k+1) = TA\hat{p}(k) + TBu(k) + Ny(k+1) + \Lambda CR(k)\beta + (I - \Lambda C)R(k)\beta. \tag{2.7}$$

By substituting $x(k+1)$ in (2.5) by (2.6), we also have

$$CR(k)\beta = y(k+1) - CNy(k+1) - CTA\hat{p}(k) - CTBu(k) + F\alpha.$$

And then by replacing $CR(k)\beta$ in (2.7), we have

$$x(k+1) = (I - \Lambda C)TA\hat{p}(k) + (I - \Lambda C)TBu(k) \\ + (N + \Lambda - \Lambda CN)\,y(k+1) + \left[(I - \Lambda C)\,R(k),\ \Lambda F\right]\begin{bmatrix}\beta \\ \alpha\end{bmatrix}.$$

Thus, we obtain $\hat{p}(k+1)$ and $\hat{H}(k+1)$ as in (2.4). □

Due to the intersection zonotope bounding uncertain states including propagated estimation errors and uncertainties, we would like to find a suitable (time-varying or time-invariant) correction matrix minimizing the effects of estimation errors and uncertainties by reducing the size of the intersection zonotope. To measure the size of a zonotope, the F_W-radius and the W-radius are used as in Definitions 1.8 and 1.9.

In the following, we first compute a time-varying Kalman correction matrix based on the F_W-radius. On the other hand, with a W-radius minimization criterion, a correction matrix can be obtained by solving an off-line optimization problem. This off-line correction matrix can also be updated following an on-line updating procedure.

2.1.1.1 Computing the Correction Matrix via Kalman Filtering Procedure

From Definition 1.8, the size of the intersection zonotope $\hat{\mathcal{X}}(k+1)$ can be measured by the F_W-radius as

$$\begin{aligned} \ell_{F,W}(k+1) &= \left\| \hat{H}(k+1) \right\|_{F,W}^2 = \mathrm{tr}\left(\hat{H}(k+1)^\top W \hat{H}(k+1) \right) \\ &= \mathrm{tr}\left(W\hat{H}(k+1)\hat{H}(k+1)^\top \right) = \mathrm{tr}\Big(W P(k+1) \Big), \end{aligned} \tag{2.8}$$

where $P(k+1) = \hat{H}(k+1)\hat{H}(k+1)^\top$. As in the Kalman filtering procedure described in [2, Theorem 5], a time-varying Kalman correction matrix $\Lambda^*(k)$ can be obtained by minimizing $\ell_{F,W}(k+1)$ of the intersection zonotope $\langle \hat{p}(k+1), \hat{H}(k+1)\rangle$.

Theorem 2.2 (Kalman correction matrix) *Given the intersection zonotope $\hat{\mathcal{X}}(k+1) = \langle \hat{p}(k+1), \hat{H}(k+1)\rangle$ in (2.4) and a weighting matrix $W \in \mathbb{S}_{\succ 0}$. The optimal correction matrix $\Lambda^*(k)$ minimizes $J = \ell_{F,W}(k+1)$ and its explicit solution is given by*

$$\Lambda^*(k) = L(k)S(k)^{-1}, \tag{2.9}$$

$$L(k) = \bar{R}(k)C(k)^\top, \tag{2.10}$$

$$S(k) = C\bar{R}(k)C^\top + Q_v, \tag{2.11}$$

$$\bar{R}(k) = T\left(A\bar{P}(k)A^\top + Q_w\right)T^\top + NQ_vN^\top, \tag{2.12}$$

with $\bar{P}(k) = \bar{H}(k)\bar{H}(k)^\top$, $Q_w = DD^\top$ and $Q_v = FF^\top$.

Proof From (2.4), we have

$$\begin{aligned} P(k+1) = &(I - \Lambda C)\,TA\bar{P}(k)A^\top T^\top (I - \Lambda C)^\top + (I - \Lambda C)\,TQ_wT^\top (I - \Lambda C)^\top \\ &+ (I - \Lambda C)\,NQ_vN^\top (I - \Lambda C)^\top + \Lambda Q_v\Lambda^\top. \end{aligned}$$

The criterion $J = \ell_{F,W}(k+1)$ is convex with respect to Λ. By setting L, S and $\bar{R}$ as in (2.10), (2.11) and (2.12), we take the partial-derivative of $J = \ell_{F,W}(k+1)$ in (2.8) with respect to Λ to obtain

$$\frac{\partial}{\partial \Lambda}\mathrm{tr}\Big(WP(k+1)\Big) = \frac{\partial}{\partial \Lambda}\mathrm{tr}\left(W\Lambda S\Lambda^\top\right) - 2\frac{\partial}{\partial \Lambda}\mathrm{tr}\left(WL\Lambda^\top\right).$$

Then, $\Lambda^*(k)$ is the value of Λ such that $\frac{\partial}{\partial \Lambda}\mathrm{tr}\Big(WP(k+1)\Big) = 0$. By using (1.14a) and (1.14b), we have

$$S(k)\Lambda^*(k)^\top W + S(k)^\top \Lambda^*(k)^\top W^\top - 2L(k)^\top W^\top = 0,$$

from which, since that $S(k)$ is also symmetric, we thus obtain $W\Lambda^*(k)S(k) = WL(k)$, which leads to (2.9). □

Remark 2.1 From Theorem 2.2, the optimal correction matrix $\Lambda^*(k)$ is independent of the weighting matrix W. Hence, this weighting matrix W can be set freely and we can also use the non-weighted Frobenius norm to measure the zonotope size as the F-radius.

2.1.1.2 Computing the Correction Matrix Using Optimization-Based Methods

From Definition 1.9, the size of the intersection zonotope $\hat{\mathcal{X}}(k+1)$ can also be measured by the W-radius as

$$\begin{aligned} \ell_W(k+1) &= \max_{z \in \mathbf{B}^{(q+n_x+2n_y)}} \left\| \hat{H}(k+1)z \right\|_{2,W}^2 \\ &= \max_{z \in \mathbf{B}^{(q+n_x+2n_y)}} z^\top \hat{H}(k+1)^\top (\Lambda) W \hat{H}(k+1)z. \end{aligned} \tag{2.13}$$

We now present a W-radius minimization criterion and the corresponding linear matrix inequality (LMI) condition to find a constant correction matrix Λ in the following theorem.

Theorem 2.3 (W-radius minimization criterion) *Given the intersection zonotope $\hat{\mathcal{X}}(k+1) = \langle \hat{p}(k+1), \hat{H}(k+1)\rangle$ in (2.4), $\gamma \in (0,1)$ and $\epsilon > 0$. The zonotope minimization criterion*

$$\ell_W(k+1) \le \gamma \ell_W(k) + \epsilon, \tag{2.14}$$

holds if there exist matrices $W \in \mathbb{S}^{n_x}_{\succ 0}$, $Y \in \mathbb{R}^{n_x \times n_y}$, diagonal matrices $\Gamma \in \mathbb{S}^{n_x}_{\succ 0}$, $\Upsilon \in \mathbb{S}^{n_y}_{\succ 0}$ and $\Omega \in \mathbb{S}^{n_y}_{\succ 0}$ such that

$$\mathrm{tr}(\Gamma) + \mathrm{tr}(\Upsilon) + \mathrm{tr}(\Omega) < \epsilon, \tag{2.15a}$$

$$\begin{bmatrix} \gamma W & \star & \star & \star & \star \\ 0 & \Gamma & \star & \star & \star \\ 0 & 0 & \Upsilon & \star & \star \\ 0 & 0 & 0 & \Omega & \star \\ (W-YC)TA & (W-YC)TD & (W-YC)NF & YF & W \end{bmatrix} \succ 0. \tag{2.15b}$$

Proof By combining (2.13) and (2.14), we have

$$\max_{z \in \mathbf{B}^{(q+n_x+2n_y)}} \left\| \hat{H}(k+1)z \right\|_{2,W}^2 - \max_{\hat{z} \in \mathbf{B}^q} \gamma \left\| \bar{H}(k)\hat{z} \right\|_{2,W}^2 - \epsilon \leq 0. \tag{2.16}$$

Let us set $z = \left[\bar{z}^\top,\ b_1^\top,\ b_2^\top,\ b_3^\top\right]^\top \in \mathbf{B}^{(q+n_w+2n_v)}$ with $\bar{z} \in \mathbf{B}^q$, $b_1 \in \mathbf{B}^{n_w}$, $b_2 \in \mathbf{B}^{n_v}$ and $b_3 \in \mathbf{B}^{n_v}$. Since $\max_{\hat{z} \in \mathbf{B}^q} \left\| \bar{H}(k)\hat{z} \right\|_{2,W}^2 \geq \left\| \bar{H}(k)\bar{z} \right\|_{2,W}^2$, $\forall \bar{z} \in \mathbf{B}^q$, we obtain a sufficient condition of (2.16)

$$\max_{\bar{z} \in \mathbf{B}^q, b_1 \in \mathbf{B}^{n_w}, b_2 \in \mathbf{B}^{n_v}, b_3 \in \mathbf{B}^{n_v}} \left(\left\| \hat{H}(k+1)z \right\|_{2,W}^2 - \gamma \left\| \bar{H}(k)\bar{z} \right\|_{2,W}^2 - \epsilon \right) < 0 \tag{2.17}$$

Then, we obtain a sufficient condition of (2.17)

$$\|\hat{H}(k+1)z\|_{2,W}^2 - \gamma \left\| \bar{H}(k)\bar{z} \right\|_{2,W}^2 - \epsilon < 0, \forall \bar{z} \in \mathbf{B}^q, \forall b_1 \in \mathbf{B}^{n_w}, \forall b_2 \in \mathbf{B}^{n_v}, \forall b_3 \in \mathbf{B}^{n_v}. \tag{2.18}$$

Recall $\hat{H}(k+1)$ in (2.4b) and set $Y = W\Lambda$. Let us denote

$$\tilde{R} = \left[(W - YC)TA,\ (W - YC)TD,\ (W - YC)NF,\ YF\right]. \tag{2.19}$$

Then, (2.18) can be reformulated as

$$\begin{bmatrix} \bar{H}(k)\bar{z} \\ b_1 \\ b_2 \\ b_3 \end{bmatrix}^\top \tilde{R}^\top W^{-1} \tilde{R} \begin{bmatrix} \bar{H}(k)\bar{z} \\ b_1 \\ b_2 \\ b_3 \end{bmatrix} - \gamma \bar{z}^\top \bar{H}(k)^\top W \bar{H}(k)\bar{z} - \epsilon < 0, \tag{2.20}$$

for any $\bar{z} \in \mathbf{B}^q$, $b_1 \in \mathbf{B}^{n_w}$, $b_2 \in \mathbf{B}^{n_v}$ and $b_3 \in \mathbf{B}^{n_v}$. If Γ, Υ and Ω are diagonal positive semi-definite matrices, then we have $b_1^\top \Gamma b_1 = \sum_{i=1}^{n_w} b_1^2 \Gamma_i \leq \mathrm{tr}(\Gamma)$, $b_2^\top \Upsilon b_2 = \sum_{i=1}^{n_v} b_2^2 \Upsilon_i \leq \mathrm{tr}(\Upsilon)$, $b_3^\top \Omega b_3 = \sum_{i=1}^{n_v} b_3^2 \Omega_i \leq \mathrm{tr}(\Omega)$, for any $b_1 \in \mathbf{B}^{n_w}$, $b_2 \in \mathbf{B}^{n_v}$ and $b_3 \in \mathbf{B}^{n_v}$, where Γ_i, Υ_i and Ω_i are each diagonal element of Γ, Υ and Ω. Therefore, we obtain

$$\mathrm{tr}(\Gamma) - b_1^\top \Gamma b_1 \geq 0, \forall b_1 \in \mathbf{B}^{n_w}, \tag{2.21a}$$

$$\mathrm{tr}(\Upsilon) - b_2^\top \Upsilon b_2 \geq 0, \forall b_2 \in \mathbf{B}^{n_v}, \tag{2.21b}$$

$$\mathrm{tr}(\Omega) - b_3^\top \Omega b_3 \geq 0, \forall b_3 \in \mathbf{B}^{n_v}. \tag{2.21c}$$

By adding (2.21) to (2.20), we obtain a sufficient condition of (2.20)

$$\begin{bmatrix} \bar{H}(k)\bar{z} \\ b_1 \\ b_2 \\ b_3 \end{bmatrix}^\top \tilde{R}^\top W^{-1} \tilde{R} \begin{bmatrix} \bar{H}(k)\bar{z} \\ b_1 \\ b_2 \\ b_3 \end{bmatrix} - \gamma \bar{z}^\top \bar{H}(k)^\top W \bar{H}(k)\bar{z} + \mathrm{tr}(\Gamma) - b_1^\top \Gamma b_1$$
$$+ \mathrm{tr}(\Upsilon) - b_2^\top \Upsilon b_2 + \mathrm{tr}(\Omega) - b_3^\top \Omega b_3 - \epsilon < 0.$$

If (2.15a) holds, then we obtain

$$\begin{bmatrix} \bar{H}(k)\bar{z} \\ b_1 \\ b_2 \\ b_3 \end{bmatrix}^\top \left(\tilde{R}^\top W^{-1} \tilde{R} - \begin{bmatrix} \gamma W & 0 & 0 & 0 \\ 0 & \Gamma & 0 & 0 \\ 0 & 0 & \Upsilon & 0 \\ 0 & 0 & 0 & \Omega \end{bmatrix} \right) \begin{bmatrix} \bar{H}(k)\bar{z} \\ b_1 \\ b_2 \\ b_3 \end{bmatrix} < 0.$$

Again, from the above inequality, we have a sufficient condition

$$\begin{bmatrix} \gamma W & 0 & 0 & 0 \\ 0 & \Gamma & 0 & 0 \\ 0 & 0 & \Upsilon & 0 \\ 0 & 0 & 0 & \Omega \end{bmatrix} - \tilde{R}^\top W^{-1} \tilde{R} \succ 0.$$

By using the Schur complement and $\tilde{R}$ in (2.19), we obtain (2.15b). □

Proposition 2.1 (Ultimate bound of the W-radius) *Given the intersection zonotope* $\hat{\mathcal{X}}(k) = \langle \hat{p}(k), \hat{H}(k) \rangle$, $\forall k \in \mathbb{N}$, $\gamma \in (0, 1)$ *and* $\epsilon > 0$. *If the criterion (2.14) holds, then the W-radius of intersection zonotope* $\hat{\mathcal{X}}(k)$ *is ultimately bounded by*

$$\ell_W(\infty) \leq \frac{\epsilon}{1-\gamma}. \tag{2.22}$$

Proof Given $\gamma \in (0, 1)$ and $\epsilon > 0$, we take $k \to \infty$ in (2.14) to obtain $\ell_W(\infty) \leq \gamma \ell_W(\infty) + \epsilon$ that implies the ultimate bound (2.22) of ℓ_W. □

Since (2.22) characterizes an ellipsoid with given $\gamma \in (0, 1)$ and $\epsilon > 0$, in order to minimize the ultimate bound $\ell_W(\infty)$, we can maximize a norm of W. For instance, we choose to maximize $\mathrm{tr}(W)$. Therefore, the optimization problem to find the off-line correction matrix Λ_f can be expressed as

$$\underset{W, Y, \Gamma, \Upsilon, \Omega}{\text{maximize}} \;\; \mathrm{tr}(W), \tag{2.23}$$

subject to (2.15a) and (2.15b).

Then, the optimal solution of the optimization problem (2.23) gives

$$\Lambda_f = W^{-1}Y.$$

To tighten the size of the intersection zonotope during iterations, we also introduce an on-line method to update the correction matrix $\Lambda_o(k)$ with the weighting matrix W obtained by solving (2.23).

Theorem 2.4 *Given the intersection zonotope $\hat{\mathcal{X}}(k) = \langle \hat{p}, \hat{H} \rangle$, $\forall k \in \mathbb{N}$ and the matrix W obtained by solving (2.23). If there exists a diagonal matrix $M \in \mathbb{S}^{n_x}$ such that*

$$\begin{bmatrix} M & \star \\ W\hat{H}(k+1) & W \end{bmatrix} \succ 0, \tag{2.24}$$

then $\ell_W(k+1)$ in (2.13) is bounded by

$$\ell_W(k+1) < \max_{z \in \mathbf{B}^{(q+n_x+2n_y)}} \|Mz\|_2^2. \tag{2.25}$$

Proof According to [3], the vertices of the intersection zonotope $\hat{\mathcal{X}}(k+1)$ can be approximated by using a diagonal matrix. With a diagonal matrix $M \in \mathbb{R}^{n_x \times n_x}$, a sufficient condition of (2.25) can be obtained as

$$z^\top \hat{H}(k+1)^\top W \hat{H}(k+1) z < z^\top M z, \; \forall z \in \mathbf{B}^{(q+n_x+2n_y)}.$$

Then, from this inequality, we have a sufficient condition $M - \hat{H}(k+1)^\top W \hat{H}(k+1) \succ 0$. By applying the Schur complement, we obtain (2.24). □

At each time step, minimizing the size of the intersection zonotope measured by the W-radius ℓ_W can be implemented by minimizing the trace of the diagonal matrix M. Therefore, the on-line updating correction matrix $\Lambda_o(k)$ can be obtained by solving the following optimization problem:

$$\underset{\Lambda}{\text{minimize}} \; \operatorname{tr}(M), \tag{2.26}$$

subject to (2.24).

Then, the optimal solution of the optimization problem (2.26) gives

$$\Lambda_o(k) = \Lambda.$$

Remark 2.2 It is worth mentioning that the off-line correction matrix Λ_f could already be useful for estimating the states. Hence, sometimes Λ_o obtained through the on-line updating implementation with (2.26) does not provide significant improvements since the state estimations are already satisfactory in terms of degrees of freedom of the intersection zonotope defined in (2.4).

2.1.2 Zonotopic Kalman Observer of Discrete-Time Descriptor Systems

We now design a zonotopic Kalman observer for the descriptor system (2.1). Unlike the presented set-membership approach with an implementation of consistency test, this zonotopic observer structure is defined based on a standard Luenberger observer structure.

2.1.2.1 Zonotopic Observer Structure for Descriptor Systems

With a pair of matrices T and N satisfying (2.3), we consider the Luenberger observer structure for the descriptor system (2.1) in a prediction type [4] as

$$\begin{aligned}\hat{x}(k+1) &= TA\hat{x}(k) + TBu(k) + TDw(k) + Ny(k+1)\\ &\quad - NFv(k+1) + G(k)\left(y(k) - C\hat{x}(k) - Fv(k)\right),\end{aligned} \tag{2.27}$$

where $\hat{x} \in \mathbb{R}^{n_x}$ denotes the estimated state vector, $G \in \mathbb{R}^{n_x \times n_y}$ denotes a time-varying observer gain.

For the descriptor system (2.1), we would like to bound the uncertain system states $x(k)$, $\forall k \in \mathbb{N}$ in a zonotopic set. A suitable observer gain $G(k)$ is used to reduce the state estimation error with a measured output $y(k)$. We first recursively define the structure of the zonotopic observer.

Theorem 2.5 (Prediction-type zonotopic observer for descriptor systems) *Given the descriptor system in (2.1), measured outputs $y(k)$, $y(k+1)$, $x(0) \in \mathcal{X}(0)$, $w(k) \in \mathcal{W}$, $v(k) \in \mathcal{V}$, $\forall k \in \mathbb{N}$, $x(k) \in \langle \hat{p}(k), \hat{H}(k)\rangle \subseteq \langle \hat{p}(k), \bar{H}(k)\rangle$ with $\bar{H}(k) = \downarrow_{q,W} (\hat{H}(k))$, $T \in \mathbb{R}^{n_x \times n_x}$ and $N \in \mathbb{R}^{n_x \times n_y}$ satisfying (2.3). The zonotope bounding uncertain states can be recursively defined by $x(k+1) \in \hat{\mathcal{X}}(k+1) = \langle \hat{p}(k+1), \hat{H}(k+1)\rangle$, where*

$$\hat{p}(k+1) = (TA - G(k)C)\,\hat{p}(k) + TBu(k) + G(k)y(k) + Ny(k+1), \tag{2.28a}$$

$$\hat{H}(k+1) = \left[(TA - G(k)C)\,\bar{H}(k),\ TD,\ -NF,\ -G(k)F\right]. \tag{2.28b}$$

Proof Considering $x(k) \in \langle \hat{p}(k), \bar{H}(k)\rangle$, we set $\hat{x}(k) = x(k) \in \langle \hat{p}(k), \bar{H}(k)\rangle$. Since $w(k) \in \mathcal{W}$, $v(k) \in \mathcal{V}$, $\forall k \in \mathbb{N}$, from (2.27), we can derive that

$$\begin{aligned}x(k+1) \in \hat{\mathcal{X}}(k+1) &= \langle \hat{p}(k+1), \hat{H}(k+1)\rangle\\ &= \left((TA - G(k)C)\langle \hat{p}(k), \bar{H}(k)\rangle\right) \oplus (TB\langle u(k), 0\rangle)\\ &\quad \oplus (G(k)\langle y(k), 0\rangle) \oplus (N\langle y(k+1), 0\rangle) \oplus \left(TD\langle 0, I_{n_w}\rangle\right)\\ &\quad \oplus \left((-NF)\langle 0, I_{n_v}\rangle\right) \oplus \left((-G(k)F)\langle 0, I_{n_v}\rangle\right).\end{aligned}$$

By applying properties in (1.9) to the above equation, we obtain $\hat{p}(k+1)$ and $\hat{H}(k+1)$ as in (2.28). □

From the state bounding zonotope in (2.28), the state estimation error $\varepsilon(k+1)$ is bounded by the zonotope $\varepsilon(k+1) = x(k+1) - \hat{p}(k+1) \in \mathcal{E}_x(k+1) := \langle 0, \hat{H}(k+1)\rangle$. The objective for the zonotopic observer design is to find a time-varying observer gain $G(k)$ to minimize the estimation error, that corresponds to the size of $\mathcal{E}_x(k+1)$.

2.1.2.2 Optimal Kalman Observer Gain for Descriptor Systems

As in Theorem 2.2, the minimization criterion is based on the F_W-radius. The optimal observer gain $G^*(k)$ can be found by minimizing the F_W-radius of $\mathcal{E}_x(k+1)$, that is minimizing $\tilde{J} = \mathrm{tr}\left(W\tilde{P}(k+1)\right)$ with $\tilde{P}(k+1) = \hat{H}(k+1)\hat{H}(k+1)^\top$.

Theorem 2.6 (Optimal Kalman observer gain for descriptor systems) *Given $\mathcal{E}_x(k+1) = \langle 0, \hat{H}(k+1)\rangle$ with $\hat{H}(k+1)$ in (2.28b) and any weighting matrix $W \in \mathbb{S}_{\succ 0}$. The optimal observer gain $G^*(k)$ minimizes $\tilde{J} = \mathrm{tr}\left(W\tilde{P}(k+1)\right)$ and its explicit solution is given by*

$$G^*(k) = TAK(k), \tag{2.29}$$

$$K(k) = \tilde{L}(k)\tilde{S}(k)^{-1}, \tag{2.30}$$

$$\tilde{L}(k) = \tilde{P}(k)C^\top, \tag{2.31}$$

$$\tilde{S}(k) = C\tilde{P}(k)C^\top + Q_v, \tag{2.32}$$

with $\tilde{P}(k) = \bar{H}(k)\bar{H}(k)^\top$ and $Q_v = FF^\top$.

Proof Based on (2.28b), we can derive that

$$\begin{aligned}\tilde{P}(k+1) = (TA - G(k)C)\bar{H}(k)\bar{H}(k)^\top(TA - G(k)C)^\top + TDD^\top T^\top \\ + NFF^\top N^\top + G(k)FF^\top G(k)^\top.\end{aligned}$$

Since $\tilde{J}$ is convex with respect to $G(k)$, $G^*(k)$ is the value of $G(k)$ such that $\frac{\partial}{\partial G}\mathrm{tr}\left(W\tilde{P}(k+1)\right) = 0$. By setting $\tilde{L}(k)$ and $\tilde{S}(k)$ as in (2.31) and (2.32), we have that

$$\frac{\partial}{\partial G(k)} tr\left(WG(k)\tilde{S}(k)G(k)^\top\right) - 2\frac{\partial}{\partial G(k)} tr\left(WTA\tilde{L}(k)G(k)^\top\right) = 0.$$

Due to the symmetry of $\tilde{S}(k)$, by using (1.14a) and (1.14b), we obtain $WG^*(k)\tilde{S}(k) = WTA\tilde{L}(k)$. Set $K(k)$ as in (2.30). Thus, $G^*(k)$ can be found as in (2.29). □

From Theorem 2.6, G^* is also independent of the weighting matrix W. To make use of $\downarrow_{q,W}(\cdot)$, a weighting matrix W is required. One selection of W is proposed in the following proposition.

Proposition 2.2 *Given the nominal descriptor system*

$$\begin{aligned} Ex(k+1) &= Ax(k) + Bu(k), \\ y(k) &= Cx(k), \end{aligned}$$

with matrices $T \in \mathbb{R}^{n_x \times n_x}$ and $N \in \mathbb{R}^{n_x \times n_y}$ satisfying (2.3a). The Luenberger observer defined by

$$\hat{x}(k+1) = TA\hat{x}(k) + TBu(k) + \bar{G}\left((y(k) - C\hat{x}(k)\right) + Ny(k+1),$$

is μ-stable (stable with a decay rate μ) if there exists $W \in \mathbb{S}^{n_x}_{\succ 0}$, $Y \in \mathbb{R}^{n_x \times n_y}$, and a scalar $\mu \in (0, 1]$ such that

$$\begin{bmatrix} \mu W & \star \\ WTA - YC & W \end{bmatrix} \succ 0. \tag{2.33}$$

Proof With matrices T and N satisfying (2.3a), the nominal system dynamics can be expressed as

$$x(k+1) = TAx(k) + TBu(k) + Ny(k+1).$$

Define the state estimation error $e(k) = x(k) - \hat{x}(k)$. Then, we have the error dynamics $e(k+1) = x(k+1) - \hat{x}(k+1) = (TA - \bar{G}C)e(k)$. With $W \in \mathbb{S}^{n_x}_{\succ 0}$, the Lyapunov candidate function is chosen as $V(k) = e(k)^\top We(k)$. With $\mu \in (0, 1]$, we have that

$$\begin{aligned} \Delta V(k) &= e(k+1)^\top We(k+1) - e(k)^\top \mu We(k) \\ &= e(k)^\top (TA - \bar{G}C)^\top W(TA - \bar{G}C)e(k) - e(k)^\top \mu We(k). \end{aligned}$$

For $e(k) \neq 0$, $\Delta V(k) < 0$ gives a sufficient condition $\mu W - \left(TA - \bar{G}C\right)^\top W\left(TA - \bar{G}C\right) \succ 0$. By applying the Schur complement with $\mu W \succ 0$ and $Y = W\bar{G}$, we obtain (2.33). □

For the nominal descriptor system $Ex(k+1) = Ax(k) + Bu(k)$, a nominal observer gain without taking into account system uncertainties can also be found by satisfying (2.33) with $\bar{G} = W^{-1}Y$. We will use $\bar{G}$ with the zonotopic observer structure defined in (2.28) to compare with the optimal Kalman gain $G^*(k)$ in order to assess the state bounding performance.

2.1.3 Discussions on Set-Membership Approach and Zonotopic Kalman Observer

2.1.3.1 Relationship Between the Proposed Approaches

Comparing the parameterized intersection zonotope structure proposed in Theorem 2.1 and the zonotopic observer structure proposed in Theorem 2.5, the intersection zonotope is formulated by considering the measurement output $y(k+1)$ to implement the system consistency test while the zonotopic observer includes measurement outputs $y(k)$ and $y(k+1)$.

To find the relationship between these two approaches, we also consider a current estimation-type zonotopic observer for the descriptor system (2.1) only containing the current measurement output $y(k+1)$ as follows:

$$\begin{aligned}\hat{x}(k+1) &= TA\hat{x}(k) + TBu(k) + TDw(k) + Ny(k+1) \\ &\quad - NFv(k+1) + \hat{G}\left(y(k+1) - C\check{x}(k+1) - Fv(k+1)\right),\end{aligned} \tag{2.34}$$

where $\hat{G} \in \mathbb{R}^{n_x \times n_y}$ is an observer gain for the current estimation-type zonotopic observer. $\check{x}(k+1)$ denotes the predicted state from the previous observed state $\hat{x}(k)$ that can be defined by

$$\check{x}(k+1) = TA\hat{x}(k) + TBu(k) + TDw(k) + Ny(k+1) - NFv(k+1). \tag{2.35}$$

Theorem 2.7 *Consider the descriptor system (2.1). The proposed set-membership approach is equivalent to the current estimation-type zonotopic observer in the structure of (2.34).*

Proof In terms of the zonotopic observer in the current estimation-type, by substituting $\check{x}(k+1)$ by (2.35) to (2.34), we can derive

$$\begin{aligned}\hat{x}(k+1) &= \left(I - \hat{G}C\right)TA\hat{x}(k) + \left(I - \hat{G}C\right)TBu(k) + \left(I - \hat{G}C\right)TDw(k) \\ &\quad + \left(N + \hat{G} - \hat{G}CN\right)y(k+1) - \left(I - \hat{G}C\right)NFv(k+1) - \hat{G}Fv(k+1).\end{aligned}$$

Considering $x(k) \in \langle \hat{p}(k), \bar{H}(k)\rangle$ with $\bar{H}(k) = \downarrow_{q,W}(\hat{H}(k))$, $w(k) \in \mathcal{W}$ and $v(k+1) \in \mathcal{V}$, the uncertain state $x(k+1)$ is bounded into the zonotope $\tilde{\mathcal{X}}(k+1) = \langle \tilde{p}(k+1), \tilde{H}(k+1)\rangle$, where

$$\begin{aligned}x(k+1) \in \tilde{\mathcal{X}}(k+1) &= \langle \tilde{p}(k+1), \tilde{H}(k+1)\rangle \\ &= \left((I - \hat{G}C)TA\langle p(k), \bar{H}(k)\rangle\right) \oplus \left((I - \hat{G}C)TB\langle u(k), 0\rangle\right) \\ &\quad \oplus \left((I - \hat{G}C)TD\langle 0, I_{n_w}\rangle\right) \oplus \left((N + \hat{G} - \hat{G}CN)\langle y(k+1), 0\rangle\right)\end{aligned}$$

$$\oplus\left(\left(-(I-\hat{G}C)\right)NF\langle 0, I_{n_v}\rangle\right)\oplus\left((-\hat{G}F)\langle 0, I_{n_v}\rangle\right).$$

By using properties in (1.9), we obtain $\tilde{p}(k+1)$ and $\tilde{H}(k+1)$ as follows:

$$\tilde{p}(k+1) = (I-\hat{G}C)TAp(k) + (I-\hat{G}C)TBu(k) + (N+\hat{G}-\hat{G}CN)y(k+1), \tag{2.36a}$$

$$\tilde{H}(k+1) = \left[(I-\hat{G}C)TA\bar{H}(k),\ (I-\hat{G}C)TD,\ -(I-\hat{G}C))NF,\ -\hat{G}F\right]. \tag{2.36b}$$

By definition of the zonotope, the subtraction sign in the last two terms of (2.36b) can be removed. Therefore, (2.4) and (2.36) are equivalent with $\Lambda = \hat{G}$. □

Remark 2.3 Since the structure of $\langle \tilde{p}(k+1), \tilde{H}(k+1)\rangle$ is equivalent to the intersection zonotope $\langle \hat{p}(k+1), \hat{H}(k+1)\rangle$ in (2.4), the observer gain $\hat{G}$ can be obtained by using methods proposed for the set-membership approach in Sect. 2.1.1.

2.1.3.2 Extension to Dynamical Systems with Unknown Inputs

In the case of $\text{rank}(E) = n_x$, the system (2.1) becomes a dynamical system. The unknown input d can be decoupled by finding matrices $\bar{T} \in \mathbb{R}^{n_x \times n_x}$ and $\bar{N} \in \mathbb{R}^{n_x \times n_y}$ that satisfy

$$\bar{T} + \bar{N}C = I_{n_x}, \tag{2.37a}$$

$$\bar{T}D_d = 0. \tag{2.37b}$$

By combining (2.37a) and (2.37b), we obtain $D_d = \bar{N}CD_d$ and $\bar{T} = I_{n_x} - \bar{N}C$. Assume D_d to be full column rank. The condition to guarantee the existence of $\bar{T}$ and $\bar{N}$ is given by $\text{rank}(D_d) = \text{rank}(CD_d)$. In this case, the proposed set-membership approach and zonotopic Kalman observer in Sects. 2.1.1 and 2.1.2 can be applied to dynamical systems subject to unknown inputs, which can be considered an improvement on the methods presented in [2, 5]. Under this structure with $\bar{T}$ and $\bar{N}$, the effects of unknown inputs can be decoupled. In this way, the limitation of zonotope-based approach that requires the system disturbances to be bounded is relaxed.

2.1.4 Numerical Example

To illustrate the proposed state estimation approaches, consider a discrete-time descriptor system as defined in (2.1) with

$$E = \begin{bmatrix} 1 & 0 & 0 \\ 0 & 1 & 0 \\ 0 & 0 & 0 \end{bmatrix}, A = \begin{bmatrix} 0.5 & 0 & 0 \\ 0.8 & 0.95 & 0 \\ -1 & 0.5 & 1 \end{bmatrix}, B = \begin{bmatrix} 1 & 0 \\ 0 & 1 \\ 0 & 0 \end{bmatrix}, D_d = \begin{bmatrix} 0 \\ 0 \\ 0.8 \end{bmatrix},$$

$$C = \begin{bmatrix} 1 & 0 & 1 \\ 1 & -1 & 0 \end{bmatrix}, D = \begin{bmatrix} 0.1 & 0 & 0 \\ 0 & 1.5 & 0 \\ 0 & 0 & 0.6 \end{bmatrix}, F = \begin{bmatrix} 0.5 & 0 \\ 0 & 1.5 \end{bmatrix}.$$

and the known input signal is given by $u(k) = \begin{bmatrix} 0.5\sin(0.01\pi k) + 1 \\ -2\cos(0.01\pi k) \end{bmatrix}$, for 100 sampling steps. The system disturbances $w(k)$ and measurement noise $v(k)$ are random Gaussian white noise bounded in zonotopes: $w(k) \in \mathcal{W} = \langle 0, I_3 \rangle$ and $v(k) \in \langle 0, I_2 \rangle$, $\forall k \in \mathbb{N}$.

Since E, C and D_d satisfy the rank condition (2.2), there exists a solution of matrices T and N satisfying (2.3). Therefore, we choose one feasible solution as follows:

$$T = \begin{bmatrix} 0.6667 & 0.3333 & 0 \\ 0.3333 & 0.6667 & 0 \\ -0.6667 & -0.3333 & 0 \end{bmatrix}, N = \begin{bmatrix} 0 & 0.3333 \\ 0 & -0.3333 \\ 1 & -0.3333 \end{bmatrix}.$$

The initial state zonotope $\mathcal{X}(0)$ is given by $\mathcal{X}(0) = \langle p(0), H(0) \rangle$, where

$$p(0) = \begin{bmatrix} 0.5 \\ 0.5 \\ 0.25 \end{bmatrix}, \quad H(0) = \begin{bmatrix} 0.1 & 0 & 0 \\ 0 & 1.5 & 0 \\ 0 & 0 & 0.6 \end{bmatrix}$$

The actual initial state vector $x(0) = [0.5,\ 0.5,\ 0.25]^{\mathrm{T}}$ is considered unknown for the state estimation scheme. We choose $q = 15$ in the zonotope reduction operator to reduce the computation load and simulation time. Simulations have been carried out in a PC with the CPU of Intel (R) Core (TM) i7-5500U 2.4GHz, 12GB RAM and MATLAB R2015a. As a result, the state estimation results are shown in Figs. 2.2 and 2.3. These plots show that both the set-membership approach and the zonotopic Kalman observer are able to provide the interval-based state estimation results.

Recall that $\Lambda^*(k)$ behaves as the Kalman correction matrix, Λ_f is obtained by solving the off-line optimization problem (2.23), $\Lambda_o(k)$ is obtained by solving the on-line optimization problem (2.26), $G^*(k)$ is the optimal Kalman gain and $\bar{G}$ with $\mu = 1$ is the nominal observer gain of the prediction-type zonotopic Kalman observer, and $\hat{G}$ is the optimal Kalman gain of the current estimation type. Besides, the optimal weighting matrix W^* is obtained also by solving (2.23). The observation error is defined as $e(k) = x(k) - \hat{x}(k) = x(k) - p(k)$, where $\hat{x}(k) \in \langle p(k), H(k) \rangle$ and the subscript : represents any time instant $k \in \mathbb{N}$. The root mean square error (MSE) between the real uncertain states and observed states can be computed by

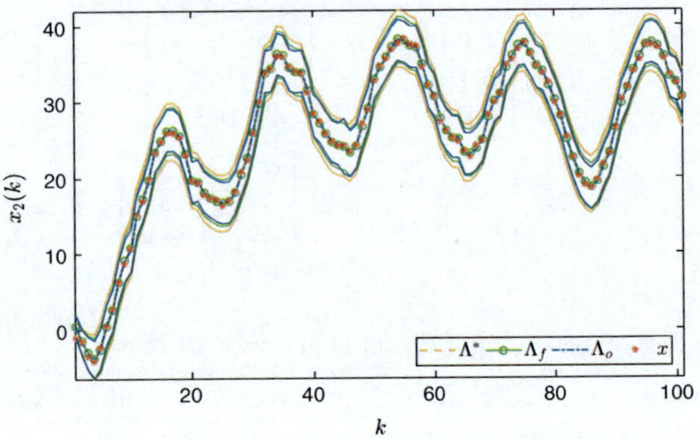

Fig. 2.2 Result of applying the set-membership approach

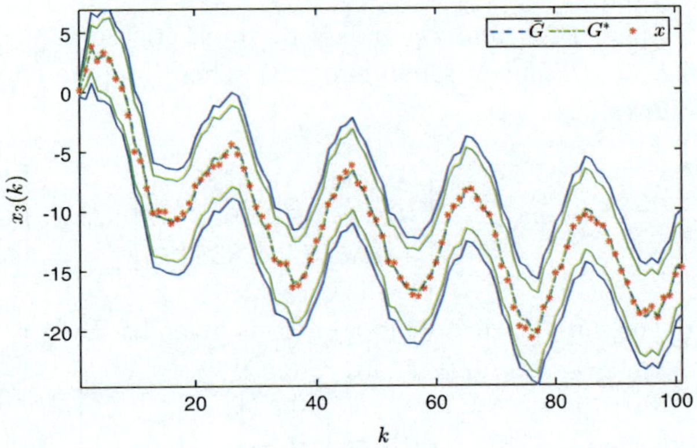

Fig. 2.3 Result of applying the prediction-type zonotopic Kalman observer

$$MSE(e) := \sqrt{\left(\frac{1}{N}\sum_{k=1}^{N}\frac{1}{n_x}\|e(k)\|_2^2\right)}.$$

Besides, we also compute the root mean squared (RMS) value of $rs(H(k))$ that is denoted by RMS($rs(H)$). The weighted and non-weighted Frobenius norm as well as the weighted 2-norm of the segment matrix of zonotopes are computed to compare the sizes of the state zonotopes for all the scenarios. Tables 2.1 and 2.2 show the comparison results of all the cases using RMS up to the step 100 with weighted and non-weighted zonotope reduction operator.

From the $MSE(e)$ results of Λ^* and $G^*(k)$ in Table 2.1, the performance of the set-membership approach is better than the zonotopic Kalman observer in prediction-type. This is because the prediction-type observer structure includes two consecutive-step measurement outputs and noise. Both the measurement noise $v(k)$ and $v(k+1)$ should be over-approximated by the terms $-NF$ and $-G(k)F$ in (2.28b). Hence, this could enlarge the size of the zonotope and gives more conservative estimation intervals. In terms of the real-time implementation of control loops, in both proposed approaches, the estimate $x(k+1)$ depends on $y(k+1)$. Hence, a state feedback

Table 2.1 Comparison with weighted zonotope reduction operator $\downarrow_{q,W}(H)$

Approach	Λ/G	$MSE(e)$	$\mathrm{RMS}(rs(H))$	$\|H\|_F$	$\|H\|_{F,W}$	$\|H\|_{2,W}$	Time (s)
Set-membership approach	Λ^*	0.0355	3.0963	1.5485	0.7070	–	–
	Λ_f	0.0326	2.5443	–	–	14.1385	3.809
	Λ_o	0.0287	2.4755	–	–	14.1373	46.238
Zonotopic observer	G^*	0.2528	3.8781	1.9694	1.9594	–	–
	$\bar{G}$	0.3027	5.0544	2.2605	2.2490	–	–
	$\hat{G}$	0.0355	3.0963	1.5485	0.7070	–	–

Table 2.2 Comparison with non-weighted zonotope reduction operator $\downarrow_q(H)$

Approach	Λ/G	$MSE(e)$	$\mathrm{RMS}(rs(H))$	$\|H\|_F$	$\|H\|_{F,W}$	$\|H\|_{2,W}$	Time (s)
Set-membership approach	Λ^*	0.0539	3.1970	1.5110	0.7070	–	–
	Λ_f	0.0404	2.5542	–	–	14.1385	2.909
	Λ_o	0.0365	2.5466	–	–	14.1373	44.158
Zonotopic observer	G^*	0.2118	3.9386	1.9737	1.2672	–	–
	$\bar{G}$	0.2335	5.0927	2.2357	1.9308	–	–
	$\hat{G}$	0.0539	3.1970	1.5110	0.7070	–	–

control like $u(k+1) = Kx(k+1)$ cannot be applied at the same time as $y(k+1)$ is acquired. However, this real-time synchronization difficulty does not exist when implementing a control loop based on the zonotopic Kalman filter in prediction-type form for dynamical systems as proposed in [2]. Hence, a real-time synchronization remains an open problem when implementing a state feedback control loop with the proposed state estimators for descriptor systems. The possible option can be realized by using a delayed state feedback control.

From the results with Λ_f and $\Lambda_o(k)$, the mean-square error and the size of the intersection zonotopes using the on-line method are smaller than the one using the off-line method. According to $\mathrm{RMS}(rs(H))$ and $\|H\|_{2,W}$ of the set-membership approach, the on-line method improves the correction matrix Λ with the weighting matrix W computed off-line. Since the optimization problem (2.26) is implemented on-line, the simulation time is longer than the off-line method. For the prediction-type zonotopic Kalman observer, the optimal Kalman gain $G^*(k)$ deals with uncertainties better than the nominal observer gain $\bar{G}$.

Besides, by comparing the first and last rows of Tables 2.1 and 2.2, it is numerically shown that the set-membership approach is equivalent to the current estimation-type zonotopic Kalman observer as the discussion in Theorem 2.7. From Table 2.2, all the approaches are run with non-weighted zonotope reduction operator $\downarrow_q(H)$. From results of $\mathrm{RMS}(rs(H))$, the size of each zonotope is larger than the case with $\downarrow_{q,W}(H)$. This is because the non-weighted zonotope reduction operator can bring more over-approximated results.

2.2 Extension of Set-Membership Approach for LPV Descriptor Systems

In this section, we extend the proposed set-membership approach for LPV descriptor systems with a new defined zonotope minimization criterion based on the $\mathcal{L}_\infty$ norm. Let us consider the following LPV descriptor system

$$Ex(k+1) = A\big(\theta(k)\big)x(k) + B\big(\theta(k)\big)u(k) + D\big(\theta(k)\big)d(k), \tag{2.38a}$$

$$y(k) = Cx(k) + Fd(k), \tag{2.38b}$$

where $x \in \mathbb{R}^{n_x}$, $u \in \mathbb{R}^{n_u}$ and $y \in \mathbb{R}^{n_y}$ denote the state, input and output vectors, respectively. $d \in \mathbb{R}^{n_d}$ denotes the system uncertainty vector. $C \in \mathbb{R}^{n_y \times n_x}$ and $F \in \mathbb{R}^{n_y \times n_d}$ are measurement matrices. Besides, $E \in \mathbb{R}^{n_x \times n_x}$ is a singular matrix corresponding to the definition of the descriptor system. As expressed in [6], $A\big(\theta(k)\big)$, $B\big(\theta(k)\big)$ and $D\big(\theta(k)\big)$ are expressed in the polytopic form

$$A\big(\theta(k)\big) = \sum_{j=1}^{h} \rho_j\left(\theta(k)\right) A_j,$$

$$B\big(\theta(k)\big) = \sum_{j=1}^{h} \rho_j\left(\theta(k)\right) B_j,$$

$$D\big(\theta(k)\big) = \sum_{j=1}^{h} \rho_j\left(\theta(k)\right) D_j,$$

where $A_j \in \mathbb{R}^{n_x \times n_x}$, $B_j \in \mathbb{R}^{n_x \times n_u}$, $D_j \in \mathbb{R}^{n_x \times n_d}$, $\forall j = 1, \ldots, h$ are known constant matrices. $\theta(k) \in \mathbb{R}^{n_\theta}$ is a scheduling vector that can be measured online and $\rho_j\left(\theta(k)\right), \forall j = 1, \ldots, h$ are weighting functions satisfying

$$\rho_j\left(\theta(k)\right) \geq 0, \quad \sum_{j=1}^{h} \rho_j\left(\theta(k)\right) = 1, \ \forall j = 1, \ldots, h. \tag{2.39}$$

Compared to (2.1), the system uncertainty vector d in (2.38) can be divided as $d := \left[w^\top, v^\top\right]^\top \in \mathbb{R}^{n_d}$ with $n_d = n_w + n_v$, where $w \in \mathbb{R}^{n_w}$ and $v \in \mathbb{R}^{n_v}$ are the vectors of system disturbances and measurement noise. Besides, $D\big(\theta(k)\big) = \left[\bar{D}\big(\theta(k)\big), 0\right]$, $F = \left[0, \bar{F}\right]$ with $\bar{D}\big(\theta(k)\big) \in \mathbb{R}^{n_x \times n_w}$ and $\bar{F} \in \mathbb{R}^{n_y \times n_v}$.

Assumption 2.4 The uncertainty vector $d(k)$ is unknown but bounded in a known centered zonotope $\mathcal{D}$ as $d(k) \in \mathcal{D} = \langle 0, H_d \rangle$, $\forall k \in \mathbb{N}$ and the initial uncertain state $x(0)$ is also bounded in the zonotope $\mathcal{X}(0) = \langle p(0), H(0) \rangle$.

Remark 2.4 Since $d(k) \in \mathcal{D}, \forall k \in \mathbb{N}$, the worst-case $d(k)$ on the boundary of $\mathcal{D} = \langle 0, H_d \rangle$ is given by $\max_{b \in \mathbf{B}^{n_d}} \|H_d b\|$. Meanwhile, by definition, the $\mathcal{L}_\infty$ norm of $d(k)$ is denoted by $\|d\|_\infty = \sup_k \|d(k)\|$, which satisfies

$$\|d\|_\infty = \sup_k \|d(k)\| = \max_{b \in \mathbf{B}^{n_d}} \|H_d b\| . \tag{2.40}$$

Definition 2.4 (*C-Observability* [7]) The LPV descriptor system (2.38) is said to be C-observable if the initial state $x(0)$ of the system can be uniquely determined by $u(k)$ and $y(k)$, $\forall k \in \mathbb{N}$.

Assumption 2.5 The LPV descriptor system (2.38) is assumed to be C-observable. Then, matrices E, A_j and C satisfy $\operatorname{rank} \begin{bmatrix} zE - A \\ C \end{bmatrix} = n_x, \forall j = 1, \ldots, h, \forall z \in \mathbb{C}$, z finite and

$$\operatorname{rank} \begin{bmatrix} E \\ C \end{bmatrix} = n_x . \tag{2.41}$$

Lemma 2.1 ([8]) *Since (2.41) holds, there exist two matrices T and N such that the following condition is satisfied*

$$TE + NC = I_{n_x} . \tag{2.42}$$

Then, the general solutions of T and N are given by

$$T = \Psi^\dagger \alpha_1 + S \left(I_{n_x + n_y} - \Psi \Psi^\dagger \right) \alpha_1, \tag{2.43a}$$
$$N = \Psi^\dagger \alpha_2 + S \left(I_{n_x + n_y} - \Psi \Psi^\dagger \right) \alpha_2, \tag{2.43b}$$

with $\Psi = \begin{bmatrix} E \\ C \end{bmatrix}$, $\alpha_1 = \begin{bmatrix} I_{n_x} \\ 0 \end{bmatrix}$ *and* $\alpha_2 = \begin{bmatrix} 0 \\ I_{n_y} \end{bmatrix}$, *where S is an arbitrary matrix of appropriate dimension.*

Based on the result in Sect. 2.1.1, since the uncertain state $x(k-1)$ is bounded in the zonotope $\mathcal{X}(k-1) = \langle p(k-1), H(k-1) \rangle$, the estimated uncertain state $x(k)$ is over-approximated by implementing three steps including prediction, measurement and correction. We now extend Definitions 2.1–2.3 for the LPV descriptor system (2.38).

Definition 2.5 (*Uncertain state set*) Given the LPV descriptor system (2.38) and Assumption 2.4 holds, the uncertain state set $\bar{\mathcal{X}}(k)$ propagated by (2.38a) is defined as

$$\bar{\mathcal{X}}(k) = \Big\{x \in \mathbb{R}^{n_x} \Big| \Big(Ex - B\big(\theta(k-1)\big)u(k-1)\Big) \in \Big(A\big(\theta(k-1)\big)\mathcal{X}(k-1) \oplus D\big(\theta(k-1)\big)\mathcal{D}\Big)\Big\}. \tag{2.44}$$

Definition 2.6 (*Measurement state set*) Given the LPV descriptor system (2.38) and a measured output vector $y(k)$, the measurement consistent state set at time instant k is defined as

$$\mathcal{X}_y(k) = \big\{x \in \mathbb{R}^{n_x} \mid (y(k) - Cx) \in F\mathcal{D}\big\}. \tag{2.45}$$

Definition 2.7 (*Exact uncertain state set*) Given the LPV descriptor system (2.38), the exact consistent uncertain state set $\mathcal{X}(k)$, that encloses uncertain states consistent with measured outputs, is defined as

$$\mathcal{X}(k) = \bar{\mathcal{X}}(k) \cap \mathcal{X}_y(k). \tag{2.46}$$

2.2.1 *Zonotopic Set-Membership Approach for Discrete-Time LPV Descriptor Systems*

We now present the set-membership state estimation approach for discrete-time LPV descriptor systems. Based on the system model (2.38), the prediction step can be implemented using the Minkowski sum and the model information through the forward set propagation. With the output data measured from the real system, the set defined in (2.45) can be obtained in the measurement step. Then, we compute the consistent state set (2.46) by a suitable approximation allowing to implement the consistency test in the correction step.

2.2.1.1 Intersection Zonotope for LPV Descriptor Systems

The set $\mathcal{X}(k)$ defined in (2.46) is a polytope obtained by an intersection between the zonotope $\bar{\mathcal{X}}(k)$ and the polytope $\mathcal{X}_y(k)$. To implement the steps of the set-membership state estimation approach in an iterative way, we first construct a parameterized intersection zonotope to over-approximate $\mathcal{X}(k)$ in the following theorem, which includes the three steps of the set-membership state estimation.

Theorem 2.8 *Consider the LPV descriptor system (2.38), $x(k-1) \in \hat{\mathcal{X}}(k-1) = \langle \hat{p}(k-1), \hat{H}(k-1)\rangle$, the measured output $y(k)$ and the measurement consistent state set $\mathcal{X}_y(k)$. Then, for a parameter-varying correction matrix $\Lambda\,(\theta(k-1)) \in \mathbb{R}^{n_x \times n_y}$, the consistent uncertain state set $\mathcal{X}(k)$ is over-approximated by the zonotope $\hat{\mathcal{X}}(k)$,*

$$\bar{\mathcal{X}}(k) \cap \mathcal{X}_y(k) \subseteq \hat{\mathcal{X}}(k) = \big\langle \hat{p}(k), \hat{H}(k)\big\rangle, \tag{2.47}$$

where

$$\begin{aligned}\hat{p}(k) = &\big(I - \Lambda\left(\theta(k-1)\right) C\big) T A\big(\theta(k-1)\big) \hat{p}(k-1) \\ &+ \big(I - \Lambda\left(\theta(k-1)\right) C\big) T B\big(\theta(k-1)\big) u(k-1) \\ &+ \big(N - \Lambda\left(\theta(k-1)\right) C N + \Lambda\left(\theta(k-1)\right)\big) y(k),\end{aligned} \tag{2.48a}$$

$$\hat{H}(k) = \Big[\big(I - \Lambda\left(\theta(k-1)\right) C\big) R(k),\ \ \Lambda\left(\theta(k-1)\right) F H_d\Big], \tag{2.48b}$$

$$R(k) = \Big[T A\big(\theta(k-1)\big) \hat{H}(k-1),\ \ T D\big(\theta(k-1)\big) H_d,\ \ N F H_d\Big]. \tag{2.48c}$$

Proof For any $\hat{x}(k)$ satisfying $\hat{x}(k) \in \bar{\mathcal{X}}(k) \cap \mathcal{X}_y(k)$, it implies $\hat{x}(k) \in \bar{\mathcal{X}}(k)$ and $\hat{x}(k) \in \mathcal{X}_y(k)$. First, in the prediction step, from $\hat{x}(k) \in \bar{\mathcal{X}}(k)$ and (2.38a), there exists an unitary vector z_1 such that

$$\begin{aligned}E\hat{x}(k) = &A\big(\theta(k-1)\big) \hat{p}(k-1) + B\big(\theta(k-1)\big) u(k-1) \\ &+ \big[A\big(\theta(k-1)\big) \hat{H}(k-1),\ D\big(\theta(k-1)\big) H_d\big] z_1.\end{aligned} \tag{2.49}$$

Therefore, in the measurement step, from $\hat{x}(k) \in \mathcal{X}_y(k)$ and (2.38b), there exists another unitary vector z_2 such that

$$C\hat{x}(k) - y(k) = F H_d z_2. \tag{2.50}$$

With a pair of T and N satisfying (2.42), (2.49) and (2.50) can be combined to obtain

$$\begin{aligned}\hat{x}(k) = &T A\big(\theta(k-1)\big) \hat{p}(k-1) + T B\big(\theta(k-1)\big) u(k-1) + N y(k) \\ &+ \big[T A\big(\theta(k-1)\big) \hat{H}(k-1),\ T D\big(\theta(k-1)\big) H_d,\ N F H_d\big] \begin{bmatrix} z_1 \\ z_2 \end{bmatrix}.\end{aligned}$$

Set $R(k)$ as in (2.48c) and $s = \big[z_1^\top, z_2^\top\big]^\top$. Finally, in the correction step, we introduce a parameter-varying correction matrix $\Lambda\left(\theta(k-1)\right) \in \mathbb{R}^{n_x \times n_y}$ and a correction term $\Lambda\left(\theta(k-1)\right) C R(k) s$ such that substituting $\hat{x}(k)$ in (2.50) by (2.51), it becomes

$$C R(k) s = (I - C N)\, y(k) - C T A\big(\theta(k-1)\big) \hat{p}(k-1) - C T B\big(\theta(k-1)\big) u(k-1) + F H_d z_2.$$

Adding and subtracting this correction term $\Lambda\left(\theta(k-1)\right) C R(k) s$ in (2.51), we can derive that

$$\begin{aligned}\hat{x}(k) &= T A\big(\theta(k-1)\big) \hat{p}(k-1) + T B\big(\theta(k-1)\big) u(k-1) + N y(k) \\ &\quad + \Lambda\left(\theta(k-1)\right) C R(k) s + \left(I - \Lambda\left(\theta(k-1)\right) C\right) R(k) s \\ &= T A\big(\theta(k-1)\big) \hat{p}(k-1) + T B\big(\theta(k-1)\big) u(k-1) + N y(k)\end{aligned}$$

$$
\begin{aligned}
&+ \Lambda\left(\theta(k-1)\right)(I - CN)\,y(k) - \Lambda\left(\theta(k-1)\right)CTA\big(\theta(k-1)\big)\hat{p}(k-1)\\
&- \Lambda\left(\theta(k-1)\right)CTB\big(\theta(k-1)\big)u(k-1) + \left(I - \Lambda\left(\theta(k-1)\right)C\right)R(k)s\\
&+ \Lambda\left(\theta(k-1)\right)FH_d z_2\\
&= \big(I - \Lambda\left(\theta(k-1)\right)C\big)\big(TA\big(\theta(k-1)\big)\hat{p}(k-1) + TB\big(\theta(k-1)\big)u(k-1)\big)\\
&+ \big(N - \Lambda\left(\theta(k-1)\right)CN + \Lambda\left(\theta(k-1)\right)\big)y(k)\\
&+ \big[\left(I - \Lambda\left(\theta(k-1)\right)C\right)R(k),\ \Lambda\left(\theta(k-1)\right)FH_d\big]\begin{bmatrix} s\\ z_2\end{bmatrix},
\end{aligned}
$$

from which we obtain the zonotope $\hat{\mathcal{X}}(k)$ with the center $\hat{p}(k)$ and the generator matrix $\hat{H}(k)$ as in (2.48). □

Remark 2.5 Along an iterative estimation procedure, the order of $\hat{\mathcal{X}}(k)$, $\forall k \in \mathbb{N}$ is growing because at each time step, the term $\Lambda\left(\theta(k-1)\right)FH_d$ is added into $\hat{H}(k)$, $\forall k \in \mathbb{N}$. From the application point of view, the order of the intersection zonotope with time should be limited. To achieve this, we use the reduction operator $\downarrow_{q,W}(\cdot)$ to fix the maximum number of columns of the intersection zonotope to preserve the inclusion property

$$
\langle \hat{p}(k-1), \hat{H}(k-1)\rangle \subseteq \langle \hat{p}(k-1), \bar{H}(k-1)\rangle,
$$

with $\bar{H}(k-1) = \downarrow_{q,W}(\hat{H}(k-1))$, where q is maximum column of $\bar{H}(k-1)$ and W denotes a weighting matrix of appropriate dimension.

Considering the polytopic form of the system (2.38), we introduce the polytopic representation of the parameterized intersection zonotope in the following corollary.

Corollary 2.1 *Consider the LPV descriptor system (2.38). If there exists a parameter-varying correction matrix $\Lambda\left(\theta(k-1)\right)$ in a polytopic form*

$$
\Lambda\left(\theta(k-1)\right) = \sum_{i=1}^{h} \rho_i\left(\theta(k-1)\right)\Lambda_i, \tag{2.51}
$$

with $\Lambda_i \in \mathbb{R}^{n_x \times n_y}$ for $i = 1, \ldots, h$, then the intersection zonotope $\hat{\mathcal{X}}(k)$ can be reformulated as follows:

$$
\begin{aligned}
\hat{p}(k) = &\sum_{i=1}^{h}\sum_{j=1}^{h} \rho_i\left(\theta(k-1)\right)\rho_j\left(\theta(k-1)\right)\Big((I - \Lambda_i C)TA_j\hat{p}(k-1) + (I - \Lambda_i C)TB_j u(k-1)\Big)\\
&+ \sum_{i=1}^{h} \rho_i\left(\theta(k-1)\right)\Big((N - \Lambda_i CN + \Lambda_i)y(k)\Big),
\end{aligned} \tag{2.52a}
$$

$$
\hat{H}(k) = \Bigg[\sum_{i=1}^{h}\sum_{j=1}^{h} \rho_i\left(\theta(k-1)\right)\rho_j\left(\theta(k-1)\right)\Big((I - \Lambda_i C)TA_j\bar{H}(k-1)\Big),
$$

$$\sum_{i=1}^{h}\sum_{j=1}^{h}\rho_i\left(\theta(k-1)\right)\rho_j\left(\theta(k-1)\right)\Big((I-\Lambda_i C)TD_jH_d\Big),$$

$$\sum_{i=1}^{h}\rho_i\left(\theta(k-1)\right)\Big((I-\Lambda_i C)NFH_d\Big),$$

$$\left.\sum_{i=1}^{h}\rho_i\left(\theta(k-1)\right)\Lambda_i FH_d\right]. \tag{2.52b}$$

Proof Based on (2.39), $A(\theta(k-1))$, $B(\theta(k-1))$ and $D(\theta(k-1))$ can be reformulated by

$$\begin{bmatrix} A(\theta(k-1)) \\ B(\theta(k-1)) \\ D(\theta(k-1)) \end{bmatrix} = \sum_{j=1}^{h}\rho_j\left(\theta(k-1)\right)\begin{bmatrix} A_j \\ B_j \\ D_j \end{bmatrix},\ j=1,\ldots,h, \tag{2.53}$$

with $\rho_j\left(\theta(k-1)\right)\geq 0$ and $\sum_{j=1}^{h}\rho_j\left(\theta(k-1)\right)=1$. By combining (2.47) with (2.51) and (2.53), we obtain (2.52). □

2.2.1.2 Computing Optimal Correction Matrix

Since all uncertain states are bounded in the intersection zonotope $\hat{\mathcal{X}}(k)$, we would like to find a suitable correction matrix $\Lambda\left(\theta(k-1)\right)$ in such a way that the size of $\hat{\mathcal{X}}(k)$ is limited. As presented in Sect. 2.1.1, the size of a zonotope can be measured by the W-radius. Based on Definitions 1.9 and 1.10, we propose the condition to limit the size of $\hat{\mathcal{X}}(k)$ in the following theorem.

Theorem 2.9 *Consider the LPV descriptor system (2.38) and $\hat{\mathcal{X}}(k)$ in (2.47). If there exists a positive scalar $\gamma>0$, a matrix $W\in\mathbb{R}^{n_x}$ and a parameter-varying correction matrix $\Lambda\left(\theta(k-1)\right)\in\mathbb{R}^{n_x\times n_y}$ such that*

$$\begin{bmatrix} \alpha W & \star & \star & \star \\ 0 & (1-\alpha)\beta I & \star & \star \\ 0 & 0 & (1-\alpha)(1-\beta)I & \star \\ \Phi_1 & \Phi_2 & \Phi_3 & W \end{bmatrix} \succ 0, \tag{2.54a}$$

$$\begin{bmatrix} I & \star \\ \gamma W & W \end{bmatrix} \preceq 0, \tag{2.54b}$$

with

$$\Phi_1 = W\left(I-\Lambda\left(\theta(k-1)\right)C\right)TA(\theta(k-1)),$$
$$\Phi_2 = W\left(I-\Lambda\left(\theta(k-1)\right)C\right)TD(\theta(k-1)),$$
$$\Phi_3 = W\left(NF-\Lambda\left(\theta(k-1)\right)CNF+\Lambda\left(\theta(k-1)\right)F\right),$$

then the parameterized intersection zonotope $\hat{\mathcal{X}}(k)$, $\forall k \in \mathbb{Z}_+$ *satisfies*

$$\ell_W(k) \leq \alpha \ell_W(k-1) + (1-\alpha)\,\ell_d, \tag{2.55a}$$

$$\ell(k) \leq \gamma^2 \ell_W(k), \tag{2.55b}$$

with $\alpha, \beta \in (0, 1)$ *and*

$$\ell_d = \max_{b_1 \in \mathbf{B}^{n_d}} \beta \, \|H_d b_1\|^2 + \max_{b_2 \in \mathbf{B}^{n_d}} (1-\beta)\, \|H_d b_2\|^2 . \tag{2.56}$$

Proof By Definition 1.9, the W-radius of the intersection zonotope $\hat{\mathcal{X}}(k)$ in (2.47) at time instant k can be formulated as

$$\ell_W(k) = \max_{\hat{z} \in \mathbf{B}^{n+2n_d}} \left\| \hat{H}(k)\big(\Lambda\left(\theta(k-1)\right)\big)\hat{z} \right\|_W^2 ,$$

where $\hat{z} \in \mathbf{B}^{n+2n_d}$ is an unitary vector. According to (2.48b), the vector $\hat{z}$ can be partitioned as $\hat{z} = \left[\bar{z}^\top, \; b_1^\top, \; b_2^\top\right]^\top$ with $\bar{z} \in \mathbf{B}^n$. By combining (2.55a) and (2.56), we obtain

$$\begin{aligned}\max_{\hat{z} \in \mathbf{B}^{n+2n_d}} \left\| \hat{H}(k)\hat{z} \right\|_W^2 \leq \max_{\bar{z} \in \mathbf{B}^n} \alpha \left\| \hat{H}(k-1)\bar{z} \right\|_W^2 + \max_{b_1 \in \mathbf{B}^{n_d}} (1-\alpha)\,\beta\,\|H_d b_1\|^2 \\ + \max_{b_2 \in \mathbf{B}^{n_d}} (1-\alpha)\,(1-\beta)\,\|H_d b_2\|^2 ,\end{aligned} \tag{2.57}$$

Then, we obtain a sufficient condition of (2.57) as

$$\begin{aligned}&\hat{z}^\top \hat{H}(k)^\top W \hat{H}(k)\hat{z} - \alpha \bar{z}^\top \hat{H}(k-1)^\top W \hat{H}(k-1)\bar{z} - (1-\alpha)\,\beta b_1^\top H_d^\top H_d b_1 \\ &\quad - (1-\alpha)\,(1-\beta)\, b_2^\top H_d^\top H_d b_2 < 0,\end{aligned} \tag{2.58}$$

for $\forall \hat{z}$, $\forall \bar{z}$, $\forall b_1$ and $\forall b_2$. Set $\xi = \hat{H}(k-1)\bar{z}$, $\phi = H_d b_1$ and $\varphi = H_d b_2$. By substituting $\hat{H}(k)$ defined (2.48b) in (2.58), it follows that

$$\begin{bmatrix} \xi \\ \phi \\ \varphi \end{bmatrix}^\top \underbrace{\begin{bmatrix} w_{11} & \star & \star \\ w_{21} & w_{22} & \star \\ w_{31} & w_{32} & w_{33} \end{bmatrix}}_{w} \begin{bmatrix} \xi \\ \phi \\ \varphi \end{bmatrix} < 0, \tag{2.59}$$

with

$$\begin{aligned}w_{11} &= A\big(\theta(k-1)\big)^\top T^\top \big(I - \Lambda\left(\theta(k-1)\right) C\big)^\top W \big(I - \Lambda\left(\theta(k-1)\right) C\big) T A\big(\theta(k-1)\big) - \alpha W, \\ w_{21} &= D\big(\theta(k-1)\big)^\top T^\top \big(I - \Lambda\left(\theta(k-1)\right) C\big)^\top W \big(I - \Lambda\left(\theta(k-1)\right) C\big) T A\big(\theta(k-1)\big), \\ w_{22} &= D\big(\theta(k-1)\big)^\top T^\top \big(I - \Lambda\left(\theta(k-1)\right) C\big)^\top W \big(I - \Lambda\left(\theta(k-1)\right) C\big) T D\big(\theta(k-1)\big) \\ &\quad - (1-\alpha)\beta I,\end{aligned}$$

$$\begin{aligned}
w_{31} &= \big(NF - \Lambda(\theta(k-1))CNF + \Lambda(\theta(k-1))F\big)^\top W\big(I - \Lambda(\theta(k-1))C\big)TA\big(\theta(k-1)\big),\\
w_{32} &= \big(NF - \Lambda(\theta(k-1))CNF + \Lambda(\theta(k-1))F\big)^\top W\big(I - \Lambda(\theta(k-1))C\big)TD\big(\theta(k-1)\big),\\
w_{33} &= \big(NF - \Lambda(\theta(k-1))CNF + \Lambda(\theta(k-1))F\big)^\top W\\
&\quad \cdot \big(NF - \Lambda(\theta(k-1))CNF + \Lambda(\theta(k-1))F\big) - (1-\alpha)(1-\beta)I.
\end{aligned}$$

By the definition of a positive definite matrix, (2.59) implies $w \prec 0$. By applying Schur complement lemma [9] to this matrix inequality, we obtain (2.54a).

On the other hand, by Definition 1.10, the radius of the intersection zonotope $\hat{\mathcal{X}}(k)$ in (2.47) at time instant k can be formulated as $\ell(k) = \max_{\hat{z} \in \mathbf{B}^{n+2n_d}} \left\| \hat{H}(k)\hat{z} \right\|^2$. From (2.55b), we derive

$$I - \gamma^2 W \preceq 0. \tag{2.60}$$

Again, by applying Schur complement lemma to (2.60), we thus obtain (2.54b). □

Considering that $A(\theta(k-1))$ and $\Lambda(\theta(k-1))$ are defined in the polytopic form, (2.54a) leads to a double-sum problem. The following result is used for the reformulation of a double-sum problem.

Lemma 2.2 ([10, 11]) *Consider the following double sum inequality condition*

$$\Gamma(\vartheta(k), \vartheta(k)) = \sum_{i=1}^{r}\sum_{j=1}^{r} \mu_i(\vartheta(k))\,\mu_j(\vartheta(k))\,\Gamma_{i,j} \succ 0. \tag{2.61}$$

Then, the condition (2.61) is fulfilled provided that the following conditions hold

$$\Gamma_{i,i} \succ 0, \quad i = 1, \ldots, r, \tag{2.62a}$$

$$\frac{2}{r-1}\Gamma_{i,i} + \Gamma_{i,j} + \Gamma_{j,i} \succeq 0, \quad , 1 \le i < j \le r. \tag{2.62b}$$

Based on Lemma 2.2, we now reformulate (2.54a) with multiple vertices in the form of (2.61) in the following corollary.

Corollary 2.2 *Consider the LPV descriptor system (2.38). If there exist matrices $W \in \mathbb{S}^{n_x}_{\succ 0}$ and $Y_i \in \mathbb{R}^{n_x \times n_y}$ for $i = 1, \ldots, h$ such that*

$$\Psi_{i,i} \succ 0, \quad i = 1, \ldots, h, \tag{2.63a}$$

$$\frac{2}{h-1}\Psi_{i,i} + \Psi_{i,j} + \Psi_{j,i} \succeq 0, \quad , 1 \le i < j \le h, \tag{2.63b}$$

with

$$\Psi_{i,j} = \begin{bmatrix} \alpha W & \star & \star & \star \\ 0 & (1-\alpha)\beta I & \star & \star \\ 0 & 0 & (1-\alpha)(1-\beta)I & \star \\ WTA_j - Y_iCTA_j & WTD_j - Y_iCTD_j & WNF - Y_iCNF + Y_iF & W \end{bmatrix}, \tag{2.64}$$

then (2.54a) is satisfied.

Proof For the polytopic representation of $A(\theta(k-1))$, $D(\theta(k-1))$ and $\Lambda(\theta(k-1))$, (2.54a) can be reformulated by the double sum as (2.61). Thus, we obtain (2.63) by means of (2.62). □

Based on the condition in Theorem 2.9, an adaptive bound, that is the upper bound of the radius of the intersection zonotope, can be obtained in the following theorem.

Theorem 2.10 *The $\mathcal{L}_\infty$ performance of the radius of the intersection zonotope $\hat{\mathcal{X}}(k)$ in (2.47) at time instant k is characterized by*

$$\ell(k) \le \gamma^2\alpha^k\ell_W(0) + \gamma^2 \|d\|_\infty^2, \tag{2.65}$$

with $\ell_W(0) = \max\limits_{b(0)\in\mathbf{B}^{n(0)}} \|H(0)b(0)\|_W^2$.

Proof From (2.40) and (2.56), we have $\ell_d = \max\limits_{b\in\mathbf{B}^{n_d}} \|H_db\|^2 = \|d\|_\infty^2$. From (2.55a), for some $\alpha \in (0, 1)$, we can derive that

$$\begin{aligned} \ell_W(k) &\le \alpha\ell_W(k-1) + (1-\alpha)\|d\|_\infty^2, \\ &\le \alpha^k\ell_W(0) + (1-\alpha)\sum_{i=0}^{k-1}\alpha^i \|d\|_\infty^2, \\ &\le \alpha^k\ell_W(0) + \|d\|_\infty^2. \end{aligned}$$

Then, from (2.55b), we obtain

$$\ell(k) \le \gamma^2\ell_W(k) \le \gamma^2\left(\alpha^k\ell_W(0) + \|d\|_\infty^2\right), \tag{2.66}$$

which gives (2.65). □

Remark 2.6 Note that Theorem 2.9 provides a procedure to obtain the most adjusted zonotope that outer-bounds the intersection of the measurement consistent state set $\bar{\mathcal{X}}(k)$ and the consistent uncertain state set $\mathcal{X}_y(k)$. The radius $\ell(k)$ (introduced in Definition 1.10) is used to measure the size of the resulting zonotope. According to Theorem 2.9, this radius satisfies (2.55b). On the other hand, Theorem 2.10 introduces a time-varying bound for this radius considering the worst-case disturbances. As shown in the proof of Theorem 2.10, the relation of this worst-case bound with the one obtained in Theorem 2.9 is given by (2.66) which leads to (2.65). This inequality establishes that the time-varying radius $\ell(k)$ is bounded by $\ell_W(0)$ (from the initial condition), the worst-case disturbance, a given scalar $\alpha \in (0, 1)$ as well as a scalar $\gamma > 0$. As the time k increases, the term α^k is going to be zero. Hence, for $k \geq k_M$ (let us denote k_M as an arbitrary large integer), a worst-case bound for $\ell(k)$ is obtained considering the worst-case disturbance as $\ell(k) \leq \gamma^2 \|d\|_\infty^2$ for $\gamma > 0$.

Based on Remark 2.6, the optimal polytopic correction matrices Λ_i for $i = 1, \ldots, h$ can be found by solving the following optimization problem

$$\underset{W, Y_i}{\text{minimize}} \ \gamma, \tag{2.67}$$

subject to (2.54b) and (2.63) that allows to obtain the least conservative worst case bound of $\ell(k)$.

Then, the optimal solutions of the optimization problem (2.67) give

$$\Lambda_i^* = W^{*-1} Y_i^*, \ i = 1, \ldots, h.$$

Remark 2.7 The constraints in (2.63) are linear and hence convex with given $\alpha, \beta \in (0, 1)$. To deal with term γW in (2.54b), the optimization problem (2.67) can be solved by a linear programming solver with a line search to find the minimum γ.

Remark 2.8 The condition (2.55a) can be replaced by the one in Theorem 2.3, which can be formulated as

$$\ell_W(k) \leq \sigma \ell_W(k-1) + \epsilon, \tag{2.68}$$

with $\sigma \in [0, 1)$ and ϵ is a scalar that can be determined by system uncertainties. From $d(k) = \begin{bmatrix} w(k)^\top, & v(k)^\top \end{bmatrix}^\top \in \mathcal{D}, \forall k \in \mathbb{N}$, we consider that the set $\mathcal{D}$ can be rewritten by the Cartesian product as $\mathcal{D} = \mathcal{W} \times \mathcal{V}$ with $w(k) \in \mathcal{W} = \langle 0, H_w \rangle$ and $v(k) \in \mathcal{V} = \langle 0, H_v \rangle$, $\forall k \in \mathbb{N}$, where H_w and H_v are the segment matrices of appropriate dimensions. Therefore, according to Theorem 2.3, ϵ can be estimated by

$$\epsilon = \max_{\bar{b}_1 \in \mathbf{B}^{n_w}} \left\| H_w \bar{b}_1 \right\|^2 + \max_{\bar{b}_2 \in \mathbf{B}^{n_v}} \left\| H_v \bar{b}_2 \right\|^2 + \max_{\bar{b}_3 \in \mathbf{B}^{n_v}} \left\| H_v \bar{b}_3 \right\|^2.$$

From (2.68), we follow the proof of Theorem 2.9 to obtain

$$\begin{bmatrix} \sigma W & \star & \star & \star & \star \\ 0 & D(\theta(k-1))^\top T^\top T D(\theta(k-1)) & \star & \star & \star \\ 0 & 0 & F^\top N^\top N F & F^\top F & \star \\ \bar{\Phi}_1 & \bar{\Phi}_2 & \bar{\Phi}_3 & \bar{\Phi}_4 & W \end{bmatrix} \succ 0, \tag{2.69}$$

with

$$\begin{aligned}
\bar{\Phi}_1 &= W\left(I - \Lambda\left(\theta(k-1)\right)C\right)TA(\theta(k-1)),\\
\bar{\Phi}_2 &= W\left(I - \Lambda\left(\theta(k-1)\right)C\right)TD(\theta(k-1)),\\
\bar{\Phi}_3 &= W\left(NF - \Lambda\left(\theta(k-1)\right)CNF\right),\\
\bar{\Phi}_4 &= W\Lambda\left(\theta(k-1)\right)F,
\end{aligned}$$

which can also be reformulated to be the polytopic form as presented in Corollary 2.2.

Besides, when time tends to infinity, (2.68) can be bounded by $\ell_W(\infty) \leq \sigma\ell_W(\infty) + \epsilon$ leading to $\ell_W(\infty) \leq \frac{\epsilon}{1-\sigma}$. To minimize the P-radius $\ell_W(\infty)$ of the intersection zonotope (2.47), we can solve an eigenvalue optimization problem with a scalar $\tau > 0$ as follows:

$$\underset{W,Y_i}{\text{maximize}} \;\; \tau, \tag{2.70}$$

subject to $\frac{(1-\sigma)W}{\epsilon} \succeq \tau I$ and the polytopic form of (2.69).

Then, the optimal solutions of the optimization problem (2.70) give

$$\bar{\Lambda}_i^* = W^{*-1}Y_i^*, \; i = 1, \ldots, h.$$

Remark 2.9 The main difference between using criteria (2.55a) and (2.68) is that although the resulting approaches compute the intersection zonotope based on the same structure in Corollary 2.1, the corresponding correction matrices Λ_i^* and $\bar{\Lambda}_i^*$ for $i = 1, \ldots, h$ are obtained using different objectives. In the case of the approach based on (2.55a) proposed in this section, the optimization problem (2.67) seeks to minimize the upper bound of the time-varying radius (Definition 1.10) of the intersection zonotope, while in the approach based on (2.68), the optimization problem (2.70) minimizes the steady W-radius (Definition 1.9) of the intersection zonotope.

Finally, we summarize the proposed set-membership state estimation approach for the discrete-time LPV descriptor system (2.38) in Algorithm 2.1.

2.2.2 *Case Study: The Truck-Trailer Model*

The truck-trailer system [12] is modeled by the system (2.38) in the polytopic form of LPV descriptor system with the following matrices

$$E = \begin{bmatrix} 1 & 0 & 0 & 0 \\ 0 & 1 & 0 & 0 \\ 0 & 0 & 1 & -1 \\ 0 & 0 & 0 & 0 \end{bmatrix}, \; A_1 = \begin{bmatrix} 1.025 & 0 & 0 & 0 \\ -0.218 & 1 & 0 & 0 \\ 0 & 0 & 1 & 1 \\ 0 & 0.06 & 0 & 1 \end{bmatrix},$$

Algorithm 2.1 Set-Membership State Estimation for LPV Descriptor Systems

1: Given $\mathcal{X}(0)$ and $\mathcal{D}$;
2: $\mathcal{X}(k-1) \Longleftarrow \mathcal{X}(0)$;
3: Solve the optimization problem (2.67) (or (2.70))
  to obtain Λ_i^* (or $\bar{\Lambda}_i^*$);
4: **for** $k := 1$: end **do**
5:   Obtain $\theta(k-1)$;
6:   Measure $y(k)$;
7:   Compute the intersection zonotope by (2.52) obtaining
  $\hat{\mathcal{X}}(k) = \langle \hat{p}(k), \hat{H}(k) \rangle$;
8:   Obtain the upper and lower bounds $x_i(k) \in \left[\underline{x}_i(k), \overline{x}_i(k)\right]$ for $i = 1, \ldots, n_x$ by
  $\overline{x}_i(k) = \hat{p}_i(k) + rs\left(\hat{H}(k)\right)_i$,
  $\underline{x}_i(k) = \hat{p}_i(k) - rs\left(\hat{H}(k)\right)_i$,
  where $\hat{p}_i(k)$ is the i-th element of $\hat{p}(k)$ and $rs\left(\hat{H}(k)\right)_i$ returns the i-th diagonal element of $rs\left(\hat{H}(k)\right)$.
9: **end for**

$$A_2 = \begin{bmatrix} 1.05 & 0 & 0 & 0 \\ -0.436 & 1 & 0 & 0 \\ 0 & 0 & 1 & 1 \\ 0 & 0.12 & 0 & 1 \end{bmatrix}, B_1 = \begin{bmatrix} -0.025 \\ 0 \\ 0 \\ 0 \end{bmatrix}, B_2 = \begin{bmatrix} -0.05 \\ 0 \\ 0 \\ 0 \end{bmatrix},$$

$$D_1 = \begin{bmatrix} 0 & 0 & 0 & 0 \\ 0 & 0 & 0 & 0 \\ -0.12 & 0 & 0 & 0 \\ 0 & 0 & 0 & 0 \end{bmatrix}, D_2 = \begin{bmatrix} 0 & 0 & 0 & 0 \\ 0 & 0 & 0 & 0 \\ -0.24 & 0 & 0 & 0 \\ 0 & 0 & 0 & 0 \end{bmatrix},$$

and the sampling time is $\Delta t = 0.2s$. The speed of backing up $\theta(k)$ varies in the range $\theta(k) \in [-1.2, -0.6]$ and the weighting functions $\rho_j(\theta(k))$ for $j = 1, 2$ are computed as $\rho_1(\theta(k)) = \frac{\theta(k)+1.2}{0.6}$ and $\rho_2(\theta(k)) = \frac{\theta(k)+0.6}{-0.6}$. Besides, the initial state is chosen as $x(0) = [0.1745,\ 0.3491,\ 3,\ -0.4189]^{\mathrm{T}}$ and the initial estimation is bounded in the zonotope $\mathcal{X}(0) = \langle p(0), H(0) \rangle$ with $p(0) = x(0)$ and $H(0) = \text{diag}([0.02,\ 0.02,\ 0.1,\ 0.02])$.

Besides, to reduce the computation time and limit the growing complexity of the resulting zonotope, we set $q = 20$ in the zonotope reduction operator $\downarrow_{q,W}(H)$ and the weighting matrix is chosen as W. $d(k) \in \mathcal{D}$, $\forall k \in \mathbb{N}$ with $H_d = \text{diag}([0.03,\ 0.004,\ 0.004,\ 0.004])$.

From Lemma 2.1, since S is an arbitrary matrix in (2.43), we take

$$S = \begin{bmatrix} 1.01 & 5.92 & 8.87 & 2.23 & 0.34 & 4.48 & 5.57 \\ 0.72 & 3.78 & 5.99 & 1.31 & 8.80 & 3.47 & 8.89 \\ 9.05 & 9.52 & 6.90 & 1.72 & 5.22 & 7.57 & 8.40 \\ 3.12 & 7.42 & 3.91 & 9.39 & 8.08 & 3.60 & 2.44 \end{bmatrix},$$

to obtain

$$T = \begin{bmatrix} 1 & 2.7864 & 3.3204 & 2.2327 \\ 0 & -2.0097 & 3.8036 & 1.3126 \\ 0 & 2.1462 & 2.9088 & 1.7248 \\ 0 & -0.3293 & 0.5831 & 9.3860 \end{bmatrix}, N = \begin{bmatrix} -2.7864 & -3.3204 & 3.3204 \\ 3.0097 & -3.8036 & 3.8036 \\ -2.1462 & -1.9088 & 2.9088 \\ 0.3293 & -0.5831 & 1.5831 \end{bmatrix}.$$

From (2.55a), the convergence rate of the W-radius $\ell_W(k)$ is described by α. By simulations, we tune $\alpha \in (0, 1)$ to find a minimum γ. Moreover, we choose $\beta = 0.5$ in (2.56). The optimization problem (2.67) is solved using the YALMIP toolbox [13] and the MOSEK solver [14]. Through the line search, we can obtain the minimum $\gamma = 11.95$ with $\alpha = 0.75$ and the optimal polytopic correction matrices Λ_i^*, $i = 1, 2$,

$$\Lambda_1^* = \begin{bmatrix} 0.8359 & -0.0031 & 0.1491 \\ 1.0755 & -0.1189 & 0.0239 \\ -0.0558 & 0.9437 & 0.0723 \\ 0.0496 & 0.0466 & 0.9378 \end{bmatrix}, \Lambda_2^* = \begin{bmatrix} 0.7413 & 0.1768 & 0.0952 \\ 1.0728 & -0.1191 & 0.0274 \\ -0.0547 & 0.9386 & 0.0743 \\ 0.0464 & 0.0519 & 0.9371 \end{bmatrix}.$$

As a comparison, we also solve the optimization problem (2.70) by a line search with $\sigma \in [0, 1)$. Then, we obtain the maximum $\tau = 0.00024$ with $\sigma = 0.8$ and the polytopic correction matrices $\bar{\Lambda}_i^*$ for $i = 1, 2$,

$$\bar{\Lambda}_1^* = \begin{bmatrix} 1.8397 & -2.3403 & 0.8747 \\ 0.9557 & 0.0102 & 0.0185 \\ 0.0098 & 0.9453 & 0.0222 \\ 0.0200 & 0.0223 & 0.9730 \end{bmatrix}, \bar{\Lambda}_2^* = \begin{bmatrix} 1.9712 & -2.4276 & 0.8667 \\ 0.9646 & -0.0022 & 0.0199 \\ -0.0020 & 0.9627 & 0.0196 \\ 0.0206 & 0.0203 & 0.9740 \end{bmatrix}.$$

By implementing Algorithm 2.1 for $N = 300$ sampling steps with Λ_i^* and $\bar{\Lambda}_i^*$ separately, the comparison results of the state estimation are shown in Fig. 2.4, where the real states are plotted by red stars as the validation. From these two figures,

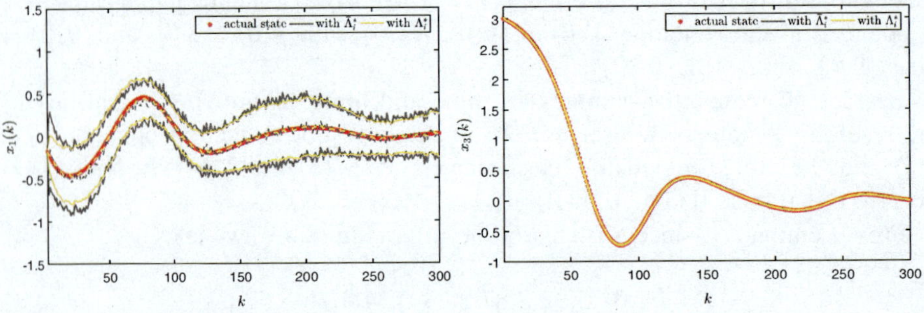

(a) x_1 (b) x_3

Fig. 2.4 State estimation results of the truck-trailer case study

Table 2.3 Comparison between Λ_i^* and $\bar{\Lambda}_i^*$

Approach	MSE	$rs(\hat{H})$	Computation time (s)
Λ_i^*	4.7362e-05	0.1332	0.0090
$\bar{\Lambda}_i^*$	5.2623e-04	0.1459	0.0099

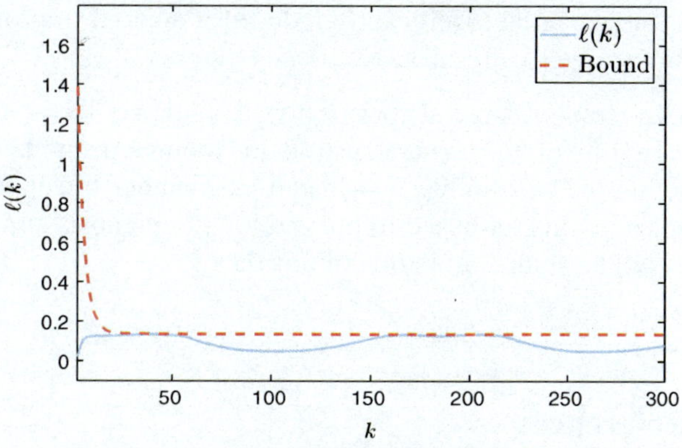

Fig. 2.5 $\mathcal{L}_\infty$ performance with Λ_i^*

the proposed approach with Λ_i^* and the comparison approach with $\bar{\Lambda}_i^*$ are able to estimate uncertain states in dash lines and propagate the estimation interval in solid lines (green ones for Λ_i^* and blue ones for $\bar{\Lambda}_i^*$).

In order to quantitatively compare the results with Λ_i^* and $\bar{\Lambda}_i^*$, we define the state estimation error between the estimated states and real states as $e(k) = x(k) - \hat{p}(k)$ and MSE. Since system uncertainties are propagated to the states during iterations, we also measure $rs(\hat{H}(k))$ to compare the size of the intersection zonotope with Λ_i^* and $\bar{\Lambda}_i^*$ bounding uncertain states. Table 2.3 shows the MSE result, the root mean square of $rs(\hat{H}(k))$ as well as the computation time. For this case study, it is clear from this table that the estimation error of the proposed approach is smaller as well as the size of intersection zonotopes. The mean computation time per one iteration for both approaches are similar and smaller than the sampling time.

Besides, with the proposed approach, the time-varying radius of the intersection zonotope is expected to be lower than the adaptive bound based on γ (as presented in Theorem 2.10), which is called the $\mathcal{L}_\infty$ performance. In Fig. 2.5, with the optimal solution γ, we can see that the radius of the intersection zonotope at each time is always constrained.

2.3 Summary

This chapter has presented a general set-based framework for discrete-time descriptor systems with application to robust state estimation. Specifically, two approaches are proposed: (i) the zonotopic set-membership approach; (ii) the zonotopic observer. It has been proved that the zonotopic observer in the current estimation type is equivalent to the set-membership approach. Several zonotope minimization criteria have been defined to find the optimal correction matrices for set-membership approach and the optimal Kalman gain for the zonotopic observer. For the set-based approaches, one weakness could be the assumption of unknown-but-bounded disturbances and noise. In this chapter, unknown inputs have been considered so that this conservativeness can be significantly reduced. Potential improvement and applications are summarized as follows:

- In terms of large-scale systems, a single set for bounding uncertain states could be difficult to be characterized due to the large number of variables. To overcome this, we will study a distributed set-membership approach in Chap. 3;
- Under this set-based framework, fault diagnosis strategies including FDI and FE will be studied in Part II of this thesis.

References

1. Wang Y, Puig V, Cembrano G (2018) Set-membership approach and Kalman observer based on zonotopes for discrete-time descriptor systems. Automatica 93:435–443
2. Combastel C (2015) Zonotopes and Kalman observers: gain optimality under distinct uncertainty paradigms and robust convergence. Automatica 55:265–273
3. Alamo T, Tempo R, Ramírez D, Camacho E (2008) A new vertex result for robustness problems with interval matrix uncertainty. Syst Control Lett 57(6):474–481
4. Wang Z, Shen Y, Zhang X, Wang Q (2012) Observer design for discrete-time descriptor systems: an LMI approach. Syst Control Lett 61(6):683–687
5. Alamo T, Bravo J, Camacho E (2005) Guaranteed state estimation by zonotopes. Automatica 41(6):1035–1043
6. Wang Z, Shi P, Lim C (2017) H_{-}/H_{∞} fault detection observer in finite frequency domain for linear parameter-varying descriptor systems. Automatica 86:38–45
7. Dai L (1989) Singular control systems. Springer, Berlin Heidelberg, Germany
8. Wang Z, Rodrigues M, Theilliol D, Shen Y (2015) Actuator fault estimation observer design for discrete-time linear parameter-varying descriptor systems. Int J Adapt Control Signal Process 29(2):242–258
9. Boyd S, El Ghaoui L, Feron E, Balakrishnan V (1994) Linear matrix inequalities in system and control theory, vol 15. Studies in applied mathematics. SIAM, Philadelphia, PA
10. Chadli M, Abdo A, Ding S (2013) Fault detection filter design for discrete-time Takagi-Sugeno fuzzy system. Automatica 49(7):1996–2005
11. Tuan H, Apkarian P, Narikiyo T, Yamamoto Y (2001) Parameterized linear matrix inequality techniques in fuzzy control system design. IEEE Trans Fuzzy Syst 9(2):324–332
12. Wang Y, Wang Z, Puig V, Cembrano G (2019) Zonotopic set-membership state estimation for discrete-time descriptor LPV systems. IEEE Trans Autom Control 64(5):2092–2099
13. Löfberg J (2004) YALMIP: a toolbox for modeling and optimization in MATLAB
14. MOSEK ApS (2015) The MOSEK optimization toolbox for MATLAB manual. Version 7.1 (Revision 28)

Chapter 3
Distributed Set-Membership Approach Based on Zonotopes

As society develops, an increasing number of large-scale systems, such as cyber-physical systems [1] and critical infrastructures (i.e. water distribution networks [2, 3] and smart grids [4]), are becoming more automatized. Such kind of systems have a large amount of states, inputs and outputs. Considering their complexity and dimension, these large- scale systems can be formulated as interconnected systems with coupled states. In the frameworks of diagnosis and optimal control of large-scale systems, a suitable distributed state estimation approach plays a significant role in the development of model-based fault diagnosis strategies [5] and the design of controllers [6, 7]. Revising the literature, different approaches have been investigated for distributed state estimation problems, as e.g. the distributed moving horizon approaches in [8, 9].

Following the research line of Chap. 2, as shown in Fig. 3.1, this chapter presents a distributed set-membership approach based on zonotopes for discrete-time systems, which improves the applicability of the set-based approach into large-scale systems. In fact, the set-based approaches [10–15] only consider the construction of single set for bounding all the uncertain states in a centralized way. Since that the dimension of large-scale systems is very high, it could be difficult to propagate a high-dimensional single set bounding all the uncertain states along iterations. As a result, a distributed set-membership approach is effective for a system with partitioned subsystems including coupled states.

Y. Wang, *Advances in State Estimation, Diagnosis and Control of Complex Systems*, Springer Theses, https://doi.org/10.1007/978-3-030-52440-1_3

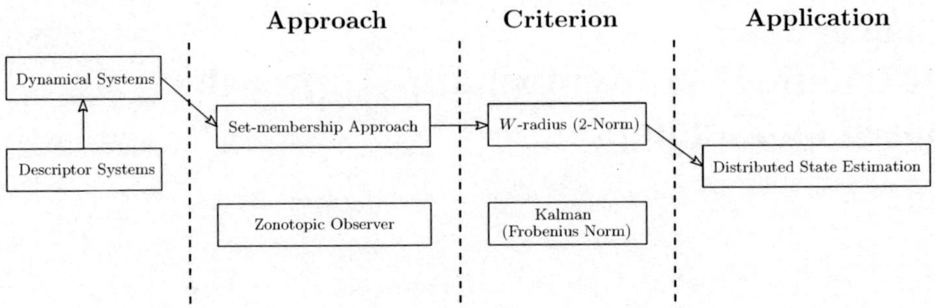

Fig. 3.1 Set-based distributed state estimation scheme

3.1 Problem Statement in Distributed Set-Membership Approach

Consider the class of discrete-time dynamical systems that can be decomposed into l interconnected subsystems (called agents) with coupled dynamics [16]. Each agent can be modeled as

$$x_i(k+1) = \sum_{j \in N_i} A_{ij} x_j(k) + B_i u_i(k) + w_i(k), \tag{3.1a}$$

$$y_i(k) = C_i x_i(k) + v_i(k), \tag{3.1b}$$

where $x_i \in \mathbb{R}^{n_{x_i}}$, $u_i \in \mathbb{R}^{n_{u_i}}$ and $y_i \in \mathbb{R}^{n_{y_i}}$ denote the state, the input and the output vectors, $w_i \in \mathbb{R}^{n_{w_i}}$ and $v_i \in \mathbb{R}^{n_{v_i}}$ denote the state disturbance and the measurement noise vectors of the i-th agent with $i = 1, \dots, l$, respectively. $A_{ii} \in \mathbb{R}^{n_{x_i} \times n_{x_i}}$, $B_i \in \mathbb{R}^{n_{x_i} \times n_{u_i}}$ and $C_i \in \mathbb{R}^{n_{y_i} \times n_{x_i}}$. Besides, N_i is defined to be the set that includes all the agents related to the agent i (i also included).

To design an iterative approach, the following assumptions are made.

Assumption 3.1 The state disturbance vector $w_i(k) \in \mathbb{R}^{n_{w_i}}$ and the measurement noise vector $v_i(k) \in \mathbb{R}^{n_{v_i}}$ are unknown but bounded by the centered zonotopes

$$w_i(k) \in \langle 0, D_{w_i} \rangle, \ v_i(k) \in \langle 0, D_{v_i} \rangle, \ \forall k \in \mathbb{N}, \tag{3.2}$$

and for $i = 1, \dots, l$, where $D_{w_i} \in \mathbb{R}^{n_{w_i} \times n_{w_i}}$ and $D_{v_i} \in \mathbb{R}^{n_{v_i} \times n_{v_i}}$.

Assumption 3.2 The initial state $x_i(0)$ is assumed to be bounded by the zonotope $x_i(0) \in \mathcal{X}_i(0) := \langle p_i(0), H_i(0) \rangle$ for $i = 1, \dots, l$.

In this chapter, the goal is to obtain robust state estimation by finding a sequence of distributed state zonotopes $\mathcal{X}_i(k)$ to independently bound the uncertain states $x_i(k)$ for $i = 1, \dots, l$, $\forall k \in \mathbb{N}$. Instead of using a centralized zonotope, these distributed state zonotopes can provide robust state estimation results.

3.2 Distributed Set-Membership Approach Based on Zonotopes

We now present a distributed set-membership approach for robust state estimation. A parameterized distributed state bounding zonotope is established for each agent considering coupled states. To determine the parameters (correction matrices) of the distributed state bounding zonotopes for robust state estimation, we propose an optimization problem based on W-radius minimization to find a set of optimal correction matrices.

3.2.1 Distributed State Bounding Zonotope

Instead of finding a centralized state bounding zonotope, we introduce the structure of the parameterized distributed state bounding zonotope $\mathcal{X}_i(k)$ for $i = 1, \dots, l$, $\forall k \in \mathbb{N}$. In this case, the coupled states are considered. Each zonotope $\mathcal{X}_i(k)$ is built to be consistent with its own measured output $y_i(k)$ of each agent. Considering that the initial states $x_i(0)$ are assumed to be bounded in an initial zonotope, the parameterized distributed state bounding zonotopes are recursively defined in the following proposition.

Proposition 3.1 (Distributed state bounding zonotope) *Given the dynamics of the distributed systems in (3.1), suppose that Assumptions 3.1 and 3.2 hold, and that* $x_i(k-1) \in \mathcal{X}_i(k-1) = \langle p_i(k-1), H_i(k-1)\rangle$, $i = 1, \dots, l$. *Then, the following inclusion holds for every correction matrix* $\Lambda_i \in \mathbb{R}^{n_{x_i} \times n_{y_i}}$

$$x_i(k) \in \mathcal{X}_i(k) = \langle p_i(k), H_i(k)\rangle, \; i = 1, \dots, l, \tag{3.3}$$

where $p_i(k)$ *and* $H_i(k)$ *are defined as*

$$p_i(k) = (I - \Lambda_i C_i)\Big(\sum_{j \in N_i} A_{ij} p_j(k-1) + B_i u_i(k-1)\Big) + \Lambda_i y_i(k), \tag{3.4a}$$

$$H_i(k) = \Big[(I - \Lambda_i C_i) \operatorname*{cat}_{j \in N_i} \big\{A_{ij} H_j(k-1)\big\}, \; (I - \Lambda_i C_i) D_{w_i}, \; \Lambda_i D_{v_i}\Big]. \tag{3.4b}$$

Proof Since $x_i(k-1) \in \langle p_i(k-1), H_i(k-1)\rangle$ for $i = 1, \dots, l$, by Definition 1.6, there exists a vector $\theta_i(k-1)$ with $\|\theta_i(k-1)\|_\infty \le 1$, $i = 1, \dots, l$ such that

$$x_i(k-1) = p_i(k-1) + H_i(k-1)\theta_i(k-1), \; i = 1, \dots, l.$$

From the dynamics in (3.1), in the prediction step, we have

$$
\begin{aligned}
x_i(k) &= \sum_{j\in N_i} A_{ij}x_j(k-1) + B_i u_i(k-1) + w_i(k-1) \\
&= \sum_{j\in N_i} A_{ij}\big(p_j(k-1) + H_j(k-1)\theta_j(k-1)\big) + B_i u_i(k-1) + w_i(k-1).
\end{aligned} \tag{3.5}
$$

From Assumption 3.1, there exists a vector $\varpi_i(k-1)$ with $\|\varpi_i(k-1)\|_\infty \leq 1$ for $i = 1, \ldots, l$ such that $w_i(k-1) = D_{w_i}\varpi_i(k-1)$. Thus, from (3.5), we derive

$$
x_i(k) = \sum_{j\in N_i} A_{ij}\big(p_j(k-1) + H_j(k-1)\theta_j(k-1)\big) + B_i u_i(k-1) + D_{w_i}\varpi_i(k-1).
$$

Set

$$
\begin{aligned}
\hat{p}_i(k) &= \sum_{j\in N_i} A_{ij}p_j(k-1) + B_i u_i(k-1), \\
R_i(k) &= \left[\operatorname*{cat}_{j\in N_i}\left\{A_{ij}H_j(k-1)\right\}, D_{w_i}\right], \\
\eta_i(k-1) &= \left[\operatorname*{cat}_{j\in N_i}\left\{\theta_j(k-1)^\top\right\}, \varpi_i(k-1)^\top\right]^\top,
\end{aligned}
$$

where $\|\eta_i(k-1)\|_\infty \leq 1$. Then, we have

$$
x_i(k) = \hat{p}_i(k) + R_i(k)\eta_i(k-1). \tag{3.6}
$$

From Assumption 3.1, there exists a vector $\sigma_i(k)$ with $\|\sigma_i(k)\|_\infty \leq 1$ for $i = 1, \ldots, l$ such that $v_i(k) = D_{v_i}\sigma_i(k)$. From the output equation of (3.1), we have

$$
y_i(k) - C_i x_i(k) - D_{v_i}\sigma_i(k) = 0. \tag{3.7}
$$

Thus, by replacing $x_i(k)$ in (3.7) with the expression in (3.6), we obtain

$$
y_i(k) - C_i\hat{p}_i(k) - C_i R_i(k)\eta_i(k-1) - D_{v_i}\sigma_i(k) = 0.
$$

Pre-multiplying by Λ_i and rearranging the terms of the above equation yields

$$
\Lambda_i y_i(k) - \Lambda_i C_i\hat{p}_i(k) - \Lambda_i C_i R_i(k)\eta_i(k-1) - \Lambda_i D_{v_i}\sigma_i(k) = 0. \tag{3.8}
$$

Finally, in the correction step, we add (3.8) to the right side of (3.6) obtaining

$$
\begin{aligned}
x_i(k) = {}& \hat{p}_i(k) + R_i(k)\eta_i(k-1) + \Lambda_i y_i(k) \\
& - \Lambda_i C_i\hat{p}_i(k) - \Lambda_i C_i R_i(k)\eta_i(k-1) - \Lambda_i D_{v_i}\sigma_i(k).
\end{aligned}
$$

By setting $p_i(k)$ and $H_i(k)$ as in (3.4), the above equation becomes

$$x_i(k) = p_i(k) + H_i(k)\begin{bmatrix}\eta_i(k-1)\\ -\sigma_i(k)\end{bmatrix}.$$

Since $\|\eta_i(k-1)\|_\infty \leq 1$ and $\|\sigma_i(k)\|_\infty \leq 1$, we thus conclude that $x_i(k) \in \langle p_i(k), H_i(k)\rangle$.

From (3.3) and (3.4), we can see that in order to find the distributed state bounding zonotope $\mathcal{X}_i(k)$ in a iterative way along the time step $k \in \mathbb{N}$, the correction matrices Λ_i for $i = 1, \ldots, l$ are required. In the following, we will investigate the way to compute Λ_i for $i = 1, \ldots, l$.

3.2.2 Computing Correction Matrices

For state estimation, the objective is to minimize the state estimation errors. Since all estimation errors and uncertainties are propagated and bounded in the distributed zonotope $\mathcal{X}_i(k)$, we would like to find Λ_i for $i = 1, \ldots, l$ to minimize the size of these distributed zonotopes. In this chapter, we also use the W-radius to measure the size of a zonotope (see Definition 1.9). In order to guarantee the global stability, we first rewrite the interconnected subsystems (3.1) as follows.

Denote $x = \left[x_1^\top, \ldots, x_l^\top\right]^\top \subset \mathbb{R}^{n_x}$ with $n_x = \sum_{i=1}^l n_{x_i}$, $u = \left[u_1^\top, \ldots, u_l^\top\right]^\top \subset \mathbb{R}^{n_u}$ with $n_u = \sum_{i=1}^l n_{u_i}$, $y = \left[y_1^\top, \ldots, y_l^\top\right]^\top \subset \mathbb{R}^{n_y}$ with $n_y = \sum_{i=1}^l n_{y_i}$, $w = \left[w_1^\top, \ldots, w_l^\top\right]^\top \subset \mathbb{R}^{n_w}$ with $n_w = \sum_{i=1}^l n_{w_i}$ and $v = \left[v_1^\top, \ldots, v_l^\top\right]^\top \subset \mathbb{R}^{n_v}$ with $n_v = \sum_{i=1}^l n_{v_i}$. The general system including l agents defined in (3.1) can be formulated as

$$x(k+1) = Ax(k) + Bu(k) + w(k), \tag{3.9a}$$
$$y(k) = Cx(k) + v(k), \tag{3.9b}$$

with

$$A = \begin{bmatrix} A_{11} & \cdots & A_{1l} \\ \vdots & \ddots & \vdots \\ A_{l1} & \cdots & A_{ll} \end{bmatrix},$$

$$B = \operatorname{diag}(B_1, \ldots, B_l), \quad C = \operatorname{diag}(C_1, \ldots, C_l),$$

where $w(k) \in \langle 0, D_w\rangle$ and $v(k) \in \langle 0, D_v\rangle, \forall k \in \mathbb{N}$, with $D_w = \operatorname{diag}\left(D_{w_1}, \ldots, D_{w_l}\right)$ and $D_v = \operatorname{diag}\left(D_{v_1}, \ldots, D_{v_l}\right)$.

Proposition 3.2 (Centralized state bounding zonotope) *Given the dynamics of the system (3.9) and suppose that $x(k-1) \in \mathcal{X}(k-1) = \langle p(k-1), H(k-1)\rangle$, for every correction matrix $\Lambda \in \mathbb{R}^{n_x \times n_y}$, the following inclusion holds*

$$x(k) \in \mathcal{X}(k) = \langle p(k), H(k)\rangle, \tag{3.10}$$

where

$$\begin{aligned} p(k) &= (I - \Lambda C)\,(Ap(k-1) + Bu(k-1)) + \Lambda y(k), \\ H(k) &= \left[(I - \Lambda C)AH(k-1),\ (I - \Lambda C)D_w,\ \Lambda D_v\right]. \end{aligned}$$

Proof Because of its similarity to Proposition 3.1, the proof is straightforward and omitted here. □

Based on the general state bounding zonotope, we present the conditions of the W-radius minimization criterion in the following theorem.

Theorem 3.1 *Given $\mathcal{X}(k) = \langle c(k), H(k)\rangle$ in (3.10), $\forall k \in \mathbb{N}$, two scalars $\gamma \in (0, 1)$ and $\epsilon > 0$. If there exist matrices $W \in \mathbb{S}^{n_x}_{\succ 0}$, $Y \in \mathbb{R}^{n_x \times n_y}$, diagonal matrices $\Gamma \in \mathbb{S}^{n_w}_{\succ 0}$ and $\Upsilon \in \mathbb{S}^{n_v}_{\succ 0}$ such that*

$$\mathrm{tr}(\Gamma) + \mathrm{tr}(\Upsilon) < \epsilon, \tag{3.11a}$$

$$\begin{bmatrix} \gamma W & \star & \star & \star \\ (W - YC)A & W & \star & \star \\ 0 & D_w^\top (W - YC)^\top & \Gamma & \star \\ 0 & D_v^\top Y^\top & 0 & \Upsilon \end{bmatrix} \succ 0, \tag{3.11b}$$

then it is guaranteed that

$$\ell_W(k) \le \gamma \ell_W(k-1) + \epsilon,\ \forall k \in \mathbb{Z}_+, \tag{3.12}$$

which leads to $\ell_W(\infty) \le \frac{\epsilon}{1-\gamma}$ when $k \to +\infty$.

Proof From (3.12), with $W \in \mathbb{S}^{n_x}_{\succ 0}$ and $\Lambda \in \mathbb{R}^{n_x \times n_y}$, for every $H(k-1)$ and $\gamma \in (0, 1)$, we have that (3.12) is equivalent to

$$\max_{\|\phi\|_\infty \le 1} \|H(k)\phi\|_W^2 - \max_{\|\theta\|_\infty \le 1} \gamma \|H(k-1)\theta\|_W^2 - \epsilon \le 0. \tag{3.13}$$

Since $\max_\theta \|H(k-1)\theta\|_W^2 \ge \|H(k-1)\theta\|_W^2$ for any $\|\theta\|_\infty \le 1$, we obtain the following sufficient condition of (3.13)

$$\max_{\|\phi\|_\infty \le 1} \|H(k)\phi\|_W^2 - \gamma \|H(k-1)\theta\|_W^2 - \epsilon \le 0. \tag{3.14}$$

Let us denote $\phi = \left[\theta^\top, \varpi^\top, \sigma^\top\right]^\top$ and $Y = R\Lambda$, then

$$R = \big[(W - YC)A,\ (W - YC)D_w,\ YD_v\big]. \tag{3.15}$$

With this notation, we have

$$H(k)\phi = W^{-1}R\big[(H(k-1)\theta)^{\top}, \varpi^{\top}, \sigma^{\top}\big]^{\top}.$$

Therefore, we rewrite (3.14) as

$$\begin{bmatrix} H(k-1)\theta \\ \varpi \\ \sigma \end{bmatrix}^{\top} R^{\top}W^{-1}R \begin{bmatrix} H(k-1)\theta \\ \varpi \\ \sigma \end{bmatrix} - \gamma(H(k-1)\theta)^{\top}WH(k-1)\theta - \epsilon < 0, \tag{3.16}$$

for any $\|\phi\|_{\infty} \leq 1$. If Γ and Υ are diagonal and positive semi-definite matrices, then we have

$$\mathrm{tr}(\Gamma) - \varpi^{\top}\Gamma\varpi \geq 0,\ \forall \|\varpi\|_{\infty} \leq 1, \tag{3.17a}$$

$$\mathrm{tr}(\Upsilon) - \sigma^{\top}\Upsilon\sigma \geq 0,\ \forall \|\sigma\|_{\infty} \leq 1. \tag{3.17b}$$

By adding (3.17) to (3.16), we can obtain a sufficient condition

$$\begin{aligned} &\begin{bmatrix} H(k-1)\theta \\ \varpi \\ \sigma \end{bmatrix}^{\top} R^{\top}W^{-1}R \begin{bmatrix} H(k-1)\theta \\ \varpi \\ \sigma \end{bmatrix} \\ &- \gamma(H(k-1)\theta)^{\top}WH(k-1)\theta - \epsilon \\ &+ \mathrm{tr}(\Gamma) - \varpi^{\top}\Gamma\varpi + \mathrm{tr}(\Upsilon) - \sigma^{\top}\Upsilon\sigma < 0. \end{aligned}$$

If (3.11a) is satisfied, then we obtain

$$\begin{bmatrix} H(k-1)\theta \\ \varpi \\ \sigma \end{bmatrix}^{\top} \left(R^{\top}W^{-1}R - \begin{bmatrix} \gamma W & \star & \star \\ 0 & \Gamma & \star \\ 0 & 0 & \Upsilon \end{bmatrix} \right) \begin{bmatrix} H(k-1)\theta \\ \varpi \\ \sigma \end{bmatrix} < 0.$$

From the above inequality, we have a sufficient condition

$$R^{\top}W^{-1}R - \begin{bmatrix} \gamma W & \star & \star \\ 0 & \Gamma & \star \\ 0 & 0 & \Upsilon \end{bmatrix} \prec 0,$$

and by changing the sign and applying the Schur complement, we obtain

$$\begin{bmatrix} \gamma W & \star & \star & \star \\ 0 & \Gamma & \star & \star \\ 0 & 0 & \Upsilon & \star \\ (W - YC)A & (W - YC)D_w & YD_v & W \end{bmatrix} \succ 0.$$

Finally, we obtain (3.11b) through a linear coordinate transformation by the matrix $T = \begin{bmatrix} I & 0 & 0 & 0 \\ 0 & 0 & I & 0 \\ 0 & 0 & 0 & I \\ 0 & I & 0 & 0 \end{bmatrix}$ applied to the above inequality. □

Remark 3.1 Denote $C_A = CA$. The inequality (3.11b) implies

$$\begin{bmatrix} \beta W & \star \\ WA - YC_A & W \end{bmatrix} > 0,$$

which is related to the design of an observer gain for the system pair (A, C_A).

From the expression of the system (3.9), it includes l agents in (3.1). Based on the definition of A, we propose the structure of matrices W and Y to be block diagonal matrices in order to find a group of Λ_i for $i = 1, \ldots, l$. Let us define the following structures [17]

$$W = \text{diag}(W_1, \ldots, W_l), \; W_i \in \mathbb{S}^{n_{x_i}}_{\succ 0}, \; i = 1, \ldots, l, \tag{3.18a}$$

$$Y = \text{diag}(Y_1, \ldots, Y_l), \; Y_i \in \mathbb{R}^{n_{x_i} \times n_{y_i}}, \; i = 1, \ldots, l. \tag{3.18b}$$

By Definition 1.9, $\ell_W(\infty) \leq \frac{\epsilon}{1-\gamma}$ leads to

$$(x(\infty) - p(\infty))^\top W (x(\infty) - p(\infty)) \leq \frac{\epsilon}{1-\gamma}, \tag{3.19}$$

which is an ellipsoid. To minimize the size of this ellipsoid in (3.19), we can maximize a norm of W, as e.g. we choose to maximize $\text{tr}(W)$. Therefore, the correction matrices $\Lambda_i \in \mathbb{R}^{n_{x_i} \times n_{y_i}}$ for $i = 1, \ldots, l$ can be obtained by solving the following optimization problem

$$\underset{W, L, \Gamma, \Upsilon}{\text{maximize}} \;\; \text{tr}(W), \tag{3.20}$$

subject to (3.11).

From the optimal solutions of (3.20), $\Lambda = W^{-1}Y$ gives $\Lambda = \text{diag}(\Lambda_1, \ldots, \Lambda_l)$ with $\Lambda_i \in \mathbb{R}^{n_{x_i} \times n_{y_i}}$ for $i = 1, \ldots, l$.

3.2.3 *Distributed Set-Membership Algorithm*

We summarize the distributed set-membership state estimation approach in the following algorithm.

Algorithm 3.1 Distributed Set-membership State Estimation

1: *(Offline procedure)* Solve the optimization problem (3.20) with the structured W and Y in (3.18) to obtain $\Lambda_i \in \mathbb{R}^{n_{x_i} \times n_{y_i}}$ for $i = 1, \ldots, l$.;
2: **for** $k := 1$: end **do**
3: Each agent i sends the state bounding zonotope $\langle p_i(k-1), H_i(k-1)\rangle$ to its neighbors for $i = 1, \ldots, l$ and $\forall k \in \mathbb{Z}_+$;
4: Receive the information $\langle p_i(k-1), H_i(k-1)\rangle, \forall j \in N_i$ from neighbors and obtain the measurement $y_i(k)$ and $\langle p_i(k-1), H_i(k-1)\rangle$. The distributed zonotope $\langle p_i(k), H_i(k)\rangle$ of the agent i is updated by (3.3) with Λ_i at time k.
5: Obtain the upper and lower bounds $x_i(k) \in [\underline{x}_i(k), \overline{x}_i(k)]$ for $i = 1, \ldots, n_x$ by

$$\overline{x}_i(k) = \hat{p}_i(k) + rs\left(\hat{H}(k)\right)_i,$$
$$\underline{x}_i(k) = \hat{p}_i(k) - rs\left(\hat{H}(k)\right)_i,$$

where $\hat{p}_i(k)$ is the i-th element of $\hat{p}(k)$ and $rs\left(\hat{H}(k)\right)_i$ returns the i-th diagonal element of $rs\left(\hat{H}(k)\right)$.
6: **end for**

3.3 Numerical Example

Given the system including two interconnected subsystems in (3.1) ($l = 2$) with system matrices

$$A_{11} = \begin{bmatrix} 0.6848 & -0.0749 & 0.1290 \\ 0.6671 & 0.9666 & -0.5852 \\ -0.2789 & -0.1119 & 1.0251 \end{bmatrix}, A_{12} = \begin{bmatrix} -0.2488 & -0.0242 \\ -0.9545 & -0.8138 \\ 0.3474 & 0.3067 \end{bmatrix},$$
$$A_{21} = \begin{bmatrix} -0.2180 & -0.0909 & 0.2027 \\ 1.1606 & 0.3804 & -0.9879 \end{bmatrix}, A_{22} = \begin{bmatrix} 0.8466 & 0.1632 \\ -1.6068 & -0.5130 \end{bmatrix},$$
$$B_1 = \begin{bmatrix} 0.8 & 0 \\ 0 & 0.58 \\ 0.6 & 0.8 \end{bmatrix}, B_2 = \begin{bmatrix} 0.8 \\ -0.75 \end{bmatrix},$$
$$C_1 = \begin{bmatrix} 1 & 1 & 0 \\ 0 & 0 & 1 \end{bmatrix}, C_2 = \begin{bmatrix} 1 & 0 \\ 0 & 1 \end{bmatrix},$$

and $w_i(k) \in \langle 0, D_{w_i}\rangle$, $v_i(k) \in \langle 0, D_{v_i}\rangle$ for $i = 1, 2$, $\forall k \in \mathbb{N}$, where

$$D_{w_1} = \begin{bmatrix} 0.1 & 0 & 0 \\ 0 & 0.15 & 0 \\ 0 & 0 & 0.25 \end{bmatrix}, D_{w_2} = \begin{bmatrix} 0.1 & 0 \\ 0 & 0.15 \end{bmatrix},$$
$$D_{v_1} = \begin{bmatrix} 0.05 & 0 \\ 0 & 0.05 \end{bmatrix}, D_{v_2} = \begin{bmatrix} 0.1 & 0 \\ 0 & 0.1 \end{bmatrix},$$

and the initial state is chosen as $x_1(0) = [0.25, 1.5, -0.5]^\top \in \langle p_1(0), H_1(0)\rangle$ and $x_2(0) = [0.8, 0]^\top \in \langle p_2(0), H_2(0)\rangle$, where $p_1(0) = x_1(0)$, $p_2(0) = x_2(0)$ and

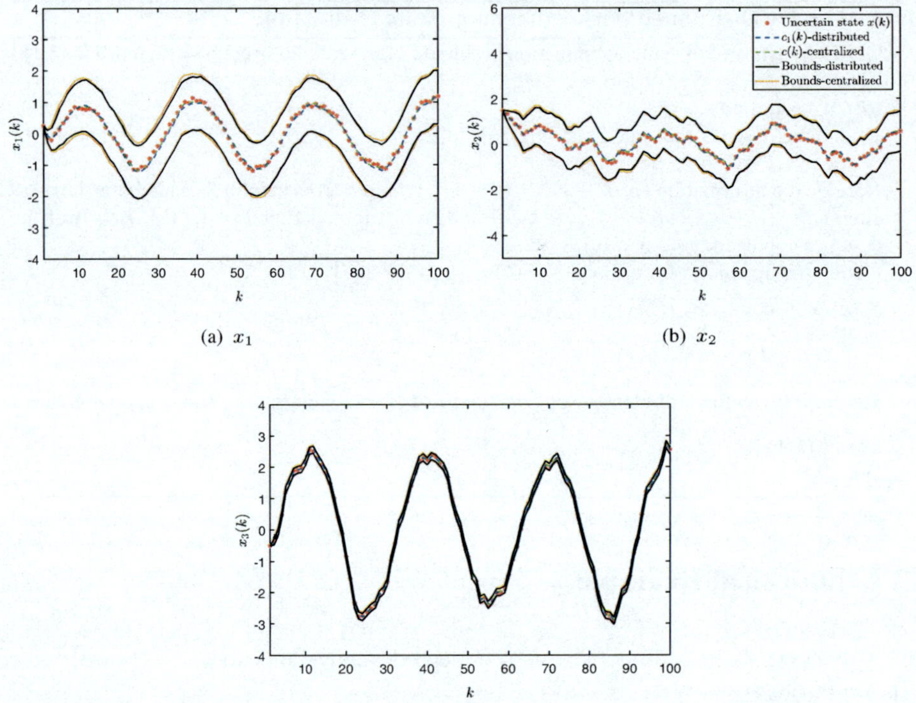

(a) x_1 (b) x_2

(c) x_3

Fig. 3.2 State estimation result of Agent 1

$$H_1(0) = \begin{bmatrix} 0.01 & 0 & 0 \\ 0 & 0.01 & 0 \\ 0 & 0 & 0.01 \end{bmatrix}, \quad H_2(0) = \begin{bmatrix} 0.01 & 0 \\ 0 & 0.01 \end{bmatrix}.$$

The simulations with this numerical example have been carried out in MATLAB and the optimization problems have been solved using the YALMIP [18] with the MOSEK solver [19]. By setting $\gamma = 0.8$, W and Y in block diagonal forms (3.18), we obtain the optimal correction matrices for two agents

$$\Lambda_1 = \begin{bmatrix} 0.2433 & 0.1841 \\ 0.7567 & -0.1841 \\ 0.0375 & 1.1077 \end{bmatrix}, \Lambda_2 = \begin{bmatrix} 1.4788 & 0.0093 \\ 0.5687 & 1.0129 \end{bmatrix}.$$

Besides, for a comparison, we also compute the centralized correction matrix Λ^c with full-dimensional W and Y obtaining

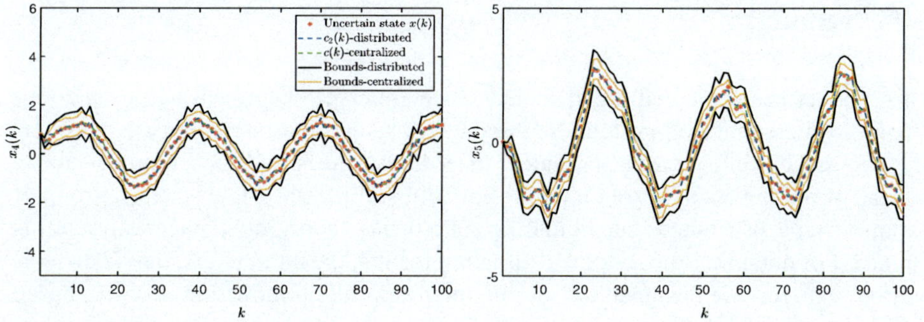

(a) x_4 (b) x_5

Fig. 3.3 State estimation result of Agent 2

Table 3.1 Comparison between the distributed and centralized approaches

	tr(W)	MSE	RMS($\sum_{k=0}^{100}(H(k))$)
Distributed approach	799.4855	0.0061	0.6174
Centralized approach	799.5274	0.0037	0.6066

$$\Lambda^c = \begin{bmatrix} 0.3273 & 0.0877 & 0.1625 & 0.0479 \\ 0.6728 & -0.0877 & -0.1625 & -0.0479 \\ 0.0920 & 1.0439 & 0.0414 & -0.0247 \\ 0.1916 & 0.1993 & 0.6911 & -0.1313 \\ 0.2677 & 0.1113 & -0.0047 & 0.7835 \end{bmatrix}.$$

Following the proposed set-membership state estimation algorithm and the corresponding centralized algorithm, robust state estimation results are shown in Figs. 3.2 and 3.3. From these plots, we can see that both approaches are able to provide state estimations with generated bounds, and the bounds of the distributed approach are not significantly larger than the centralized ones. Besides, the optimal values of the optimization problem (3.20), the MSE result and a measure of bounds are computed and shown in Table 3.1. With the unstructured W, the optimal objective $\mathrm{tr}(W)$ in the centralized approach is slightly better than the distributed one. As a result, the state estimation error and generated bounds in the centralized approach is slightly smaller than the distributed ones. However, the distributed approach uses less information and is able to get similar results as the centralized approach.

3.4 Summary

This chapter has presented a distributed set-membership approach based on zonotopes for interconnected systems with coupled states. The interconnected systems are affected by unknown-but-bounded state disturbances and measurement noise. Instead of finding a single zonotope for bounding all the uncertain states, a group of parameterized distributed state bounding zonotopes to over-bound uncertain states is defined. For obtaining robust state estimation results, the parameters, that is the correction matrices, are designed by solving the proposed optimization problem based on the W-radius minimization. The proposed approach is tested by a numerical example and compared with the centralized approach. From the simulation results, it can be seen that the distributed approach is not much worse than the centralized one since less information of measured outputs is used to correct the predicted state set for each agent. As future research, a customized geometrical set could be defined and a suitable communication strategy can be applied to improve the performance of distributed set-membership approach.

References

1. Fawzi H, Tabuada P, Diggavi S (2014) Secure estimation and control for cyber-physical systems under adversarial attacks. IEEE Trans Autom Control 59(6):1454–1467
2. Wang Y, Alamo T, Puig V, Cembrano G (2018) Economic model predictive control with nonlinear constraint relaxation for the operational management of water distribution networks. Energies 11(4):991
3. Wang Y, Puig V, Cembrano G (2017) Non-linear economic model predictive control of water distribution networks. J Process Control 56:23–34
4. Antoniadou-Plytaria KE, Kouveliotis-Lysikatos IN, Georgilakis PS, Hatziargyriou ND (2017) Distributed and decentralized voltage control of smart distribution networks: models, methods, and future research. IEEE Trans Smart Grid 8(6):2999–3008
5. Blanke M, Kinnaert M, Lunze J, Staroswiecki M (2016) Diagnosis and fault-tolerant control. Springer, Berlin Heidelberg
6. Barreiro-Gomez J, Quijano N, Ocampo-Martinez C (2016) Constrained distributed optimization: a population dynamics approach. Automatica 69:101–116
7. Christofides P, Scattolini R, Muñoz de la Peña D, Liu J (2013) Distributed model predictive control: a tutorial review and future research directions. Comput Chem Eng 51:21–41
8. Yin X, Liu J (2017) Distributed moving horizon state estimation of two-time-scale nonlinear systems. Automatica 79:152–161
9. Zeng J, Liu J (2015) Distributed moving horizon state estimation: simultaneously handling communication delays and data losses. Syst Control Lett 75:56–68
10. Alamo T, Bravo J, Camacho E (2005) Guaranteed state estimation by zonotopes. Automatica 41(6):1035–1043
11. Alamo T, Bravo J, Redondo M, Camacho E (2008) A set-membership state estimation algorithm based on DC programming. Automatica 44(1):216–224
12. Puig V, Cuguero P, Quevedo J (2001) Worst-case state estimation and simulation of uncertain discrete-time systems using zonotopes. In: European control conference (ECC), pp 1691–1697
13. Wang Y, Puig V (2016) Zonotopic extended Kalman filter and fault detection of discrete-time nonlinear systems applied to a quadrotor helicopter. In: 3rd international conference on control and fault tolerant systems (SysTol), Barcelona, Spain, pp 367–372

14. Wang Y, Puig V, Cembrano G (2018) Set-membership approach and Kalman observer based on zonotopes for discrete-time descriptor systems. Automatica 93:435–443
15. Wang Y, Wang Z, Puig V, Cembrano G (2019) Zonotopic set-membership state estimation for discrete-time descriptor LPV systems. IEEE Trans Autom Control 64(5):2092–2099
16. Wang Y, Alamo T, Puig V, Cembrano G (2018) A distributed setmembership approach based on zonotopes for interconnected systems. In: 57th IEEE conference on decision and control (IEEE-CDC), Miami, USA, pp 668–673
17. Maestre J, Muñoz de la Peña D, Camacho E, Alamo T (2011) Distributed model predictive control based on agent negotiation. J Process Control 21(5):685–697
18. Löfberg J (2004) YALMIP: a toolbox for modeling and optimization in MATLAB
19. MOSEK ApS (2015) The MOSEK optimization toolbox for MATLAB manual. Version 7.1 (Revision 28)

Part II
Diagnosis

Chapter 4
Set-Based Fault Detection and Isolation for Descriptor Systems

This chapter presents set-based FDI strategies for discrete-time descriptor systems. In this chapter, as shown in Fig. 4.1, we apply the set-based framework proposed in Chap. 2 into FDI [1, 2] for discrete-time descriptor systems. In particular, fault sensitivity should be taken into account for implementing an FD strategy [3, 4]. In this chapter, we will show two different criteria for achieving fault sensitivity: (i) the one based on a $\mathcal{H}_-$ index and therefore the condition is transformed as an LMI; (ii) the other based on a new defined criterion and algebraic solution is explicitly presented. In the first method, the effects of occurred faults are propagated in the center of state bounding zonotopes while in the second method, those effects are bounded and propagated in the segment matrix of state bounding zonotopes. Besides, the FI strategy is implemented by adopting a bank of zonotopic UIOs.

4.1 Zonotopic FD Observer for Descriptor Systems Considering the $\mathcal{H}_-$ Fault Sensitivity

Consider the following discrete-time descriptor system with additive actuator faults as

$$Ex(k+1) = Ax(k) + Bu(k) + D_w w(k) + Ff(k), \tag{4.1a}$$

$$y(k) = Cx(k) + D_v v(k), \tag{4.1b}$$

where $x \in \mathbb{R}^{n_x}$, $u \in \mathbb{R}^{n_u}$ and $y \in \mathbb{R}^{n_y}$ denote the state, the known input and the output vectors, $w \in \mathbb{R}^{n_w}$, $v \in \mathbb{R}^{n_v}$ and $f \in \mathbb{R}^{q}$ denote the state disturbance, the measurement noise and the additive fault vectors. $A \in \mathbb{R}^{n_x \times n_x}$, $B \in \mathbb{R}^{n_x \times n_u}$, $D_w \in \mathbb{R}^{n_x \times n_w}$, $F \in \mathbb{R}^{n_x \times q}$, $C \in \mathbb{R}^{n_y \times n_x}$ and $D_v \in \mathbb{R}^{n_y \times n_v}$ are system matrices. Besides, the matrix $E \in \mathbb{R}^{n_x \times n_x}$ may be singular, that is, $\text{rank}(E) \leq n_x$.

Y. Wang, *Advances in State Estimation, Diagnosis and Control of Complex Systems*, Springer Theses, https://doi.org/10.1007/978-3-030-52440-1_4

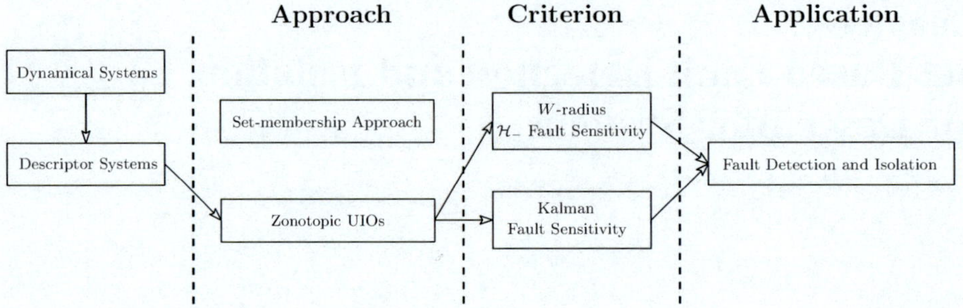

Fig. 4.1 Set-based FDI scheme

For the system (4.1), we consider Assumptions 2.1 and 2.2 hold. Besides, the descriptor system (4.1) is also assumed to be C-observable. Then, matrices E, C satisfy the rank condition (2.41). Thus, there exists two non-zero matrices $T \in \mathbb{R}^{n_x \times n_x}$ and $N \in \mathbb{R}^{n_x \times n_y}$ that can be obtained by Lemma 2.1.

Based on the result in Sect. 2.1.2, we use a zonotopic observer in Theorem 2.5 to implement the FD strategy. Recall the result of Theorem 2.5, considering a state observation $\hat{x}(k) \in \langle \hat{p}(k), \hat{H}(k) \rangle \subseteq \mathbb{R}^{n_x}$ at time step $k \in \mathbb{N}$, which also satisfies $\hat{x}(0) = x(0) \in \langle p(0), H(0) \rangle$ when $k = 0$ and no faults occurred. Then, at time step $k+1$, the state observation is recursively defined by $\hat{x}(k+1) \in \langle \hat{p}(k+1), \hat{H}(k+1) \rangle$ with

$$\hat{p}(k+1) = (TA - GC)\,\hat{p}(k) + TBu(k) + Gy(k) + Ny(k+1), \tag{4.2a}$$

$$\hat{H}(k+1) = \left[(TA - GC)\,\bar{H}(k),\ TD_w,\ -GD_v,\ -ND_v\right], \tag{4.2b}$$

where $\bar{H}(k) = \downarrow_{q,W} (\hat{H}(k))$ and $G \in \mathbb{R}^{n_x \times n_y}$ is an observer gain. Then, from the state zonotope $\hat{x}(k) \in \langle \hat{p}(k), \hat{H}(k) \rangle$, $\forall k \in \mathbb{N}$, we define the zonotopic FD observer for descriptor system (4.1) by the residual zonotope $r(k) \in \langle p_r(k), H_r(k) \rangle \subseteq \mathbb{R}^{n_y}$ with

$$p_r(k) = y(k) - C\hat{p}(k), \tag{4.3a}$$

$$H_r(k) = \left[C\hat{H}(k),\ D_v\right], \tag{4.3b}$$

We would like to design an FD observer gain G that minimizes the effects of uncertainties and meanwhile maximizes the sensitivity to faults based on the $\mathcal{H}_-$ index. Thus, the design of this gain G relies on two LMI conditions.

4.1.1 Zonotopic Observer Decomposition

For the descriptor system (4.1), with matrices T and N, descriptor dynamics can be reformulated as

$$\begin{aligned} x(k+1) = &TAx(k) + TBu(k) + TD_w w(k) + TFf(k) \\ &+ Ny(k+1) - ND_v v(k+1). \end{aligned} \tag{4.4}$$

Considering a state observation $\hat{x}(k) \in \langle \hat{p}(k), \hat{H}(k) \rangle$, we define the state estimation error as $e(k) = x(k) - \hat{p}(k)$. Then, with (4.4), the state estimation error dynamics can be formulated as

$$\begin{aligned} e(k+1) = &(TA - GC)e(k) + TD_w w(k) + TFf(k) \\ &- GD_v v(k) - ND_v v(k+1). \end{aligned} \tag{4.5}$$

From (4.5), the effects of faults appear in the error dynamics. Consider that $x(k) \in \langle p(k), H(k) \rangle = \{\langle \hat{p}(k), \hat{H}(k) \rangle \oplus \langle p_f(k), 0 \rangle\}$ is the uncertain state of the descriptor system (4.1) at time step k, where $p_f(k) \in \mathbb{R}^{n_x}$ is the center of the zonotope only affected by faults and $p_f(0) = 0$ at time step $k = 0$. We recursively provide a zonotope that bounds the uncertain state $x(k+1)$ in the following theorem.

Theorem 4.1 (State bounding zonotope decomposition) *Given the descriptor system (4.1), $w(k) \in \langle 0, I_{n_w} \rangle$ and $v(k) \in \langle 0, I_{n_v} \rangle$, $\forall k \in \mathbb{N}$ and $x(k) \in \langle p(k), H(k) \rangle = \{\langle \hat{p}(k), \hat{H}(k) \rangle \oplus \langle p_f(k), 0 \rangle\}$, $\forall k \in \mathbb{N}$. The uncertain state $x(k+1)$ is bounded by the zonotope in the decomposition form: $x(k+1) \in \langle p(k+1), H(k+1) \rangle = \{\langle \hat{p}(k+1), \hat{H}(k+1) \rangle \oplus \langle p_f(k+1), 0 \rangle\}$ where $\hat{p}(k+1)$ and $\hat{H}(k+1)$ are defined in (4.2) and*

$$p_f(k+1) = (TA - GC)p_f(k) + TFf(k). \tag{4.6}$$

Proof From $x(k) \in \langle p(k), H(k) \rangle = \{\langle \hat{p}(k), \hat{H}(k) \rangle \oplus \langle p_f(k), 0 \rangle\}$, we know that $p(k) = \hat{p}(k) + p_f(k)$ and $e(k) = x(k) - \hat{p}(k) \in \langle p_f(k), \hat{H}(k) \rangle$. Then, we have $x(k+1) = \hat{c}(k+1) + e(k+1)$. Considering $w(k) \in \langle 0, I_{n_w} \rangle, v(k) \in \langle 0, I_{n_v} \rangle, \forall k \in \mathbb{N}$ and $e(k+1)$ in (4.5), we can derive that

$$\begin{aligned} x(k+1) \in \langle p(k+1), H(k+1) \rangle &= \{\langle \hat{p}(k+1), \hat{H}(k+1) \rangle \oplus \langle p_f(k+1), 0 \rangle\} \\ &= \langle \hat{p}(k+1), 0 \rangle \oplus ((TA - GC)\langle p_f(k), \hat{H} \rangle) \oplus \left(TD_w \langle 0, I_{n_w} \rangle\right) \\ &\oplus TF\langle f(k), 0 \rangle \oplus \left(-GD_v \langle 0, I_{n_v} \rangle\right) \oplus \left(-ND_v \langle 0, I_{n_v} \rangle\right). \end{aligned}$$

By applying properties in (1.9), we thus obtain $\hat{p}(k+1)$ and $\hat{H}(k+1)$ defined as in (4.2) and $p_f(k+1)$ as in (4.6). □

From Theorem 4.1, it can be seen that the uncertain state $x(k)$ of the descriptor system (4.1) is bounded in the zonotope $\{\langle \hat{p}(k), \hat{H}(k) \rangle \oplus \langle p_f(k), 0 \rangle\}$ and $\langle \hat{p}(k), \hat{H}(k) \rangle$

is only affected by state disturbances and measurement noise while $\langle p_f(k), 0\rangle$ is only affected by additive actuator faults. Besides, with $x(k) \in \langle p(k), H(k)\rangle = \{\langle \hat{p}(k), \hat{H}(k)\rangle \oplus \langle p_f(k), 0\rangle\}$, $v(k) \in \langle 0, I_{n_v}\rangle$ and $y(k)$ in (4.1b), we define the zonotopic FD observer $r(k) \in \langle p_r(k), H_r(k)\rangle$ in the following decomposition form

$$p_r(k) = Cp_f(k), \tag{4.7a}$$

$$H_r(k) = \left[C\hat{H}(k),\ D_v\right], \tag{4.7b}$$

from which the effects of occurred faults are characterized at the center of the zonotopic FD observer while uncertainties are propagated in the zonotope segment matrix.

4.1.2 Observer Gain Design Considering $\mathcal{H}_-$ Fault Sensitivity

We now present the LMI results that allow achieving robustness against bounded uncertainties and sensitivity to faults for descriptor system (4.1). From the analysis in (4.7), we formulate the robustness condition by minimizing the size of the zonotope $\langle p_r(k), H_r(k)\rangle$. We use the W-radius to measure the size of the zonotope. From (4.7b), to minimize the size of $\langle p_r(k), H_r(k)\rangle$ is equivalent to minimize the size of $\langle \hat{p}(k), \hat{H}(k)\rangle$. Besides, we derive the $\mathcal{H}_-$ fault sensitivity condition [4] for the center $p_r(k)$ as presented in (4.7a).

4.1.2.1 Robustness Condition

According to Definition 1.9, with a matrix $W \in \mathbb{S}^{n_x}_{\succ 0}$, we recall the W-radius of the zonotope $\langle \hat{p}(k), \hat{H}(k)\rangle$ as

$$\ell_W(k) = \max_{\bar{z}\in\mathbf{B}^h} \left\| \hat{H}(k)\bar{z} \right\|^2_{2,W}. \tag{4.8}$$

Considering uncertainties (state estimation error, disturbances and noise) are propagated and bounded in the zonotope $\langle \hat{p}(k), \hat{H}(k)\rangle$, to reduce the effects of uncertainties, the size of this zonotope should be minimized. Based on Sect. 2.1.1, we implement the result of Theorem 2.3 to find the gain of zonotopic observer.

Proposition 4.1 *Given the descriptor system (4.1), the zonotope $\langle \hat{p}(k), \hat{H}(k)\rangle$, $\forall k \in \mathbb{N}$ and its W-radius in (4.8), two scalars $\gamma \in (0, 1)$ and $\epsilon > 0$. If there exists a matrix $W \in \mathbb{S}^{n_x}_{\succ 0}$ such that a minimization criterion is defined as*

$$\ell_W(k+1) \le \gamma \ell_W(k) + \epsilon,\ \forall k \in \mathbb{N}, \tag{4.9}$$

then the W-radius is ultimately bounded by $\ell_W(\infty) \leq \frac{\epsilon}{1-\gamma}$.

Proof The proof can be found in Theorem 2.3 and omitted here. □

Based on (4.9), we now formulate the robustness condition of the zonotopic FD observer.

Theorem 4.2 (Robustness condition) *Given the descriptor system (4.1), matrices $T \in \mathbb{R}^{n_x \times n_x}$ and $N \in \mathbb{R}^{n_x \times n_y}$ satisfying (2.42), two scalars $\gamma \in (0, 1)$ and $\epsilon > 0$. If there exist matrices $W \in \mathbb{S}^n_{\succ 0}$, $Y \in \mathbb{R}^{n_x \times n_y}$, and diagonal matrices $\Gamma \in \mathbb{S}^{n_w}_{\succ 0}$, $\Upsilon \in \mathbb{S}^{n_v}_{\succ 0}$ and $\Omega \in \mathbb{S}^{n_v}_{\succ 0}$ such that*

$$\operatorname{tr}(\Gamma) + \operatorname{tr}(\Upsilon) + \operatorname{tr}(\Omega) < \epsilon, \tag{4.10a}$$

$$\begin{bmatrix} \gamma W & \star & \star & \star & \star \\ 0 & \Gamma & \star & \star & \star \\ 0 & 0 & \Upsilon & \star & \star \\ 0 & 0 & 0 & \Omega & \star \\ WTA - YC & WTD_w & YD_v & WND_v & W \end{bmatrix} \succ 0, \tag{4.10b}$$

then the dynamics of $\hat{H}(k)$ in (4.2b) is stable and the W-radius minimization criterion in (4.9) holds.

Proof As $\ell_W(k)$ in (4.8), we reformulate (4.9) as follows:

$$\max_{z \in \mathbf{B}^{(h+n_w+2n_v)}} \left\| \hat{H}(k+1)z \right\|^2_{2,W} \leq \max_{\bar{z} \in \mathbf{B}^h} \gamma \left\| \hat{H}(k)\bar{z} \right\|^2_{2,W} + \epsilon.$$

Set $z = \left[\bar{z}^\top, b_1^\top, b_2^\top, b_3^\top\right]^\top$. Based on the proof of Theorem 2.3, for any $z \in \mathbf{B}^{(h+n_w+2n_v)}$, we obtain a sufficient condition

$$\left\| \hat{H}(k+1)z \right\|^2_{2,W} - \gamma \left\| \hat{H}(k)\bar{z} \right\|^2_{2,W} - \epsilon < 0. \tag{4.11}$$

By setting $Y = WG$ and recalling $\hat{H}(k+1)$ in (4.2b), we denote

$$R = [WTA - YC,\ WTD_w,\ YD_v,\ WND_v]. \tag{4.12}$$

Then, (4.11) becomes

$$\begin{bmatrix} \bar{H}(k)\bar{z} \\ b_1 \\ b_2 \\ b_3 \end{bmatrix}^\top R^\top W^{-1} R \begin{bmatrix} \bar{H}(k)\bar{z} \\ b_1 \\ b_2 \\ b_3 \end{bmatrix} - \bar{z}^\top \bar{H}^\top \gamma W \bar{H} \bar{z} - \epsilon < 0. \tag{4.13}$$

If Γ, Υ and Ω are diagonal and positive semi-definite matrices, then we have following conditions:

$$\mathrm{tr}(\Gamma) \geq b_1^\top \Gamma b_1, \ \forall b_1 \in \mathbf{B}^{n_w}, \tag{4.14a}$$

$$\mathrm{tr}(\Upsilon) \geq b_2^\top \Upsilon b_2, \ \forall b_2 \in \mathbf{B}^{n_v}, \tag{4.14b}$$

$$\mathrm{tr}(\Omega) \geq b_3^\top \Omega b_3, \ \forall b_3 \in \mathbf{B}^{n_v}. \tag{4.14c}$$

With (4.14), we can obtain a sufficient condition of (4.13) as

$$\begin{bmatrix} \bar{H}\bar{z} \\ b_1 \\ b_2 \\ b_3 \end{bmatrix}^\top R^\top W^{-1} R \begin{bmatrix} \bar{H}(k)\bar{z} \\ b_1 \\ b_2 \\ b_3 \end{bmatrix} - \bar{z}^\top \bar{H}^\top \gamma W \bar{H}\bar{z} + \mathrm{tr}(\Gamma) - b_1^\top \Gamma b_1$$
$$+ \mathrm{tr}(\Upsilon) - b_2^\top \Upsilon b_2 + \mathrm{tr}(\Omega) - b_3^\top \Omega b_3 - \epsilon < 0.$$

If (4.10a) is satisfied, then from the above condition we obtain

$$\begin{bmatrix} \bar{H}(k)\bar{z} \\ b_1 \\ b_2 \\ b_3 \end{bmatrix}^\top \left(\begin{bmatrix} \gamma W & 0 & 0 & 0 \\ 0 & \Gamma & 0 & 0 \\ 0 & 0 & \Upsilon & 0 \\ 0 & 0 & 0 & \Omega \end{bmatrix} - R^\top W^{-1} R \right) \begin{bmatrix} \bar{H}(k)\bar{z} \\ b_1 \\ b_2 \\ b_3 \end{bmatrix} > 0.$$

From the above inequality, a sufficient condition can be obtained

$$\begin{bmatrix} \gamma W & 0 & 0 & 0 \\ 0 & \Gamma & 0 & 0 \\ 0 & 0 & \Upsilon & 0 \\ 0 & 0 & 0 & \Omega \end{bmatrix} - R^\top W^{-1} R \succ 0, \ \forall \bar{H}(k)\bar{z}, \forall b_1, \forall b_2, \forall b_3,$$

from which, by applying the Schur complement, we thus obtain (4.10b). □

4.1.2.2 $\mathcal{H}_-$ Fault Sensitivity Condition

With $p_r(k)$ in (4.7a) with the propagation of $p_f(k)$ in (4.6), the $\mathcal{H}_-$ performance index β between the signals $p_r(k)$ and $f(k)$ satisfies

$$\sum_{k=0}^{\infty} p_r(k)^\top p_r(k) \geq \beta^2 \sum_{k=0}^{\infty} f(k)^\top f(k). \tag{4.15}$$

Based on the generalized KYP lemma (see Lemma 1.6), a relaxation of (4.15) is given in the following lemma.

Lemma 4.1 *Consider the fault frequency contents $\theta \in \Theta$ with Θ defined in Table 1.1, the dynamics of $p_f(k)$ in (4.6) and suppose $(TA - GC)$ to be Schur. If there exist matrices $P \in \mathbb{S}^{n_x}$ and $Q \in \mathbb{S}^{n_x}_{\succ 0}$, and a scalar $\beta > 0$ such that*

$$
\begin{aligned}
&\begin{bmatrix} TA - GC & TF \\ I_{n_x} & 0 \end{bmatrix}^{\top} \Xi \begin{bmatrix} TA - GC & TF \\ I_{n_x} & 0 \end{bmatrix} \\
&+ \begin{bmatrix} C & 0 \\ 0 & I_q \end{bmatrix}^{\top} \begin{bmatrix} -I_{n_y} & 0 \\ 0 & \beta^2 I_q \end{bmatrix} \begin{bmatrix} C & 0 \\ 0 & I_q \end{bmatrix} \prec 0,
\end{aligned} \tag{4.16}
$$

where Ξ is chosen as in Table 1.1 with respect to $\theta \in \Theta$, then the $\mathcal{H}_-$ performance index β between the signals $p_r(k)$ and $f(k)$ satisfies (4.15).

Proof Without loss of generality, let us consider $\forall \theta \in \Theta$ in the middle-frequency domain. Recall $\theta_c = \frac{\theta_1+\theta_2}{2}$ and from Table 1.1, Π is chosen as

$$
\Pi = \begin{bmatrix} -P & e^{j\theta_c} Q \\ e^{-j\theta_c} Q & P - 2\cos(\theta_w) Q \end{bmatrix}.
$$

By pre-multiplying $\begin{bmatrix} p_f^{\top}, & f^{\top} \end{bmatrix}$ and post-multiplying its transpose to both sides of (4.16), we obtain a sufficient condition

$$
\begin{aligned}
&p_f(k)^{\top} P p_f(k) - p_f(k+1)^{\top} P p_f(k+1) - p_r(k)^{\top} p_r(k) + \beta^2 f(k)^{\top} f(k) \\
&+ \mathbf{He}\left(p_f(k+1)^{\top} e^{j\theta_c} Q p_f(k)\right) - p_f(k)^{\top} 2\cos(\theta_w) Q p_f(k) \leq 0.
\end{aligned} \tag{4.17}
$$

Since the term $\mathbf{He}\left(p_f(k+1)^{\top} e^{j\theta_c} Q p_f(k)\right) - p_f(k)^{\top} 2\cos(\theta_w) Q p_f(k)$ is a number, we have that

$$
\begin{aligned}
&\mathbf{He}\left(p_f(k+1)^{\top} e^{j\theta_c} Q p_f(k)\right) - p_f(k)^{\top} 2\cos(\theta_w) Q p_f(k) \\
&= \mathrm{tr}\left(Q\left(\mathbf{He}\left(e^{j\theta_c} p_f(k) p_f(k+1)^{\top}\right) - 2\cos(\theta_w) p_f(k) p_f(k)^{\top}\right)\right).
\end{aligned}
$$

Therefore, with $p_f(0) = 0$ and $p_f(\infty) = 0$, we sum (4.17) from $k = 0$ to ∞ obtaining

$$
-\sum_{k=0}^{\infty} p_r(k)^{\top} p_r(k) + \beta^2 \sum_{k=0}^{\infty} f(k)^{\top} f(k) + \mathrm{tr}(QS) \leq 0,
$$

where $S = \sum_{k=0}^{\infty}\left(\mathbf{He}\left(e^{j\theta_c} p_f(k) p_f(k+1)^{\top}\right) - 2\cos(\theta_w) p_f(k) p_f(k)^{\top}\right)$. Based on the result of [5, Theorem 4] and considering the frequency range $\theta \in [\theta_1, \theta_2]$ of occurred faults, with (4.6), we assume that the following condition holds

$$
e^{j\theta_w} \sum_{k=0}^{\infty} \left(p_f(k+1) - e^{j\theta_1} p_f(k)\right) \left(p_f(k+1) - e^{j\theta_2} p_f(k)\right)^* \leq 0, \tag{4.18}
$$

with $\theta_w = \frac{\theta_2-\theta_1}{2}$. If (4.18) holds, then we have $\mathrm{tr}(QS) \geq 0$ and $-\sum_{k=0}^{\infty} p_r(k)^{\top} p_r(k) + \beta^2 \sum_{k=0}^{\infty} f(k)^{\top} f(k) \leq 0$ that implies (4.15).

Furthermore, following the above proof, by choosing $\theta_1 = -\theta_l$ and $\theta_2 = \theta_l$ for the low-frequency case or $\theta_1 = -\theta_h$ and $\theta_2 = 2\pi - \theta_h$ for the high-frequency case, we can obtain that (4.15) is satisfied. □

Based on the result in Lemma 4.1, we then propose the $\mathcal{H}_-$ fault sensitivity condition for designing the observer gain in the following theorem.

Theorem 4.3 ($\mathcal{H}_-$ fault sensitivity condition) *Given the zonotope $\langle p_r(k), H_r(k)\rangle$ in (4.3), $\forall k \in \mathbb{N}$ with the fault $f(k)$ in a finite-frequency domain $\theta_1 \leq \theta \leq \theta_2$, a scalar α, $L \in \mathbb{R}^{q\times n_x}$, and $T \in \mathbb{R}^{n_x\times n_x}$ and $N \in \mathbb{R}^{n_x\times n_y}$ satisfying (4.28). If there exist matrices $W \in \mathbb{S}^{n_x}_{\succ 0}$, $Y \in \mathbb{R}^{n_x\times n_y}$, $P \in \mathbb{S}^{n_x}$ and $Q \in \mathbb{S}^{n_x}_{\succ 0}$, and a scalar $\beta > 0$ such that*

$$\begin{bmatrix} \Phi_f & \star & \star \\ \alpha F^\top T^\top W^\top + LWTA - LYC & \star & \star \\ \alpha W^\top + e^{j\theta_c} Q + WTA - YC & W^\top L^\top + WTF & P + W + W^\top \end{bmatrix} \succ 0, \tag{4.19}$$

with $\Phi_f = C^\top C - P + 2\cos(\theta_w)Q + \mathbf{He}\,(\alpha WTA - \alpha YC)$, then the zonotopic FD observer in (4.3) guarantees the $\mathcal{H}_-$ performance in (4.15).

Proof For $\theta_1 \leq \theta \leq \theta_2$ in any finite-frequency domain, from (4.16), we derive

$$\begin{bmatrix} \Phi_1 & \Phi_2 \\ \Phi_3 & \Phi_4 \end{bmatrix} \prec 0, \tag{4.20}$$

where

$$\begin{aligned} \Phi_1 &= P - 2\cos(\theta_w)Q - C^\top C - (TA - GC)^\top P(TA - GC) \\ &\quad + \mathbf{He}\left((TA - GC)^\top e^{j\theta} Q\right), \\ \Phi_2 &= e^{-j\theta} QTF - (TA - GC)^\top PTF, \\ \Phi_3 &= (TF)^\top e^{j\theta} Q - (TF)^\top P(TA - GC), \\ \Phi_4 &= \beta^2 I_q - (TF)^\top PTF. \end{aligned}$$

Set $\bar{\Phi} = \begin{bmatrix} C^\top C - P + 2\cos(\theta_w)Q & 0 \\ 0 & -\beta^2 I_q \end{bmatrix}$, $Q_f = \begin{bmatrix} Q \\ 0 \end{bmatrix}$ and $A_f = [TA - GC, TF]$. Then, (4.20) is equivalent to

$$\begin{bmatrix} I_n \\ A_f \end{bmatrix}^\top \begin{bmatrix} \bar{\Phi} & \star \\ e^{j\theta_c} Q_f^\top & P \end{bmatrix} \begin{bmatrix} I_n \\ A_f \end{bmatrix} \succ 0.$$

By using the Finsler's lemma to above inequality, we obtain

$$\begin{bmatrix} \bar{\Phi} & \star \\ e^{j\theta_c} Q_f^\top & P \end{bmatrix} + \mathbf{He}\,(RU) \succ 0, \tag{4.21}$$

where $U = \begin{bmatrix} A_f, & I_n \end{bmatrix}$ and $R \in \mathbb{R}^{(2n_x+q)\times n_x}$ is an arbitrary matrix (called multiplier). Therefore, with given α and $L \in \mathbb{R}^{q\times n}$, we define a structure of the multiplier as $R = \begin{bmatrix} \alpha W \\ LW \\ W \end{bmatrix}$. By substituting R in (4.21), we thus obtain (4.19). □

Remark 4.1 From (4.21), we can see that the multiplier R is chosen arbitrarily. The defined structures of R are based on parameters α and L that can be tuned to find feasible solutions of (4.19).

4.1.2.3 Optimization Problem Setup

The objective of designing the FD observer gain is to minimize the effects of uncertainties and maximize the sensitivity to occurred faults. On the one hand, for given $\gamma \in (0, 1)$ and $\epsilon > 0$, we have the ultimate bound of the W-radius $\ell_W(\infty) \leq \frac{\epsilon}{1-\gamma}$ that corresponds to an ellipsoidal set. To minimize the size of this set, we can maximize a measure of the matrix W, for instance we choose to maximize $\mathrm{tr}(W)$. On the other hand, we can maximize the H_- fault sensitivity index β. In general, given $\gamma \in (0, 1)$, $\epsilon > 0$, α, $L \in \mathbb{R}^{q\times n_x}$, $T \in \mathbb{R}^{n_x\times n_x}$ and $N \in \mathbb{R}^{n_x\times n_y}$ satisfying (4.28), and two prioritization weights λ_r and λ_f, the optimization problem for designing the FD observer gain is expressed as follows:

$$\underset{\substack{W,Y,\Gamma,\Upsilon,\Omega,\\ P,Q,\beta^2}}{\text{maximize}} \; \lambda_r \mathrm{tr}(W) + \lambda_f \beta^2, \tag{4.22}$$

subject to (4.10b), (4.10a) and (4.19).

Then, the optimal solution of (4.22) gives the optimal FD observer gain $G = W^{-1}Y$.

Remark 4.2 The weights λ_r and λ_f are set for obtaining a trade-off between robustness and fault sensitivity conditions. For instance, the fault sensitivity objective can be enhanced by choosing $\lambda_f > \lambda_r$.

4.1.3 Zonotopic FD Algorithm

From the output equation (4.1b), we have $0 = y(k) - Cx(k) - D_v v(k)$. Hence, if $f(k) = 0$, we know $0 \in \langle p_r(k), H_r(k)\rangle$, which can be used to determine the FD alarm χ. The logic of the determination of $\chi(k)$, $\forall k \in \mathbb{N}$ is formulated as follows:

$$\chi(k) = \begin{cases} 0 & \text{if } 0 \in \langle p_r(k), H_r\rangle \\ 1 & \text{if } 0 \notin \langle p_r(k), H_r\rangle \end{cases} \tag{4.23}$$

where $\chi = 0$ means that no fault is detected and $\chi = 1$ means that a fault is detected. In general, the robust FD strategy is summarized in Algorithm 4.1.

Algorithm 4.1 Zonotopic FD for Descriptor Systems

1: Given the descriptor system (4.1), $\gamma \in (0, 1)$, $\epsilon > 0$, α, L, $x(0) \in \langle p(0), H(0) \rangle$ and suppose the faults in a finite-frequency domain $\theta \in \Theta$;
2: Obtain a pair of T and N satisfying (2.42);
3: Solve the optimization problem (4.22) to obtain $G = W^{-1}Y$;
4: **while** $k > 0$ **do**
5: Compute the state zonotope $\hat{x}(k) \in \langle \hat{p}(k), \hat{H}(k) \rangle$ using the recursive form of (4.2);
6: Compute the zonotopic FD observer $\langle p_r(k), H_r(k) \rangle$ using (4.3);
7: Determine the FD alarm $\chi(k)$ using the logic in (4.23).
8: **end while**

4.1.4 Case Study: The Chemical Mixing System

Consider the chemical mixing system in [6]. By using the Euler discretization method with the sampling time $t_s = 0.1$ s, we obtain a discrete-time descriptor model of the chemical mixing system as in (4.1) with the following matrices

$$E = \begin{bmatrix} 1 & 0 & 0 & 0 \\ 0 & 0 & 0 & 0 \\ 0 & 0 & 1 & 0 \\ 0 & 0 & 0 & 0 \end{bmatrix}, A = \begin{bmatrix} 0.9625 & 0.0067 & 0 & 0 \\ 0 & -0.1 & 0 & 0 \\ 0.03 & 0.0533 & 0.95 & -0.004 \\ 0 & 0.1 & 0 & -0.1 \end{bmatrix},$$

$$B = \begin{bmatrix} 0.01 & 0 \\ 0.1 & 0 \\ 0 & 0.002 \\ 0 & 0.1 \end{bmatrix}, D_w = \begin{bmatrix} 0.005 & 0 & 0 & 0 \\ 0 & 0.005 & 0 & 0 \\ 0 & 0 & 0.005 & 0 \\ 0 & 0 & 0 & 0.005 \end{bmatrix},$$

$$C = \begin{bmatrix} 0 & 1 & 0 & 0 \\ 0 & 0 & 1 & 0 \\ 0 & 0 & 0 & 1 \end{bmatrix}, D_v = \begin{bmatrix} 0.005 & 0 & 0 \\ 0 & 0.005 & 0 \\ 0 & 0 & 0.005 \end{bmatrix},$$

and $F = 3B$. The initial state $x(0) = [0.5, 0, 0.5, 0]^\top$ is assumed to be bounded by the zonotope $x(0) \in \langle p(0), H(0) \rangle$, where $c_0 = x_0$ and $H_0 = 0.001 I_4$. The input signal u is set as $u = 4\sin(0.3k) + 5$, $\forall k \in \mathbb{N}$. For the reduction operator $\downarrow_{q,W}(\cdot)$, $q = 20$ and W is chosen as the optimal solution of (4.22). Since the rank condition (2.41) is satisfied, we choose one solution as follows:

$$T = \begin{bmatrix} 1 & 1 & 0 & 1 \\ 0 & 1 & 0 & 1 \\ 0 & 1 & 0.5 & 1 \\ 0 & 1 & 0 & 1 \end{bmatrix}, N = \begin{bmatrix} 0 & 0 & 0 \\ 1 & 0 & 0 \\ 0 & 0.5 & 0 \\ 0 & 0 & 1 \end{bmatrix}.$$

Assume that the step fault signal $f(k) = [0.2, 0.2]^\top$, $k \geq 50$ otherwise $f(k) = 0$ in the finite-frequency domain $|\theta| \leq 0.1$. We choose $\gamma = 0.7$, $\epsilon = 1$, $\alpha = 0.3$, $L = 10F^\top$, $\lambda_r = 1$ and $\lambda_f = 10$. By solving the optimization problem (4.22), we thus obtain the optimal FD observer gain G as

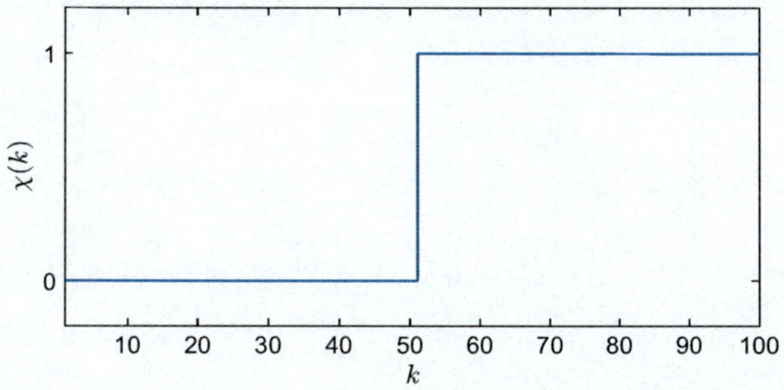

Fig. 4.2 FD alarm result

Table 4.1 Comparison of the objectives in different scenarios

	Weights	Robustness tr(W)	Fault sensitivity β
Scenario 1	$\lambda_r = 1$ and $\lambda_f = 10$	21034.9	0.0045
Scenario 2	$\lambda_r = 10$ and $\lambda_f = 1$	12348.9	0.0031
Scenario 3	$\lambda_r = 1$ and $\lambda_f = 1$	12348.9	0.0033
Scenario 4	$\lambda_r = 1$ and $\lambda_f = 0$	12348.7	0.0032

$$G = \begin{bmatrix} -1.2194 & 121.2034 & -3.7687 \\ -0.7457 & -6.0439 & 0.0837 \\ 0.0793 & 1.6789 & -0.0904 \\ -0.2943 & 19.9514 & -1.6032 \end{bmatrix},$$

and $\beta = 0.0045$ and $\text{tr}(W) = 21034.9$. By applying Algorithm 4.1, we obtain the FD alarm result as shown in Fig. 4.2. The assumed actuator faults can be detected from time $k = 51$.

To compare the performance of the optimization problem with different selections of weights λ_r and λ_f, we have carried out the simulations and the comparison results are presented in Table 4.1 to the trade-off between the robustness against uncertainties and the sensitivity to faults. In order to obtain a better $\mathcal{H}_-$ fault sensitivity performance, we can set $\lambda_f > \lambda_r$ in Scenario 1. In this case, the $\mathcal{H}_-$ fault sensitivity index β increases and the robustness objective $\text{tr}(W)$ also increases compared to the other scenarios. From Scenario 2–3, with the same $\lambda_f = 1$, $\text{tr}(W)$ reaches a locally stationary value because the fault sensitivity LMI depends on the parameters α and L, which subsequently leads to a suboptimal solution of the optimization problem (4.22). The similar suboptimal solution can also be found in Scenario 4.

In order to compare the performance of the FD observer gain, a time-varying observer gain denoted by $\bar{G}(k)$ for state estimation is considered as presented in Sect. 2.1.2, where the objective is only to minimize the effects of bounded uncertainties by reducing the size of the uncertain state zonotope. The comparison result of the generated residual bounds with two observer gains is shown in Fig. 4.3. The residual bounds $r_i(k) \in \left[\underline{r}_i(k), \overline{r}_i(k)\right]$ for $i = 1, \ldots, n_y$ and $\forall k \in \mathbb{N}$ are computed

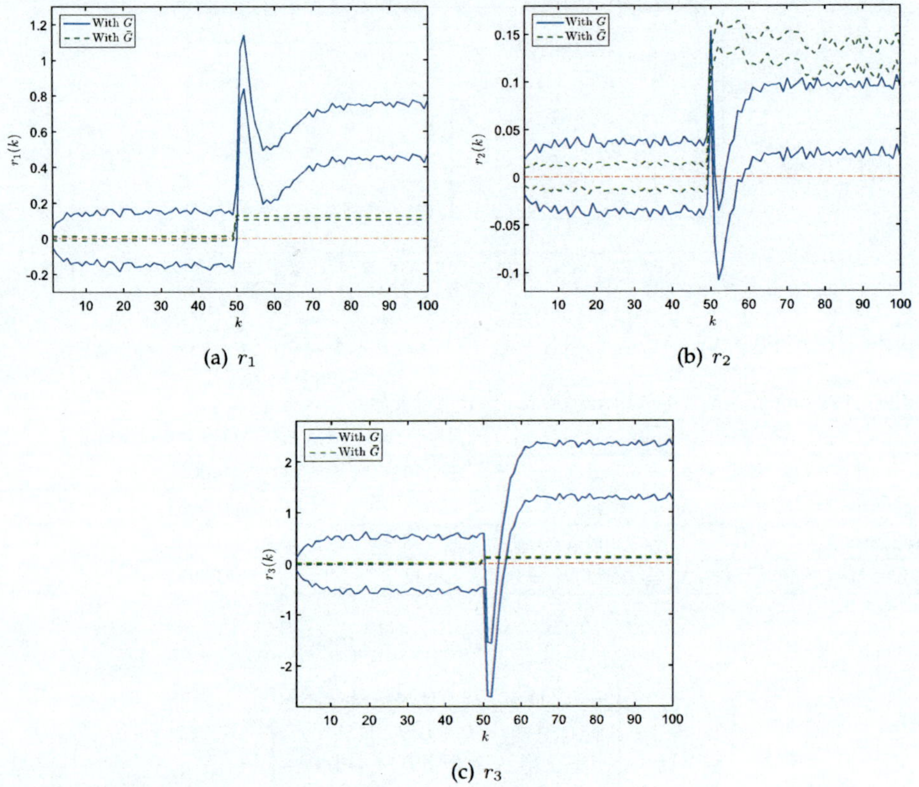

Fig. 4.3 Generated residual bounds

by using the interval hull of the zonotope (see Definition 1.7) by

$$\underline{r_i}(k) = p_{r,i}(k) - rs(H_r(k))_{i,i},$$
$$\overline{r_i}(k) = p_{r,i}(k) + rs(H_r(k))_{i,i}.$$

From Fig. 4.3, we can see that the residual bounds with G are more sensitive to the fault than the others and staying far away from the coordinate origin. Since the considered uncertainty sets are centered and the coordinate origin is used in the decision-making of the FD module, the proposed method is more effective.

4.2 Robust FDI Based on Zonotopic UIOs for LTV Descriptor Systems

In the following, we propose a robust FDI based on zonotopic UIOs for linear time-varying (LTV) descriptor systems [3, 7]. The FD observer gain is designed based on a new defined fault sensitivity criterion. Consider the discrete-time LTV descriptor systems with additive actuator faults as

$$E(k+1)x(k+1) = A(k)x(k) + B(k)u(k) + D_w(k)w(k) + F(k)f(k), \quad (4.24a)$$

$$y(k) = C(k)x(k) + D_v(k)v(k), \quad (4.24b)$$

where $x \in \mathbb{R}^{n_x}$, $u \in \mathbb{R}^m$ and $y \in \mathbb{R}^{n_y}$ denote the state, the known input and the output vectors, respectively. $w \in \mathbb{R}^{n_w}$, $v \in \mathbb{R}^{n_v}$ denote the state disturbance vector and the measurement noise vector. $f \in \mathbb{R}^m$ denotes the normalized additive fault vector. $A(k) \in \mathbb{R}^{n_x \times n_x}$, $B(k) \in \mathbb{R}^{n_x \times m}$, $C(k) \in \mathbb{R}^{n_y \times n_x}$, $D_w(k) \in \mathbb{R}^{n_x \times n_w}$, $D_v(k) \in \mathbb{R}^{n_y \times n_v}$ and $F(k) \in \mathbb{R}^{n_x \times m}$, $\forall k \in \mathbb{N}$ are known time-varying system matrices. Besides, $E(k) \in \mathbb{R}^{n_x \times n_x}$ satisfies $\text{rank}(E(k)) \leq n_x, \forall k \in \mathbb{N}$. In particular, when $\text{rank}(E(k)) = n_x$, (4.24) is equivalent to a dynamical system.

Following the basic FDI framework in [8, Sect. 6.2] and [2, Sect. 3.5], the actuator fault f is modeled in an additive form with the input u. To develop a robust FDI strategy, the fault vector $f(k)$, $\forall k \in \mathbb{N}$ can be rewritten in an element-wise form as

$$f(k) = [f_1(k), \ldots, f_i(k), \ldots, f_m(k)]^\top, \quad \forall k \in \mathbb{N}, \quad (4.25)$$

where the element $f_i(k)$ with $i = 1, \ldots, m$ in the fault vector $f(k)$ corresponds to the i-th actuator fault at time step k. Then, the descriptor system (4.24) can be rewritten as

$$E(k+1)x(k+1) = A(k)x(k) + B(k)u(k) + D_w(k)w(k) + F_i(k)f_i(k) + \bar{F}_i(k)\bar{f}_i(k), \quad (4.26a)$$

$$y(k) = C(k)x(k) + D_v(k)v(k), \quad (4.26b)$$

where $F_i(k)$ denotes the corresponding fault magnitude matrix on the i-th actuator, $\bar{f}_i(k) = f(k) \setminus f_i(k)$ is the fault vector $f(k)$ excluding the i-th element and $\bar{F}_i(k) = F(k) \setminus F_i(k)$ is the matrix obtained by removing i-th column from the fault magnitude matrix $F(k)$ at time step k.

Assumption 4.1 Matrices $E(k)$ and $C(k)$ satisfy the following rank condition:

$$\text{rank} \begin{bmatrix} E(k) \\ C(k) \end{bmatrix} = n_x, \quad \forall k \in \mathbb{N}. \quad (4.27)$$

Thus, from the condition (4.27), there always exist two time-varying matrices $T(k) \in \mathbb{R}^{n_x \times n_x}$ and $N(k) \in \mathbb{R}^{n_x \times n_y}$ such that

$$T(k)E(k) + N(k)C(k) = I, \quad \forall k \in \mathbb{N}. \quad (4.28)$$

Assumption 4.2 The initial state vector is assumed to be bounded in the initial zonotope $x(0) \in \mathcal{X}(0) = \langle p(0), H(0) \rangle$ and the system disturbances and measurement noise are assumed to be unknown but bounded by the centered zonotopes:

$$w(k) \in \mathcal{W} = \langle 0, I_{n_w} \rangle, \ v(k) \in \mathcal{V} = \langle 0, I_{n_v} \rangle, \ \forall k \in \mathbb{N}. \tag{4.29}$$

Assumption 4.3 The normalized fault vector $f(k)$ is assumed to be unknown but bounded by the centered zonotope $f \in \mathcal{F} = \langle 0, I_m \rangle$, $\forall k \in \mathbb{N}$ and its magnitude is given by the distribution matrix $F(k)$.

The uncertain states are estimated by a zonotope considering that all the uncertainties are also bounded by zonotopes. Based on a recursive procedure, estimation errors and uncertainties are also propagated using set operations. We would like to design a zonotopic UIO of the descriptor LTV system (4.24) and (4.26) to implement robust FDI. The objectives for the zonotopic UIO design are summarized as follows:

(i) **Robust FD:** For the LTV descriptor system (4.24), a zonotopic UIO with an observer gain for robust FD is designed to minimize the effects of bounded uncertainties and meanwhile to maximize the fault sensitivity on actuator faults.
(ii) **Robust FI:** For the LTV descriptor system (4.24) in the representation of (4.26), a bank of zonotopic UIOs for robust FI are designed. The observer gain of the i-th zonotopic UIO is designed to remove the effect of the corresponding actuator fault f_i, to maximize the fault sensitivity on the remaining faults $\bar{f}_i$, and meanwhile to minimize the effects of bounded uncertainties.

4.2.1 Zonotopic UIO Structure of LTV Descriptor Systems

Considering Assumption 4.1, we can always find a pair of matrices $T(k+1) \in \mathbb{R}^{n_x \times n_x}$ and $N(k+1) \in \mathbb{R}^{n_x \times n_y}$ such that (4.28) holds at the time step $k+1, \forall k \in \mathbb{N}$. From the system (4.24), the descriptor dynamics can be transformed into

$$\begin{aligned} x(k+1) = & \ T(k+1)A(k)x(k) + T(k+1)B(k)u(k) \\ & + T(k+1)D_w(k)w(k) + T(k+1)F(k)f(k) \\ & + N(k+1)y(k+1) - N(k+1)D_v(k+1)v(k+1). \end{aligned} \tag{4.30}$$

According to [9–11], we consider a basic UIO structure as

$$z(k+1) = M(k)z(k) + K(k)u(k) + G(k)y(k), \tag{4.31a}$$

$$\hat{x}(k) = z(k) + N(k)y(k), \tag{4.31b}$$

$$\hat{y}(k) = C(k)\hat{x}(k), \tag{4.31c}$$

where $z \in \mathbb{R}^{n_x}$, $\hat{x} \in \mathbb{R}^{n_x}$ and $\hat{y} \in \mathbb{R}^{n_y}$ denote vectors of the observer state, the estimated state and output. Besides, $M(k) \in \mathbb{R}^{n_x \times n_x}$, $K(k) \in \mathbb{R}^{n_x \times n_u}$, $N(k) \in \mathbb{R}^{n_x \times n_y}$ and $G(k) \in \mathbb{R}^{n_x \times n_y}$ are time-varying matrices to be designed. In particular, $G(k)$ is the time-varying observer gain of the UIO (4.31).

Let us define the state estimation error as $e(k) = x(k) - \hat{x}(k)$. From (4.31b), we have $e(k) = x(k) - \hat{x}(k) = x(k) - z(k) - N(k)y(k)$. From (4.31) at time step $k + 1$, we can derive

$$\begin{aligned} e(k+1) &= x(k+1) - \hat{x}(k+1) \\ &= x(k+1) - z(k+1) - N(k+1)y(k+1). \end{aligned}$$

By substituting $x(k+1)$ by (4.30) and introducing $e(k)$ in the above equation, we obtain the state estimation error dynamics as

$$\begin{aligned} e(k+1) = M(k)e(k) &+ \Big(T(k+1)A(k) - G(k)C(k) - M(k)\Big)x(k) \\ &+ \Big(T(k+1)B(k) - K(k)\Big)u(k) + M(k)N(k)y(k) \\ &+ T(k+1)D_w(k)w(k) + T(k+1)F(k)f(k) \\ &- G(k)D_v(k)v(k) - N(k+1)D_v(k+1)v(k+1). \end{aligned} \tag{4.32}$$

We first define recursively the zonotopic UIO of the descriptor LTV system (4.24). In this case, the fault vector $f(k)$ is considered as the unknown input. For decoupling the unknown input, it is desired that matrices $E(k)$, $C(k)$ and $D_d = F(k)$ satisfy the condition in (2.2), $\forall k \in \mathbb{N}$. Therefore, there always exist two time-varying matrices $T(k) \in \mathbb{R}^{n_x \times n_x}$ and $N(k) \in \mathbb{R}^{n_x \times n_y}$ such that (4.28) and $T(k)F(k) = 0, \forall k \in \mathbb{N}$. Suppose that the state vector $x(k)$ of the descriptor LTV system (4.24) satisfies the inclusion $x(k) \in \mathcal{X}(k) = \langle p(k), H(k)\rangle$ at time step $k \in \mathbb{N}$, which also satisfies the initial state vector $x(0) \in \langle p(0), H(0)\rangle$ at time step $k = 0$.

Theorem 4.4 (Zonotopic UIO structure of LTV descriptor systems) *Consider the admissible LTV descriptor system (4.24) and $x(k) \in \mathcal{X}(k) = \langle p(k), H(k)\rangle$ at time step $k \in \mathbb{N}$. The zonotopic UIO of the descriptor system (4.24) can be recursively defined by $x(k+1) \in \mathcal{X}(k+1) = \langle p(k+1), H(k+1)\rangle$, where*

$$\begin{cases} p(k+1) &= (T(k+1)A(k) - G(k)C(k))\,p(k) + T(k+1)B(k)u(k) \\ &\quad + G(k)y(k) + N(k+1)y(k+1), \\ H(k+1) &= [R(k), T(k+1)D_w(k), -G(k)D_v(k), -N(k+1)D_v(k+1)], \end{cases} \tag{4.33}$$

with $R(k) = (T(k+1)A(k) - G(k)C(k))\,\bar{H}(k)$ and $\bar{H}(k) = \downarrow_{q,W}(H(k))$.

Proof Consider $x(k) \in \langle p(k), H(k)\rangle$ at time step $k \in \mathbb{N}$ and $\langle p(k), H(k)\rangle \subset \langle p(k), \bar{H}(k)\rangle$ holds. By setting $\hat{x}(k) = p(k)$, we have $e(k) = x(k) - \hat{x}(k) \in \langle 0, H(k)\rangle \subset \langle 0, \bar{H}(k)\rangle$. Therefore, at time step $k+1$, we have $x(k+1) = e(k+1) + \hat{x}(k+1)$.

From $e(k+1)$ in (4.32), let us choose

$$M(k) = T(k+1)A(k) - G(k)C(k), \tag{4.34a}$$

$$K(k) = T(k+1)B(k). \tag{4.34b}$$

Taking into account $f \equiv 0$, (4.32) becomes

$$\begin{aligned} e(k+1) = & \,(T(k+1)A(k) - G(k)C(k))\, e(k) \\ & + (T(k+1)A(k) - G(k)C(k))\, N(k)y(k) \\ & + T(k+1)D_w(k)w(k) - G(k)D_v(k)v(k) \\ & - N(k+1)D_v(k+1)v(k+1). \end{aligned}$$

From (4.31), we can derive

$$\begin{aligned} \hat{x}(k+1) = & \,(T(k+1)A(k) - G(k)C(k))\, p(k) \\ & + T(k+1)B(k)u(k) + \Big(G(k) - (T(k+1)A(k) - G(k)C(k))\, N(k)\Big)y(k) \\ & + N(k+1)y(k+1). \end{aligned}$$

Considering $e(k) \in \langle 0, \bar{H}(k)\rangle$, $w(k) \in \mathcal{W} = \langle 0, I_{n_w}\rangle$ and $v(k), v(k+1) \in \mathcal{V} = \langle 0, I_{n_v}\rangle$, from $x(k+1) = e(k+1) + \hat{x}(k+1)$, we derive

$$\begin{aligned} x(k+1) \in & \,\langle p(k+1), H(k+1)\rangle \\ = & \Big((T(k+1)A(k) - G(k)C(k))\langle 0, \bar{H}(k)\rangle\Big) \\ & \oplus \langle (T(k+1)A(k) - G(k)C(k))\, N(k)y(k), 0\rangle \\ & \oplus \Big(T(k+1)D_w(k)\langle 0, I_{n_w}\rangle\Big) \oplus \Big(-G(k)D_v(k)\langle 0, I_{n_v}\rangle\Big) \\ & \oplus \Big(-N(k+1)D_v(k+1)\langle 0, I_{n_v}\rangle\Big) \oplus \langle \hat{x}(k+1), 0\rangle. \end{aligned}$$

Thus, using the properties in (1.9), we obtain $p(k+1)$ and $H(k+1)$ as in (4.33). □

Remark 4.3 Note that the zonotope $\mathcal{X}(k) = \langle p(k), H(k)\rangle$ is used for bounding $x(k)$, $\forall k \in \mathbb{N}$ while the estimated state $\hat{x}(k)$ in (4.31) only determines the nominal value and the estimation error is omitted in the formulation of (4.31). According to the proof of Theorem 4.4, from $x(k) \in \langle p(k), H(k)\rangle$, we know $p(k) = \hat{x}(k)$ and the state estimation error $e(k) = x(k) - \hat{x}(k) \in \langle 0, H(k)\rangle$.

Remark 4.4 Considering Assumption 4.2, from the output equation (4.24b), for $x(k) \in \mathcal{X}(k) = \langle p(k), H(k)\rangle$, $\forall k \in \mathbb{N}$, we can derive the output zonotope $\mathcal{Y}(k) = \langle p_y(k), H_y(k)\rangle$, where

$$\begin{aligned} y(k) \in & \,\langle p_y(k), H_y(k)\rangle \\ = & \,(C(k)\langle p(k), H(k)\rangle) \oplus \Big(D_v(k)\langle 0, I_{n_v}\rangle\Big) \\ = & \,\langle C(k)p(k), [C(k)H(k),\ D_v(k)]\rangle. \end{aligned}$$

Since $p(k) = \hat{x}(k)$, from the output zonotope $\mathcal{Y}(k) = \langle p_y(k), H_y(k)\rangle$, we also know $\hat{y}(k) = p_y(k) = C(k)p(k)$ and the output estimation error $\epsilon(k) = y(k) - \hat{y}(k) \in \langle 0, H_y(k)\rangle$.

To implement a FDI strategy, let us define the residual zonotope

$$\mathcal{R}(k) = y(k) \oplus (-\mathcal{Y}(k)) \,. \tag{4.35}$$

We present the explicit computational result of this residual zonotope in the following.

Corollary 4.1 *Consider the admissible LTV descriptor system (4.24) and* $x(k) \in \mathcal{X}(k) = \langle p(k), H(k)\rangle$, $\forall k \in \mathbb{N}$. *The residual zonotope is given by* $\mathcal{R}(k) = \langle p_r(k), H_r(k)\rangle$, *where*

$$\begin{cases} p_r(k) &= y(k) - C(k)p(k), \\ H_r(k) &= \begin{bmatrix} -C(k)H(k), & -D_v(k) \end{bmatrix}. \end{cases} \tag{4.36}$$

Proof Based on Theorem 4.4, $x(k) \in \langle p(k), H(k)\rangle$ can be computed recursively, $\forall k \in \mathbb{N}$. According to the definition of $\mathcal{R}(k)$, it follows

$$\begin{aligned} \mathcal{R}(k) &= \langle p_r(k), H_r(k)\rangle \\ &= y(k) \oplus \langle -C(k)p(k), [-C(k)H(k), \ -D_v(k)]\rangle. \end{aligned}$$

By applying the properties in (1.9), we thus obtain (4.36). □

The output equation (4.24b) can be rewritten as $0 = y(k) - C(k)x(k) - D_v(k)v(k)$. Taking into account that $v(k) \in \langle 0, I_{n_v}\rangle$, if no fault occurred until time step k in the zonotope $x(k) \in \mathcal{X}(k) = \langle p(k), H(k)\rangle$, then the following condition holds:

$$0 \in \mathcal{R}(k). \tag{4.37}$$

To analyze the effects of occurred actuator faults in the defined state or residual zonotope above, we consider the normalized fault vector $f(k) \in \mathcal{F}$, $\forall k \in \mathbb{N}$, i.e., the magnitude of the fault vector $f(k)$ is stored in the matrix $F(k)$. Therefore, we present the decomposed zonotopic UIO structure for the descriptor system in the presence of faults considering $f(k) \in \mathcal{F}$, $\forall k \in \mathbb{N}$ in the following theorem.

Theorem 4.5 (Zonotopic UIO decomposition of LTV descriptor systems) *Consider the admissible LTV descriptor system (4.24) with* $f(k) \in \mathcal{F}$ *and* $x(k) \in \{\langle p_e(k), H_e(k)\rangle \oplus \langle p_f(k), H_f(k)\rangle\}$, $\forall k \in \mathbb{N}$. *The zonotopic UIO affected by actuator faults can be recursively defined in the decomposition form as* $x(k+1) \in \{\langle p_e(k+1), H_e(k+1)\rangle \oplus \langle p_f(k+1), H_f(k+1)\rangle\}$, *where*

$$\begin{cases} p_e(k+1) &= (T(k+1)A(k) - G(k)C(k))\, p_e(k) + T(k+1)B(k)u(k) \\ & \quad + G(k)y(k) + N(k+1)y(k+1), \\ H_e(k+1) &= \begin{bmatrix} R_e(k), & T(k+1)D_w(k), & -G(k)D_v(k), & -N(k+1)D_v(k+1) \end{bmatrix}, \end{cases} \tag{4.38}$$

and

$$\begin{cases} p_f(k+1) &= (T(k+1)A(k) - G(k)C(k))\, p_f(k), \\ H_f(k+1) &= \left[(T(k+1)A(k) - G(k)C(k))\, \bar{H}_f(k),\; T(k+1)F(k)\right], \end{cases} \tag{4.39}$$

with $R_e(k) = (T(k+1)A(k) - G(k)C(k))\, \bar{H}_e(k)$, $\bar{H}_e(k) = \downarrow_{q,W} (H_e(k))$, $\bar{H}_f(k) = \downarrow_{q,W} (H_f(k))$, $H_e(k+1) \in \mathbb{R}^{n_x \times n_e}$, *and* $H_f(k+1) \in \mathbb{R}^{n_x \times n_f}$.

Proof Consider $x(k) \in \{\langle p_e(k), H_e(k)\rangle \oplus \langle p_f(k), H_f(k)\rangle\} \subset \{\langle p_e(k), \bar{H}_e(k)\rangle \oplus \langle p_f(k), \bar{H}_f(k)\rangle\} = \langle \left(p_e(k) + p_f(k)\right), \left[\bar{H}_e(k), \bar{H}_f(k)\right]\rangle$. By setting $\hat{x}(k) = p_e(k) + p_f(k)$, we have $e(k) = x(k) - \hat{x}(k) \in \langle 0, \left[\bar{H}_e(k), \bar{H}_f(k)\right]\rangle$.

Let us choose the matrices $M(k)$ and $K(k)$ as in (4.34). With $w(k) \in \mathcal{W} = \langle 0, I_{n_w}\rangle$, $v(k), v(k+1) \in \mathcal{V} = \langle 0, I_{n_v}\rangle$ and $f(k) \in \mathcal{F} = \langle 0, I_m\rangle$, we derive $x(k+1) = e(k+1) + \hat{x}(k+1)$ to obtain

$$\begin{aligned} x(k+1) \in\ & \{\langle p_e(k+1), H_e(k+1)\rangle \oplus \langle p_f(k+1), H_f(k+1)\rangle\} \\ =\ & \left((T(k+1)A(k) - G(k)C(k))\,\langle p_e(k) + p_f(k), \left[\bar{H}_e(k),\ \bar{H}_f(k)\right]\rangle\right) \\ & \oplus \langle T(k+1)B(k)u(k), 0\rangle \oplus \langle G(k)y(k), 0\rangle \oplus \langle N(k+1)y(k+1), 0\rangle \\ & \oplus \left(T(k+1)D_w(k)\langle 0, I_{n_w}\rangle\right) \oplus \left(-G(k)D_v(k)\langle 0, I_{n_v}\rangle\right) \\ & \oplus \left(-N(k+1)D_v(k+1)\langle 0, I_{n_v}\rangle\right) \oplus \left(T(k+1)F(k)\langle 0, I_m\rangle\right). \end{aligned}$$

Then, the zonotope $\langle p_e(k+1), H_e(k+1)\rangle$ is only affected by uncertainties while the zonotope $\langle p_f(k+1), H_f(k+1)\rangle$ is only affected by faults if they are chosen as in (4.38) and (4.39). □

Corollary 4.2 *Consider the admissible LTV descriptor system (4.24) with* $f(k) \in \mathcal{F}$ *and* $x(k) \in \{\langle p_e(k), H_e(k)\rangle \oplus \langle p_f(k), H_f(k)\rangle\}$, $\forall k \in \mathbb{N}$. *The residual zonotope* $\mathcal{R}(k) = \langle p_r(k), H_r(k)\rangle$ *can be decomposed as* $\mathcal{R}(k) = \{\langle p_{re}(k), H_{re}(k)\rangle \oplus \langle p_{rf}(k), H_{rf}(k)\rangle\}$, *where*

$$\begin{cases} p_{re}(k) &= y(k) - C(k)p_e(k), \\ H_{re}(k) &= \left[-C(k)H_e(k),\ -D_v(k)\right]. \end{cases} \tag{4.40}$$

and

$$\begin{cases} p_{rf}(k) &= -C(k)p_f(k), \\ H_{rf}(k) &= -C(k)H_f(k). \end{cases} \tag{4.41}$$

Proof The proof is straightforward based on the zonotope properties and therefore is omitted here. □

From Theorem 4.5 and Corollary 4.2, we have divided the effects of system uncertainties and actuator faults. Specifically, the effects of uncertainties (disturbances and noise) are propagated to the zonotope $\langle p_e(k+1), H_e(k+1)\rangle$ while the effects of actuator faults are constrained in the zonotope $\langle p_f(k+1), H_f(k+1)\rangle$. Hence, for the FD observer gain G design, we use the decomposed zonotopic UIO structure defined in Theorem 4.5 to discuss robustness to uncertainties and sensitivity to actuator faults.

4.2.2 *Observer Gain Designs*

4.2.2.1 Optimal Kalman Gain for LTV Descriptor Systems

As discussed in Theorem 4.5, we can characterize the effects of uncertainties and faults with the zonotopes $\langle p_e(k+1), H_e(k+1)\rangle$ and $\langle p_f(k+1), H_f(k+1)\rangle$ separately. Hence, the problem of designing an FD observer gain to be robust against uncertainties and to be sensitive to faults is transformed to minimizing or maximizing the size of these zonotopes. Following the result in Sect. 2.1.2, the size of a zonotope can be measured by the F_W-radius. For state estimation, the objective of the observer gain design is only to minimize the effects of uncertainties. According to Theorem 2.6, the optimal Kalman observer gain for the admissible LTV descriptor system (4.24) in the fault-free case ($f = 0$) can be computed in the following corollary.

Corollary 4.3 (Optimal Kalman gain for LTV descriptor systems) *Given the zonotopic UIO structure in (4.33) of the admissible LTV descriptor system (4.24) with $f \equiv 0$ and a matrix $W \in \mathbb{S}^{n_x}_{\succ 0}$, the optimal time-varying Kalman gain $\bar{G}(k) = \arg\min_{G(k)} J_s$, where $J_s = \|H(k+1)\|^2_{F,W}$ is computed by the following procedure*

$$\bar{G}(k) = T(k+1)A(k)\bar{K}(k), \tag{4.42a}$$
$$\bar{K}(k) = L(k)S(k)^{-1}, \tag{4.42b}$$
$$L(k) = \bar{P}(k)C(k)^{\top}, \tag{4.42c}$$
$$S(k) = C(k)\bar{P}(k)C(k)^{\top} + D_v(k)D_v(k)^{\top}, \tag{4.42d}$$

with $\bar{P}(k) = \bar{H}(k)\bar{H}(k)^{\top}$.

Proof For $x(k+1) \in \langle p(k+1), H(k+1)\rangle$ in (4.33), the criterion

$$J_s = \|H(k+1)\|^2_{F,W} = \mathrm{tr}(WP(k+1))$$

with $P(k+1) = H(k+1)(H(k+1))^{\top}$ is convex with respect to $G(k)$. The optimal Kalman gain $\bar{G}(k)$ satisfies

$$\frac{d}{dG(k)}\mathrm{tr}\,(WP(k+1)) = 0.$$

Hence, we compute derivative of J_s with respect to $G(k)$. Selecting $L(k)$ and $S(k)$ as in (4.42), we have

$$\frac{d}{dG(k)}\text{tr}\left(WG(k)S(k)G(k)^\top\right) - 2\frac{d}{dG(k)}\text{tr}\left(WT(k+1)A(k)L(k)G(k)^\top\right) = 0.$$

From the above equation, we obtain the optimal Kalman gain $\bar{G}(k)$ as in (4.42).
□

4.2.2.2 FD Observer Gain

To design an FD observer gain, in addition to guarantee robustness to uncertainties, we would like to maximize the fault sensitivity on actuator faults, which can be realized by maximizing the F_W-radius of the zonotope $\langle p_f(k+1), H_f(k+1)\rangle$. Assume that there exist matrices $T(k+1)$, $N(k+1)$ and $G(k)$ such that the zonotopic UIO in (4.33) is stable. The objectives of the FD observer gain $G(k)$ can be implemented by solving the following optimization problem

$$\underset{G}{\text{minimize}}\|H_e(k+1)\|_{F,W_1}^2 \;\text{ and simultaneously }\; \underset{G}{\text{maximize}}\|H_f(k+1)\|_{F,W_2}^2, \tag{4.43}$$

with matrices $W_1, W_2 \in \mathbb{S}_{\succ 0}^{n_x}$.

To implement the optimization problem above, we define a performance criterion as

$$J_{e/f} = \frac{\|H_f(k+1)\|_{F,W_1}^2}{\|H_e(k+1)\|_{F,W_2}^2}. \tag{4.44}$$

Therefore, the optimization problem (4.43) is converted to maximize $J_{e/f}$. In order to find the solution of the FD observer gain, we first reformulate $\|H_f(k+1)\|_{F,W_1}^2$ and $\|H_e(k+1)\|_{F,W_2}^2$ using the properties in (1.9) as follows. From Definition 1.8, for $\|H_f(k+1)\|_{F,W_1}^2$, we have

$$\begin{aligned}
\|H_f(k+1)\|_{F,W_1}^2 &= \text{tr}\left((H_f(k+1))^\top W_1 H_f(k+1)\right) = \text{tr}\left(W_1 H_f(k+1)(H_f(k+1))^\top\right) \\
&= \mathbf{vec}\left(H_f(k+1)\right)^\top \mathbf{vec}\left(W_1 H_f(k+1)\right) \\
&= \mathbf{vec}\left(H_f(k+1)\right)^\top \left(I_{n_f} \otimes W_1\right) \mathbf{vec}\left(H_f(k+1)\right),
\end{aligned}$$

and from (4.39), we have

$$\mathbf{vec}\left(H_f(k+1)\right) = \begin{bmatrix} -(C(k)\bar{H}_f(k)^\top \otimes I_{n_x} \\ 0 \end{bmatrix} \mathbf{vec}(G(k)) + \begin{bmatrix} \mathbf{vec}\left(T(k+1)A(k)\bar{H}_f(k)\right) \\ \mathbf{vec}\left(T(k+1)F(k)\right) \end{bmatrix}.$$

Selecting

$$\theta(k) = \mathbf{vec}(G(k)),$$

and

$$S_f(k) = \begin{bmatrix} -(C(k)\bar{H}_f(k))^\top \otimes I_{n_x} \\ 0 \end{bmatrix}, \ b_f(k) = \begin{bmatrix} \mathbf{vec}\left(T(k+1)A(k)\bar{H}_f(k)\right) \\ \mathbf{vec}\left(T(k+1)F(k)\right) \end{bmatrix},$$

we have

$$\begin{aligned} \|H_f(k+1)\|_{F,W_1}^2 &= (S_f(k)\theta(k) + b_f(k))^\top \left(I_{n_f} \otimes W_1\right)(S_f(k)\theta(k) + b_f(k)) \\ &= \begin{bmatrix} \theta(k) \\ 1 \end{bmatrix}^\top \left[S_f(k),\, b_f(k)\right]^\top \left(I_{n_f} \otimes W_1\right)\left[S_f(k),\, b_f(k)\right] \begin{bmatrix} \theta(k) \\ 1 \end{bmatrix} \\ &= \bar{\theta}(k)^\top Q_f(k)\bar{\theta}(k), \end{aligned}$$

where

$$\bar{\theta}(k) = \begin{bmatrix} \theta(k) \\ 1 \end{bmatrix},$$

and

$$Q_f(k) = \left[S_f(k),\, b_f(k)\right]^\top \left(I_{n_f} \otimes W_1\right)\left[S_f(k),\, b_f(k)\right]. \tag{4.45}$$

Similarly, $\|H_e(k+1)\|_F^2$ can be reformulated as

$$\|H_e(k+1)\|_{F,W_2}^2 = \bar{\theta}(k)^\top Q_e(k)\bar{\theta}(k),$$

where

$$Q_e(k) = \left[S_e(k),\, b_e(k)\right]^\top \left(I_{n_e} \otimes W_2\right)\left[S_e(k),\, b_e(k)\right], \tag{4.46}$$

and

$$S_e(k) = \begin{bmatrix} -(C(k)\bar{H}_e(k))^\top \otimes I_{n_x} \\ -D_v(k)^\top \otimes I_{n_x} \\ 0 \\ 0 \end{bmatrix}, \ b_e(k) = \begin{bmatrix} \mathbf{vec}\left(T(k+1)A(k)\bar{H}_e(k)\right) \\ 0 \\ \mathbf{vec}\left(T(k+1)D_w(k)\right) \\ -\mathbf{vec}\left(N(k+1)D_v(k+1)\right) \end{bmatrix}.$$

Then, the performance criterion $J_{e/f}$ defined in (4.44) can be rewritten as

$$J_{e/f} = \frac{\bar{\theta}(k)^\top Q_f(k)\bar{\theta}(k)}{\bar{\theta}(k)^\top Q_e(k)\bar{\theta}(k)}. \tag{4.47}$$

Due to change of variables $\bar{\theta}(k) = \begin{bmatrix} \theta(k) \\ 1 \end{bmatrix}$ and $\theta(k) = \mathbf{vec}(G(k))$, finding an FD observer gain G by maximizing $J_{e/f}$ in (4.44) is equivalent to finding $\bar{\theta}^*(k)$ such that $J_{e/f}$ in (4.47) is maximum. Then, we provide the explicit solutions of the optimal $\bar{\theta}^*(k)$ corresponding to maximum $J^*_{e/f}$ in the following theorem.

Theorem 4.6 *Given the criterion $J_{e/f}$ defined in (4.47) with respect to $\bar{\theta}(k)$ and matrices $W_1, W_2 \in \mathbb{S}^{n_x}_{>0}$, the maximum $J_{e/f}$ is the maximum generalized eigenvalue of $\left(Q_f(k), Q_e(k)\right)$ with $Q_f(k)$ as in (4.45) and $Q_e(k)$ as in (4.46), that is denoted by $J^*_{e/f} = \lambda_{\max}\left(Q_f(k), Q_e(k)\right)$, and the optimal $\bar{\theta}^*(k)$ belongs to the null space of $\left(Q_f(k) - J^*_{e/f} Q_e(k)\right)$, that is also the generalized eigenvector of $\left(Q_f(k), Q_e(k)\right)$ corresponding to its maximum generalized eigenvalue.*

Proof To find the optimal $\bar{\theta}^*(k)$ corresponding to the maximum $J^*_{e/f}$, we take the derivative of $J_{e/f}$ in (4.47) with respect to $\bar{\theta}(k)$ as

$$\frac{d}{d\bar{\theta}(k)} J_{e/f} = \frac{2Q_f(k)\bar{\theta}(k)\left(\bar{\theta}(k)^\top Q_e(k)\bar{\theta}(k)\right) - 2Q_e(k)\bar{\theta}(k)\left(\bar{\theta}(k)^\top Q_f(k)\bar{\theta}(k)\right)}{\left(\bar{\theta}(k)^\top Q_e(k)\bar{\theta}(k)\right)^2},$$

By setting $\frac{d}{d\bar{\theta}(k)} J_{e/f} = 0$, we obtain

$$2Q_f(k)\bar{\theta}^*(k)\left(\bar{\theta}^*(k)^\top Q_e\bar{\theta}^*(k)\right) - 2Q_e(k)\bar{\theta}^*(k)\left(\bar{\theta}^*(k)^\top Q_f(k)\bar{\theta}^*(k)\right) = 0,$$

which can be simplified to be

$$Q_f(k)\bar{\theta}^*(k) = \frac{\bar{\theta}^*(k)^\top Q_f(k)\bar{\theta}^*(k)}{\bar{\theta}^*(k)^\top Q_e(k)\bar{\theta}^*(k)} Q_e(k)\bar{\theta}^*(k).$$

From (4.47) and $\bar{\theta}^*(k)$ corresponding to the maximum $J^*_{e/f}$, we have

$$Q_f(k)\bar{\theta}^*(k) = J^*_{e/f} Q_e(k)\bar{\theta}^*(k). \tag{4.48}$$

Then, (4.48) leads to a generalized eigenvalue problem with $J^*_{e/f} = \lambda_{\max}\left(Q_f(k), Q_e(k)\right)$ being the maximum generalized eigenvalue and $\bar{\theta}^*(k)$ being the corresponding eigenvector. Besides, from (4.48), we can also derive

$$\left(Q_f(k) - J^*_{e/f} Q_e(k)\right)\bar{\theta}^*(k) = 0.$$

Hence, $\bar{\theta}^*(k)$ also belongs to the null space of $\left(Q_f(k) - J^*_{e/f} Q_e(k)\right)$. □

Based on the optimal solution $\bar{\theta}^*(k)$, we derive the optimal FD observer gain in the following theorem.

Theorem 4.7 (Optimal FD observer gain for LTV descriptor systems) *Given the optimal solution $\bar{\theta}^*(k)$ from Theorem 4.6 as $\bar{\theta}^*(k) = \begin{bmatrix} \tilde{\theta}^*(k) \\ \breve{\theta}^*(k) \end{bmatrix}$, $\tilde{\theta}^*(k) \in \mathbb{R}^{(n_x \cdot n_y) \times 1}$ and $\breve{\theta}^*(k) \in \mathbb{R}$, the optimal FD observer gain $G^*(k)$ can be computed by*

$$G^*(k) = \mathbf{vec}^{-1}\left(\frac{\tilde{\theta}^*(k)}{\breve{\theta}^*(k)}\right). \tag{4.49}$$

Proof By dividing $\breve{\theta}^*(k)$ in both sides of (4.48), we have

$$Q_f(k)\begin{bmatrix} \frac{\tilde{\theta}^*(k)}{\breve{\theta}^*(k)} \\ 1 \end{bmatrix} = J^*_{e/f} Q_e(k) \begin{bmatrix} \frac{\tilde{\theta}^*(k)}{\breve{\theta}^*(k)} \\ 1 \end{bmatrix}.$$

Based on the structure of $\bar{\theta}(k) = \begin{bmatrix} \theta(k) \\ 1 \end{bmatrix}$, we thus obtain $G^*(k)$ as in (4.49). □

4.2.3 Robust FDI Using Zonotopic UIOs

To include robust FI, the idea is to design a bank of zonotopic UIOs for identifying the effect from each actuator fault. From (4.25), the single fault is considered as an unknown input to be decoupled for the corresponding zonotopic UIO. The general robust FDI scheme is presented in Fig. 4.4. For the LTV descriptor system (4.24) with m actuators, we would like to design m zonotopic UIOs. By checking the residual zonotopes obtained by m zonotopic UIOs, the FDI alarm can be determined by the FDI module.

From the LTV descriptor representation in (4.26), we treat $f_i(k)$, $i = 1, \ldots, m$ as an unknown input of the LTV descriptor system (4.26). With $f_i(k)$ and $\bar{f}_i(k)$, $i = 1, \ldots, m$, the descriptor dynamics can be reformulated as

$$\begin{aligned} x(k+1) = {} & T_i(k+1)A(k)x(k) + T_i(k+1)B(k)u(k) + N_i(k+1)y(k+1) \\ & + T_i(k+1)D_w(k)w(k) - N_i(k+1)D_v(k+1)v(k+1) \\ & + T_i(k+1)F_i(k)f_i(k) + T_i(k+1)\bar{F}_i(k)\bar{f}_i(k), \end{aligned} \tag{4.50}$$

and from (4.50), state estimation error dynamics can also be reformulated as

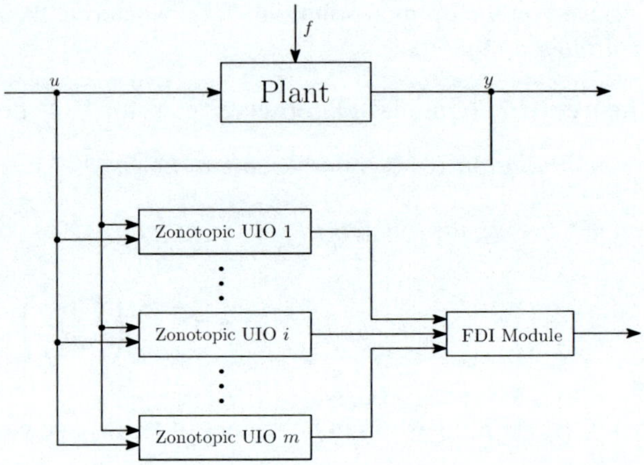

Fig. 4.4 Zonotopic UIO-based robust FDI scheme

$$
\begin{aligned}
e(k+1) = & \left(T_i(k+1)A(k) - G(k)C(k)\right) e(k) \\
& + \left(T_i(k+1)A(k) - G(k)C(k)\right) N_i(k)y(k) \\
& + T_i(k+1)D_w(k)w(k) - G(k)D_v(k)v(k) \\
& - N_i(k+1)D_v(k+1)v(k+1) \\
& + T_i(k+1)F_i(k)f_i(k) + T_i(k+1)\bar{F}_i(k)\bar{f}_i(k).
\end{aligned} \tag{4.51}
$$

To remove the effect of $f_i(k)$ and preserve the effect of $\bar{f}_i(k)$ in (4.39) and (4.51), a pair of matrices $T_i(k+1) \in \mathbb{R}^{n_x \times n_x}$ and $N_i(k+1) \in \mathbb{R}^{n_x \times n_y}$ for the i-th zonotopic UIO also satisfies (4.28) and

$$T_i(k+1)F_i(k) = 0, \tag{4.52a}$$

$$T_i(k+1)\bar{F}_i(k) \neq 0. \tag{4.52b}$$

Based on the condition (2.2), we present the condition for the existence of matrices $T_i(k+1)$ and $N_i(k+1)$ satisfying (4.28) and (4.52a) in the following.

Assumption 4.4 For the LTV descriptor system (4.24), matrices $E(k+1)$, $C(k+1)$ and $F_i(k)$ satisfy the following rank condition

$$\text{rank}\begin{bmatrix} I_{n_x} \otimes \begin{bmatrix} E(k+1) & F_i(k) \\ C(k+1) & 0 \end{bmatrix} \\ \mathbf{vec}\left(\begin{bmatrix} I_{n_x} \\ 0 \end{bmatrix}\right)^{\top} \end{bmatrix} = n_x \cdot \begin{bmatrix} E(k+1) & F_i(k) \\ C(k+1) & 0 \end{bmatrix}, \ \forall k \in \mathbb{N}. \tag{4.53}$$

Therefore, from the proof of Theorem 4.4, we know $x(k+1) = e(k+1) + \hat{x}(k+1)$. In (4.51), the effect of f_i is removed in $e(k+1)$ by using the matrix $T_i(k+1)$ with $T_i(k+1)F_i(k) = 0$ and meanwhile the effect of $\bar{f}_i(k)$ is preserved. Besides, for designing the i-th observer gain, considering $\bar{f}_i(k) \in \mathcal{F}_i = \langle 0, I_{m-1}\rangle$, we replace

$T(k+1)$ by $T_i(k+1)$ and $F(k)$ by $\bar{F}_i(k)$ in (4.39). Following the design procedure of FD gain, the optimal observer gain $G_i^*(k)$ for robust FDI can be obtained.

After having m zonotopic UIOs, at each time step, a sequence of residual zonotopes $\langle p_{r,i}(k), H_{r,i}(k)\rangle$, $i = 1, \ldots, m$ can be generated based on Corollary 4.1. Then, a fault can be determined in the FDI module. The logic of the FDI module are proposed as follows.

The logic of the FDI module:

$$\begin{cases} 0 \in \langle p_{r,i}(k), H_{r,i}(k)\rangle \text{ and } 0 \notin \langle p_{r,j}(k), H_{r,j}(k)\rangle, \; i \neq j = 1, \ldots, m & \textbf{The } i\textbf{-th actuator fault is detected} \\ 0 \in \langle p_{r,i}(k), H_{r,i}(k)\rangle, \; i = 1, \ldots, m & \textbf{No fault is detected} \end{cases}$$

Remark 4.5 Note that we can also decouple $m-1$ actuator faults as unknown inputs by finding suitable $T_i(k)$ and N_i, $\forall k \in \mathbb{N}$. For the remaining fault that is not decoupled, if $0 \notin \langle p_{r,i}, H_{r,i}\rangle$, $i = 1, \ldots, m$, then it can be detected.

Algorithm 4.2 Robust FDI based on Zonotopic UIOs

```
Given the discrete-time LTV descriptor system (4.24) with system matrices E(k+1), A(k),
  B(k), C(k), D_w(k), D_v(k), F(k) and x(0) ∈ ⟨p(0), H(0)⟩, w(k) ∈ ⟨0, I_{n_w}⟩, v(k) ∈ ⟨0, I_{n_v}⟩,
  f(k) ∈ ⟨0, I_m⟩, ∀k ∈ N;
p(k) ⟵ p(0), H(k) ⟵ H(0);
p_e(k) ⟵ p(0), H_e(k) ⟵ H(0);
p_f(k) ⟵ 0, H_f(k) ⟵ 0;
for k = 1 : Γ do
  Obtain the residual zonotope ⟨p_{r,i}(k), H_{r,i}(k)⟩ in (4.36);
  Determine the FDI alarm by the logic of the FDI module;
  for i = 1 : end do
    Reformulate F(k) to find F̄_i(k) = F(k) \ F_i(k);
    Obtain matrices T_i(k+1) and N_i(k+1) for i = 1, ..., m by satisfying (4.28) and (4.52);
    F ⟵ F̄_i, T(k+1) ⟵ T_i(k+1), N(k+1) ⟵ N_i(k+1);
    Compute the zonotopes ⟨p_e(k+1), H_e(k+1)⟩ by (4.38) and ⟨p_f(k+1), H_f(k+1)⟩
      by (4.39) for the i-th zonotopic UIO;
    Compute the observer gain G_i^*(k) for (4.33) following the proposed computation steps
      presented above;
    Gather the system outputs y(k) and y(k+1);
    Update the state zonotope x(k+1) ∈ ⟨p(k+1), H(k+1)⟩ in (4.33);
  end for
end for
```

We now summarize the robust FDI strategy in Algorithm 4.2 considering a simulation horizon of Γ. Note that this robust FDI strategy based on zonotopic UIO can also be applied to standard dynamical systems, that is when $\text{rank}(E) = n_x$, $\forall k \in \mathbb{N}$.

4.2.4 Case Studies

4.2.4.1 Numerical Example

To compare the FD observer gain obtained with the proposed approach with zonotopic Kalman observer gain, we consider the LTV descriptor system (4.24) with

$$E = \begin{bmatrix} 1&0&0&0 \\ 0&1&0&0 \\ 0&0&1&0 \\ 0&0&0&0 \end{bmatrix},\ A(k) = \begin{bmatrix} 0.5 & 0.3\sin(0.4k) & 0 & 0 \\ 0 & 0.3 & 0 & 0 \\ 0 & 0 & 0.6 & 0 \\ 0 & -0.5 & -0.5 & 0.8 \end{bmatrix},\ B = \begin{bmatrix} 0.1 \\ 1 \\ -0.1 \\ -1 \end{bmatrix},$$

$$C = \begin{bmatrix} 0&1&0&0 \\ 0&0&1&0 \\ 0&0&0&1 \end{bmatrix},\ D_w = 0.005I_4,\ D_v = 0.01I_3,\ F = 2B,$$

and the initial state $x(0) = [2, 2, 3, 3.125]^{\mathrm{T}}$ is assumed to be bounded by the zonotope $x(0) \in \langle p(0), H(0)\rangle$, where $p(0) = x(0)$ and $H(0) = 0.1I_4$. The weighting matrices W_1 and W_2 for designing the FD observer gain are chosen to be identity matrices of appropriate dimensions. The input signal u is set as $u(k) = 2, \forall k \in \mathbb{N}$. For the reduction operator $\downarrow_{q,W}(\cdot)$, q and W are set respectively as $q = 20$ and $W = I$. With constant matrices E and C, by satisfying the condition (4.28), we consider one solution of constant matrices T and N as follows:

$$T = \begin{bmatrix} 1&0&0&0 \\ 0&0.5&0&0 \\ 0&0&0.5&0 \\ 0&0&0&1 \end{bmatrix},\ N = \begin{bmatrix} 0&0&0 \\ 0.5&0&0 \\ 0&0.5&0 \\ 0&0&1 \end{bmatrix}.$$

The simulation has been carried out in MATLAB for 100 sampling time steps. With this example, we compute the time-varying Kalman gain $\bar{G}(k)$ (following Corollary 4.3) and the designed FD observer gain $G^*(k)$ (following Theorem 4.7) at each time step. Since the system has three measurement outputs, the residual zonotope $\langle p_r(k), H_r(k)\rangle, \forall k \in \mathbb{N}$ is in a 3-dimensional space. Therefore, the interval hull (see Definition 1.7) of the residual zonotope is used to plot the individual residual bounds $r_i(k) \in \left[\underline{r_i}(k), \overline{r_i}(k)\right]$ for $r_i(k) \in \mathbb{R}^{n_y}$, where

$$\underline{r_i}(k) = p_{r,i}(k) - rs(H_r(k))_{i,i},\ i = 1, \ldots, n_y, \forall k \in \mathbb{N},$$
$$\overline{r_i}(k) = p_{r,i}(k) + rs(H_r(k))_{i,i},\ i = 1, \ldots, n_y, \forall k \in \mathbb{N}.$$

Consider a step actuator fault $f(k) = 0.3, k \geq 30$. The comparative results of the residuals and their lower and upper bounds are shown in Fig. 4.5. From these plots, it is shown that when no fault occurred ($f(k) = 0, k < 30$), the coordinate origin is inside all the residual bounds, which means it is also inside the residual zonotope. Besides, the bounds with the Kalman gain $\bar{G}(k)$ are tighter than those obtained with

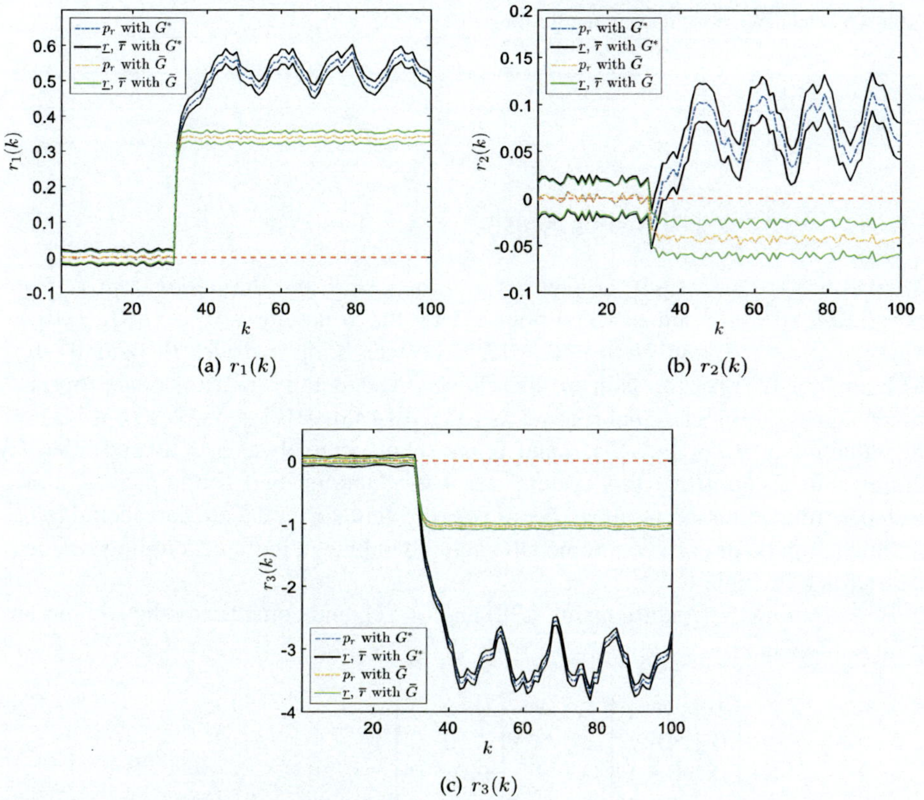

Fig. 4.5 Comparison of generated residuals

Table 4.2 Minimum detectable fault using optimal Kalman and FD gains

	$\bar{G}(k)$	$G^*(k)$	Improvement
Minimal detectable fault	0.0135	0.0089	51.69%

the FD observer gain $G^*(k)$. This is because the objective of the Kalman gain design is to minimize the effects of uncertainties. On the other hand, when the system is affected by actuator fault $f(k) = 0.3$, $k \geq 30$, the residual bounds obtained with $G^*(k)$ are moved further away from the coordinate origin which means they are more sensitive with respect to the occurred fault. This fault sensitivity will be useful when faults with small magnitude occur.

Table 4.2 presents the minimal detectable faults of this example with two observer gains obtained in simulation. It is shown that the observer with G^* is able to detect smaller faults when the fault sensitivity is considered. The trade-off between robustness to uncertainties and sensitivity to faults is improved using $G^*(k)$.

Table 4.3 Unknown input decoupling for robust FDI strategy

	T, N	f_1	f_2
Zonotopic UIO 1	T_1, N_1	×	×
Zonotopic UIO 2	T_2, N_2	×	

4.2.4.2 The Chemical Mixing System

We also use the case study in Sect. 4.1.4 with $F = B$ and the initial state $x(0) = [0.5, 0, 0.5, 0]^\top$ is assumed to be bounded by the zonotope $x(0) \in \langle p(0), H(0) \rangle$, where $p(0) = x(0)$ and $H(0) = 0.001 I_4$. The weighting matrices W_1 and W_2 for designing the FD observer gain are also chosen to be identity matrices of appropriate dimensions. The input signal u is set as $u(k) = [4\sin(0.3k) + 5, 5]^\top$, $\forall k \in \mathbb{N}$. For the reduction operator $\downarrow_{q,W}(\cdot)$, q and W are set respectively as $q = 20$ and $W = I$. Taking into account that this system has two actuators, two zonotopic UIOs are used. For implementing robust FDI strategy, the actuator faults are considered to be unknown inputs for each zonotopic UIO and the unknown input decoupling strategy is described in Table 4.3.

By satisfying the conditions in (4.28) and (4.52), and considering the strategy in Table 4.3, we have

$$
T_1 = \begin{bmatrix} 1 & 1 & 0 & 1 \\ 0 & 1 & 0 & 1 \\ 0 & 1 & 0.5 & 1 \\ 0 & 1 & 0 & 1 \end{bmatrix}, \quad N_1 = \begin{bmatrix} 0 & 0 & 0 \\ 1 & 0 & 0 \\ 0 & 0.5 & 0 \\ 0 & 0 & 1 \end{bmatrix},
$$

$$
T_2 = \begin{bmatrix} 1 & 23.4588 & 0.1777 & 0.0032 \\ 0 & 23.4588 & 0.1777 & 0.0032 \\ 0 & 23.4588 & 0.1777 & 0.0032 \\ 0 & 23.4588 & 0.1777 & 0.0032 \end{bmatrix}, \quad N_2 = \begin{bmatrix} 0 & -0.1777 & 0 \\ 1 & -0.1777 & 0 \\ 0 & 0.8223 & 0 \\ 0 & -0.1777 & 1 \end{bmatrix}.
$$

The zonotopic UIO 1 is designed with T_1 and N_1. Besides, the sensitivity to the second actuator fault is considered such that $\bar{F}_1(k) = [0, 0, 0.02, 1]^\top$. The zonotopic UIO 2 is designed with T_2 and N_2. Because the effect of the second actuator fault is removed by using the unknown input decoupling, this observer is designed to be sensitive to the first actuator fault.

From Fig. 4.6, it is shown that the coordinate origin is inside all the residual bounds of zonotopic UIO 1 and 2, that is inside the residual zonotopes corresponding to zonotopic UIO 1 and 2, which implies that there is no fault occurrence. From Fig. 4.7, it can be seen that for both zonotopic UIOs, the coordinate origin is not inside all the residual bounds after 20 sampling time steps. Based on Table 4.3 and the designed FDI strategy, the first actuator fault is detected at time step 21. Figure 4.8 shows that the coordinate origin is always inside the residual bounds of zonotopic UIO 2. Besides, after 20 sampling time steps the coordinate origin is not inside the residual bounds of zonotopic UIO 1. According to the proposed FDI strategy, the second actuator fault is detected at time step 21.

Fig. 4.6 FDI result of the chemical mixing system without faults

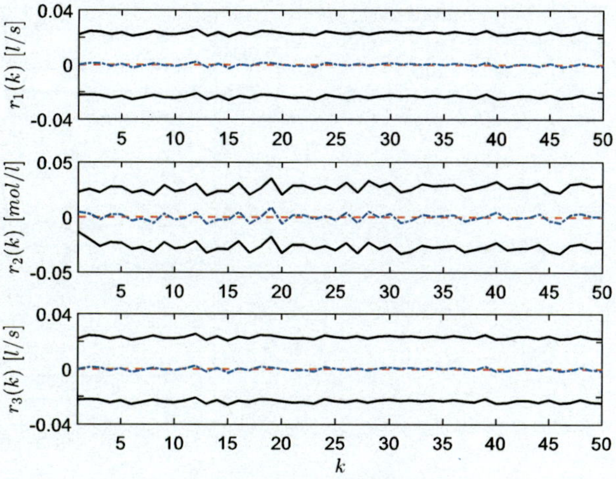

(a) Zonotopic UIO 1

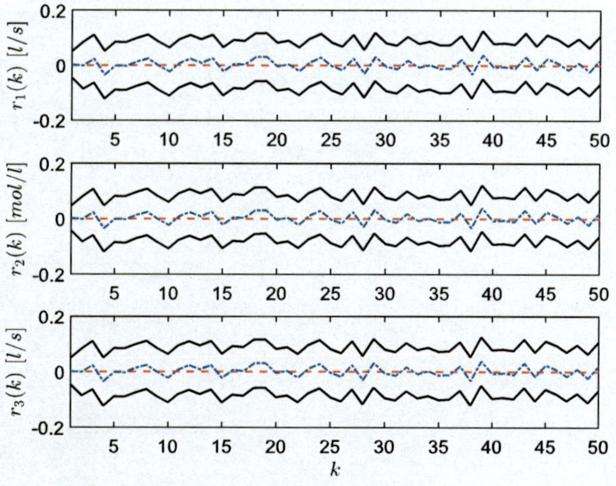

(b) Zonotopic UIO 2

Fig. 4.7 FDI result of the chemical mixing system with the first actuator fault

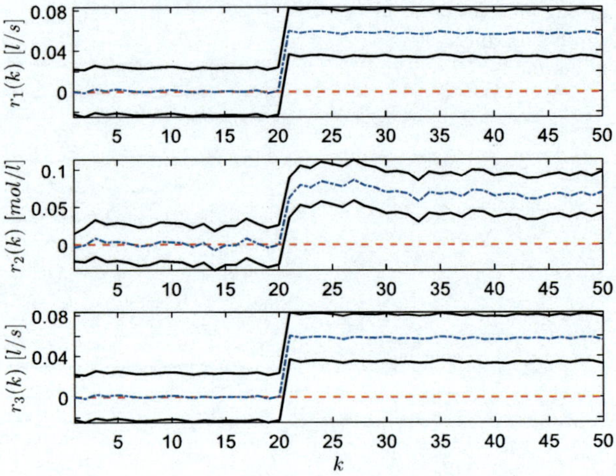

(a) Zonotopic UIO 1

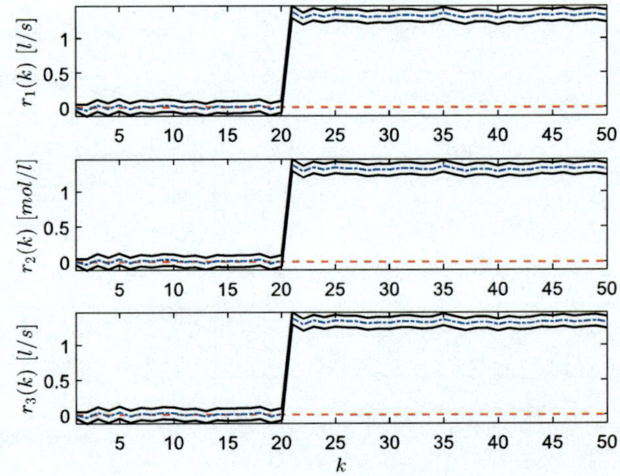

(b) Zonotopic UIO 2

Fig. 4.8 FDI result of the chemical mixing system with the second actuator fault

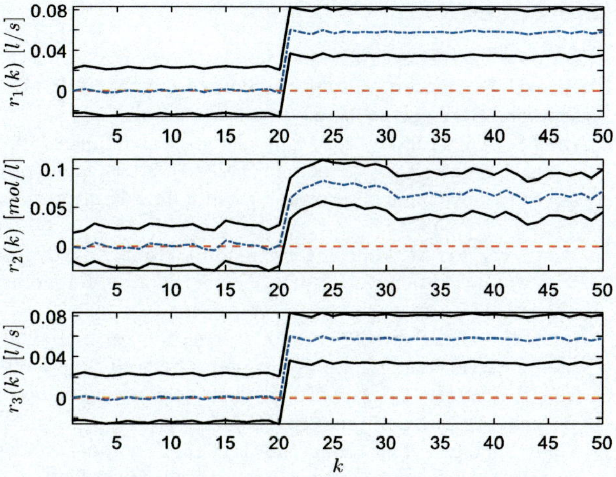

(a) Zonotopic UIO 1

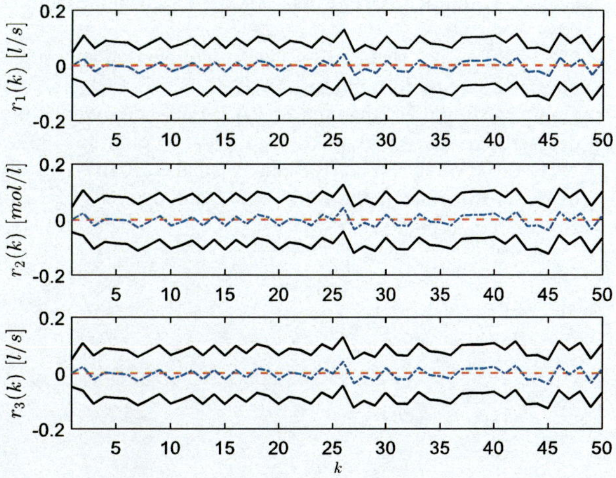

(b) Zonotopic UIO 2

4.3 Summary

This chapter has presented two robust FD methods and an application of robust FI based on zonotopic UIOs. For the two proposed FD methods, the first one based on LMI conditions to find a constant FD observer gain, where the $\mathcal{H}_-$ fault sensitivity is considered. The second FD method seeks for a time-varying FD observer gain with minimizing a defined fault sensitivity criterion. The advantage of the first method is that the stability of zonotopic FD observer can be guaranteed. However, for the second robust FD method, the stability with a time-varying FD observer gain for LTV descriptor systems deserves to be investigated as a future research.

References

1. Blanke M, Kinnaert M, Lunze J, Staroswiecki M (2016) Diagnosis and fault-tolerant control. Springer, Berlin Heidelberg
2. Ding S (2013) Model-based fault diagnosis techniques. Springer, London, UK
3. Wang Y, Puig V, Xu F, Cembrano G (2019) Robust fault detection and isolation based on zonotopic unknown input observer for discrete-time descriptor systems. J Frankl Inst 356(10):5293–5314
4. Wang Y, Zhou M, Puig V, Cembrano G, Wang Z (2017) Zonotopic fault detection observer with $\mathcal{H}_-$ performance. In: 36th Chinese control conference (CCC), Dalian, P.R. China, pp 7230–7235
5. Iwasaki T, Hara S, Fradkov A (2005) Time domain interpretations of frequency domain inequalities on (semi)finite ranges. Syst Control Lett 54(7):681–691
6. Yeu T, Kim H, Kawaji S (2005) Fault detection, isolation and reconstruction for descriptor systems. Asian J Control 7(4):356–367
7. Wang Y, Puig V, Xu F, Cembrano G (2018) Zonotopic unknown input observer of discrete-time descriptor systems for state estimation and robust fault detection. In: 10th IFAC symposium on fault detection. Supervision and safety for technical processes (IFAC-SAFEPROCESS). Warsaw, Poland, pp 307–313
8. Chen J, Patton R (1999) Robust model-based fault diagnosis for dynamic systems. Springer, New York, USA
9. Chang S, You W, Hsu P (1997) Design of general structured observers for linear systems with unknown inputs. J Frankl Inst 334(2):213–232
10. Gupta M, Tomar N, Bhaumik S (2015) Full- and reduced-order observer design for rectangular descriptor systems with unknown inputs. J Frankl Inst 352(3):1250–1264
11. Xu F, Tan J, Wang X, Puig V, Liang B, Yuan B (2016) A novel design of unknown input observers using set-theoretic methods for robust fault detection. In: American control conference (ACC), Boston, USA, pp 5957–5961

Chapter 5
Set-Based Fault Estimation for Descriptor Systems

Fault estimation, as a significant stage of fault diagnosis, aims to estimate the magnitude of occurred faults in a system. The problem of fault estimation has been studied using a large amount of approaches during the past decades [1–4]. A suitable fault estimation with robust performance against system uncertainties is very useful for implementing an active fault-tolerant control system [5]. Based on the robust control techniques, robust fault estimations are implemented in a variety of systems, such as [6–8], where the effects of uncertainties are bounded, and as a result, fault estimation results are obtained with the minimum estimation error.

The chapter presents a robust FE based on zonotopic Kalman filter for discrete-time descriptor systems subject to unknown-but-bounded uncertainties and additive actuator faults [9, 10] following the research line shown in Fig. 5.1. In Contrast to traditional FE methods, the FE results by using the proposed method provide not only a punctual value but also a deterministic set bounding the propagated uncertainties. Following the set-based framework for descriptor systems in Sect. 2.1.2, we first define the structure of the zonotopic FE filter based on fault detectability indices and matrix proposed in [11]. The zonotopic FE filter gain is formulated in a parametrized form. The optimal filter gain is designed to achieve the robustness against uncertainties and meanwhile the identification of occurred actuator faults. Furthermore, we discuss the boundedness of the propagated zonotopic FE.

5.1 Problem Statement in FE

Consider the discrete-time LTI descriptor system with additive actuator faults as follows:

Y. Wang, *Advances in State Estimation, Diagnosis and Control of Complex Systems*, Springer Theses, https://doi.org/10.1007/978-3-030-52440-1_5

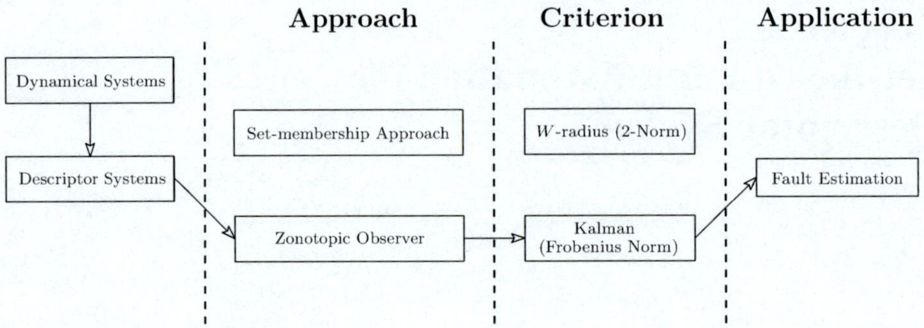

Fig. 5.1 Set-based FE scheme

$$Ex(k+1) = Ax(k) + Bu(k) + D_w w(k) + Ff(k), \tag{5.1a}$$
$$y(k) = Cx(k) + D_v v(k), \tag{5.1b}$$

where $x \in \mathbb{R}^{n_x}$ and $u \in \mathbb{R}^{n_u}$ denote the system state and the known input vectors, $w \in \mathbb{R}^{n_w}$ and $v \in \mathbb{R}^{n_v}$ denote the state disturbance vector and measurement noise vector, $y \in \mathbb{R}^{n_y}$ denotes the measurement output vector, $f \in \mathbb{R}^{q}$ denotes the actuator fault vector. $A \in \mathbb{R}^{n_x \times n_x}$, $B \in \mathbb{R}^{n_x \times n_u}$, $C \in \mathbb{R}^{n_y \times n_x}$, $D_w \in \mathbb{R}^{n_x \times n_w}$ and $D_v \in \mathbb{R}^{n_y \times n_v}$ are the system matrices. Besides, $F \in \mathbb{R}^{n_x \times q}$ denotes the fault distribution matrix describing the directions of the fault vector. In terms of the descriptor system (5.1), the matrix $E \in \mathbb{R}^{n_x \times n_x}$ might be singular, that is $\text{rank}(E) \leq n_x$.

For the system (5.1), we consider that Assumptions 2.1 and 2.2 hold. Besides, based on [11], we assume that $\text{rank}(C) = r$ and $\text{rank}(F) = q$ with $q \leq r$ and the system (5.1) is C-observable. Then, matrices E and C satisfy (2.41). Thus, there always exist two non-empty matrices $T \in \mathbb{R}^{n_x \times n_x}$ (T also non-singular) and $N \in \mathbb{R}^{n_x \times n_y}$ that can be obtained by Lemma 2.1.

In the following, we will design a set-based robust FE filter for the discrete-time descriptor system (5.1) to estimate the actuator fault magnitude f. The FE filter is built in a zonotopic framework considering unknown-but-bounded disturbances and measurement noise. Using this framework, robustness against uncertainties can be achieved by minimizing the size of the zonotope bounding estimation errors, disturbances and noise. The FE results are bounded using a zonotopic set.

5.2 Zonotopic FE Filter for Descriptor Systems

We now propose a zonotopic FE filter for the descriptor system (5.1). By means of fault detectability indices and matrix, we analyze and construct the FE zonotope to estimate occurred actuator faults. Therefore, the optimal FE filter gain is computed. Besides, we discuss boundedness of zonotopic FE.

5.2.1 Fault Detectability Indices and Matrix

Denote the fault distribution matrix $F = \left[F_1, \ldots, F_q\right]$ and the fault vector $f(k) = \left[f_1(k), \ldots, f_q(k)\right]^\top$, $\forall k \in \mathbb{N}$, where F_i is the i -th column of F and $f_i(k)$ is the i-th element of $f(k)$ for $i = 1, \ldots, q$, $\forall k \in \mathbb{N}$. We recall definitions of fault detectability indices and matrix first introduced in [11, 12] and extended for descriptor systems in [7] as follows.

Definition 5.1 (*Fault detectability indices* [7]) The discrete-time descriptor system (5.1) is said to have fault detectability indices $\rho = \left\{\rho_1, \rho_2, \ldots, \rho_q\right\}$ if

$$\rho_i = \min\left\{\sigma \mid C(TA)^{\sigma-1}TF_i \neq 0, i = 1, 2, \ldots\right\}. \tag{5.2}$$

and $s = \max\left\{\rho_1, \rho_2, \ldots, \rho_q\right\}$ denotes the maximum of fault detectability indices.

Assumption 5.1 Without loss of generality, the discrete-time descriptor system (5.1) is assumed with finite fault detectability indices.

Definition 5.2 (*Fault detectability matrix* [7]) With the fault detectability indices of the descriptor system (5.1) defined as $\rho = \left\{\rho_1, \rho_2, \ldots, \rho_q\right\}$, the fault detectability matrix is given by

$$\Upsilon = C\Psi, \tag{5.3}$$

with

$$\Psi = \left[(TA)^{\rho_1-1}TF_1,\ (TA)^{\rho_2-1}TF_2,\ \ldots,\ (TA)^{\rho_q-1}TF_q\right]. \tag{5.4}$$

5.2.2 Zonotopic FE Filter

When the condition (2.41) is fulfilled, there exists matrices T and N satisfying (2.42). We consider a state estimation filter for the discrete-time descriptor system (5.1) as

$$\begin{cases} z(k+1) &= TA\hat{x}(k) + TBu(k) + G(k)\left(y(k) - C\hat{x}(k)\right) \\ \hat{x}(k) &= z(k) + Ny(k), \end{cases} \tag{5.5}$$

where $\hat{x} \in \mathbb{R}^{n_x}$ denotes the estimated state vector and $z \in \mathbb{R}^{n_x}$ denotes the filter state vector.

Let us define the state estimation error $e(k) = x(k) - \hat{x}(k)$ and the output estimation error $\varepsilon(k) = y(k) - C\hat{x}(k)$. Then, the error dynamics of e and ε can be written as follows:

$$\begin{cases} e(k+1) &= (TA - G(k)C)e(k) + TFf(k) + TD_w w(k) \\ &\quad -G(k)D_v v(k) - ND_v v(k+1), \\ \varepsilon(k) &= Ce(k) + D_v v(k). \end{cases}$$

In order to analyze the effects of uncertainties and faults, we split e and ε into two parts: $e = e_f + e_w$ and $\varepsilon = \varepsilon_f + \varepsilon_w$, where e_f and ε_f are the errors only affected by actuator faults ($w(k) = 0$ and $v(k) = 0$, $\forall k \in \mathbb{N}$), and e_w and ε_w are the errors only affected by disturbances and noise ($f(k) = 0$, $\forall k \in \mathbb{N}$).

$$\begin{cases} e_f(k+1) & = (TA - G(k)C)e_f(k) + TFf(k), \\ \varepsilon_f(k) & = Ce_f(k), \end{cases} \tag{5.6}$$

and

$$\begin{cases} e_w(k+1) & = (TA - G(k)C)e_w(k) + TD_w w(k) \\ & \quad -G(k)D_v v(k) - ND_v v(k+1), \\ \varepsilon_w(k) & = Ce_w(k) + D_v v(k), \end{cases} \tag{5.7}$$

with the following initial conditions $e_f(k) = 0$ and $e_w(0) = e(0)$. Therefore, we know $\varepsilon_f(k) = 0$, $\forall k \in \mathbb{N}$.

We now analyze the effects of occurred actuator faults and uncertainties in the estimation errors using the fault detectability indices and matrix in Definitions 5.1 and 5.2 in the following theorem.

Theorem 5.1 (FE condition) *Consider the descriptor system (5.1). If there exists the gain $G(k) \in \mathbb{R}^{n_x \times n_y}$ such that*

$$(TA - G(k)C)\Psi = 0, \tag{5.8}$$

then the effect of the faults on $\varepsilon(k)$ can be expressed as

$$\varepsilon(k) = C\Psi \left[f_1(k-\rho_1)^\top, \ldots, f_q(k-\rho_q)^\top \right]^\top + \varepsilon_w(k). \tag{5.9}$$

Proof By merging (5.6), we can derive from the time instant $k = 0$ that

$$\varepsilon_f(k) = C\Phi^k e_f(0) + C\Phi^{k-1}TFf(0) + \cdots + C\Phi_1 TFf(k-1), \tag{5.10}$$

where $\Phi^k = \prod_{j=1}^{k} (TA - G_j C)$. According to [7, Theorem 1], we obtain

$$C\Phi^j TF_i = \begin{cases} C(TA)^{\rho_i - 1} TF_i, & j = \rho_i, \\ 0, & j \neq \rho_i. \end{cases} \tag{5.11}$$

Substituting (5.11) into (5.10) yields

$$\begin{aligned} \varepsilon_f(k) &= C\Phi^k e_f(0) + C(TA)^{\rho_1 - 1} TF_1 f_1(k-\rho_1) \\ &\quad + \cdots + C(TA)^{\rho_q - 1} TF_q f_q(k-\rho_q) \\ &= C\Phi^k e_f(0) + C\Psi \left[f_1(k-\rho_1), \ldots, f_q(k-\rho_q) \right]^\top. \end{aligned} \tag{5.12}$$

Since $e_f(0) = 0$, (5.12) becomes $\varepsilon_f(k) = C\Psi\left[f_1(k-\rho_1), \ldots, f_q(k-\rho_q)\right]^\top$. Therefore, from $\varepsilon(k) = \varepsilon_f(k) + \varepsilon_w(k)$, we obtain (5.9). □

From Theorem 5.1, we can see that the effects of faults and uncertainties can be separated in (5.9). Therefore, we define the zonotopic FE filter for the descriptor system (5.1) in the following theorem.

Theorem 5.2 (Zonotopic FE filter for descriptor systems) *Given the descriptor system (5.1) with $w(k) \in \langle 0, I_{n_w}\rangle$ and $v(k) \in \langle 0, I_{n_v}\rangle$, $\forall k \in \mathbb{N}$, matrices $T \in \mathbb{R}^{n_x \times n_x}$, $N \in \mathbb{R}^{n_x \times n_y}$ satisfying (2.42). Consider the state bounding zonotope $x_w(k-1) \in \langle p(k-1), H(k-1)\rangle \subseteq \langle p(k-1), \bar{H}(k-1)\rangle$ with $\bar{H}(k-1) = \downarrow_{\ell,W}(H(k-1))$, the state bounding zonotope $x_w(k) \in \langle p(k), H(k)\rangle$, $\forall k \in \mathbb{N}$ is recursively defined by*

$$
\begin{aligned}
p(k) &= (TA - G(k-1)C)\,p(k-1) + TBu(k-1) \\
&\quad + G(k-1)y(k-1) + Ny(k), && (5.13a)\\
H(k) &= \left[(TA - G(k-1)C)\,\bar{H}(k-1),\ TD_w,\ -G(k-1)D_v,\ -ND_v\right]. && (5.13b)
\end{aligned}
$$

If there exist matrices $G(k-1) \in \mathbb{R}^{n_x \times n_y}$ satisfying (5.8) and $M \in \mathbb{R}^{q \times n_y}$ satisfying

$$
M = (C\Psi)^\dagger = \Upsilon^\dagger, \tag{5.14}
$$

then the actuator faults is bounded by $\hat{f}(k) = \left[f_1(k-\rho_1), \ldots, f_q(k-\rho_q)\right]^\top \in \langle p_f(k), H_f(k)\rangle$, where

$$
\begin{aligned}
p_f(k) &= My(k) - MCp(k), && (5.15a)\\
H_f(k) &= \left[-MCH(k),\ -MD_v\right]. && (5.15b)
\end{aligned}
$$

Proof From the analysis of effects of occurred actuator faults and uncertainties in (5.9), we can build state bounding zonotope and FE zonotope in the following.

(*State bounding zonotope*) With a filter gain $G(k-1)$, from (5.5), we can derive

$$
\begin{aligned}
\hat{x}(k) &= (TA + G(k-1)C)\,\hat{x}(k-1) + TBu(k-1) \\
&\quad + G(k-1)y(k-1) + Ny(k).
\end{aligned}
$$

For $x_w(k-1) \in \langle p(k-1), \bar{H}(k-1)\rangle$, we set $\hat{x}(k-1) = p(k-1)$ and we know $e_w(k-1) = x_w(k-1) - p(k-1) \in \langle 0, \bar{H}(k-1)\rangle$. From (5.7), with $w(k) \in \langle 0, I_{n_w}\rangle$, $v(k) \in \langle 0, I_{n_v}\rangle$, $\forall k \in \mathbb{N}$, we derive $x_w(k) = \hat{x}(k) + e_w(k)$ obtaining

$$\begin{aligned} x_w(k) &\in \langle p(k), H(k)\rangle \\ &= ((TA - G(k-1)C)\langle p(k-1), 0\rangle) \oplus (TB\langle u(k-1), 0\rangle) \\ &\quad \oplus (G(k-1)\langle y(k-1), 0\rangle) \oplus (N\langle y(k), 0\rangle) \\ &\quad \oplus \big((TA - G(k-1)C)\langle 0, \bar{H}(k-1)\rangle\big) \oplus \big(TD_w\langle 0, I_{n_w}\rangle\big) \\ &\quad \oplus \big((-G(k-1)D_v)\langle 0, I_{n_v}\rangle\big) \oplus \big((-ND_v)\langle 0, I_{n_v}\rangle\big). \end{aligned}$$

By using the properties in (1.9), we obtain $p(k)$ and $H(k)$ in (5.13).

(*FE zonotope*) From $x_w(k) \in \langle p(k), H(k)\rangle$ and $\hat{x}(k) = p(k)$, we know $e_w(k) \in \langle 0, H(k)\rangle$. By definition, we also have the output estimation error $\varepsilon(k) = y(k) - Cp(k)$. On the other hand, by pre-multiplying $M \in \mathbb{R}^{q \times n_y}$ on both sides of (5.9), we obtain

$$M\varepsilon(k) = MC\Psi\left[f_1(k-\rho_1), \ldots, f_q(k-\rho_q)\right]^\top + M\varepsilon_w(k). \tag{5.16}$$

Denote $\hat{f}(k) = \left[f_1(k-\rho_1), \ldots, f_q(k-\rho_q)\right]^\top$. Taking into account M satisfying (5.14), we know $MC\Psi = I$. Therefore, from (5.16), we obtain

$$\hat{f}(k) = M\varepsilon(k) - M\varepsilon_w(k) = M\varepsilon(k) - M\left(Ce_w(k) + D_v v(k)\right). \tag{5.17}$$

Recall $\varepsilon(k) = y(k) - Cp(k)$, $e_w(k) \in \langle 0, H(k)\rangle$ and $v(k) \in \langle 0, I_{n_v}\rangle$. From (5.17), we can derive that

$$\begin{aligned} \hat{f}(k) &\in \langle p_f(k), H_f(k)\rangle \\ &= (M\langle y(k) - Cp(k), 0\rangle) \oplus (-MC\langle 0, H(k)\rangle) \oplus \big(-MD_v\langle 0, I_{n_v}\rangle\big). \end{aligned}$$

Again, by using the properties in (1.9), we obtain $p_f(k)$ and $H_f(k)$ as in (5.15).

□

Remark 1: From the structure of the zonotopic FE filter proposed in Theorem 5.2, it is clear that the estimated fault $\hat{f}(k)$ has delays for each element and the delays are determined by the fault detectability indices ρ_i for $i = 1, \ldots, q$.

5.2.3 *Optimal FE Filter Gain*

We now present the results of optimal FE filter gain. For designing the gain of the zonotopic FE filter, the following criteria are taken into account:

- $G(k)$ satisfies the algebraic condition (5.8);
- $G(k)$ minimizes the estimation error $e_w(k+1)$, that reduces the size of the zonotope $\langle p(k+1), H(k+1)\rangle$.

Following the zonotopic Kalman gain in Sect. 2.1.2, the size of a zonotope can be measured by the F_W-radius. The objective of the zonotope minimization can be defined by $J = \mathrm{tr}(WP(k+1))$ with a weighting matrix $W \in \mathbb{S}^{n_x}_{>0}$ and the covariation

$$P(k+1) = H(k+1)H(k+1)^\top. \tag{5.18}$$

Theorem 5.3 (Optimal FE filter gain) *Given $H(k+1)$, a weighting matrix $W \in \mathbb{S}^{n_x}_{>0}$, the fault detectability matrix Υ in (5.3) with* $\mathrm{rank}(\Upsilon) = q$. *The optimal gain $G^*(k)$ can be computed by the parametrized form*

$$G^*(k) = \Phi M + \bar{G}^*(k)\Omega, \tag{5.19}$$

with

$$\Phi = TA\Psi,\ M = \Upsilon^\dagger,\ \Omega = \alpha\left(I_m - \Upsilon M\right), \tag{5.20}$$

where $\alpha \in \mathbb{R}^{(p-q)\times p}$ is an arbitrary matrix guaranteeing that Ω has full-row rank and $\bar{G}(k) \in \mathbb{R}^{n\times(p-q)}$. Besides, $\bar{G}(k) = \bar{G}^*(k)$ minimizes $J = \mathrm{tr}(WP(k+1))$ with $P(k+1)$ in (5.18), which is computed through the following procedure

$$\bar{G}^*(k) = \tilde{L}(k)\tilde{S}(k)^{-1}, \tag{5.21}$$

$$\tilde{L}(k) = (TA - \Phi MC)\bar{P}(k)C^\top\Omega^\top - \Phi MV\Omega^\top, \tag{5.22}$$

$$\tilde{S}(k) = \Omega\left(C\bar{P}(k)C^\top + V\right)\Omega^\top, \tag{5.23}$$

with $\bar{P}(k) = \bar{H}(k)\bar{H}(k)^\top$ and $V = D_v D_v^\top$.

Proof From $M = \Upsilon^\dagger$ and $\mathrm{rank}(\Upsilon) = q$, we have $M\Upsilon = I_q$. Since $\mathrm{rank}(\Upsilon) = q$, we can obtain a matrix $\Omega \in \mathbb{R}^{(n_y-q)\times n_y}$ such that $\Omega\Upsilon = 0$.

Therefore, with $G(k)$ defined in (5.19), we derive

$$\begin{aligned}(TA - G(k)C)\Psi &= \left(TA - \left(\Phi M + \bar{G}(k)\Omega\right)C\right)\Psi \\ &= TA\Psi - TA\Psi MC\Psi - \bar{G}(k)\Omega C\Psi \\ &= TA\Psi - TA\Psi M\Upsilon - \bar{G}(k)\Omega\Upsilon.\end{aligned}$$

Since $M\Upsilon = I_q$ and $\Omega\Upsilon = 0$, the above equation leads to $TA\Psi - TA\Psi M\Upsilon - \bar{G}(k)\Omega\Upsilon = 0$. Thus, (5.8) is satisfied with $G(k)$ parametrized as in (5.19).

Then, the problem is converted to find $\bar{G}(k)$ minimizing $J = \mathrm{tr}(WP(k+1))$. By definition, J is convex with respect to $\bar{G}(k)$. Thus, $\bar{G}^*(k)$ is a value of $\bar{G}(k)$ such that $\frac{\partial J}{\partial \bar{G}(k)} = 0$.

Set $\tilde{L}(k)$ and $\tilde{S}(k)$ as in (5.22) and (5.23). Evaluating $\frac{\partial J}{\partial \bar{G}(k)} = 0$, we have that

$$\frac{\partial \mathrm{tr}}{\partial \bar{G}(k)}\left(W\bar{G}(k)\tilde{S}(k)\bar{G}(k)^\top\right) - 2\frac{\partial \mathrm{tr}}{\partial \bar{G}(k)}\left(W\tilde{L}(k)\bar{G}(k)^\top\right) = 0. \tag{5.24}$$

Algorithm 5.1 Zonotopic FE algorithm for descriptor systems

```
Given the system matrices E, A, B, C, D_w, D_v and F and the initial state bounded
  in x_0 ∈ ⟨c_0, H_0⟩;
Solve the equation (2.42) by using  to obtain T and N;
Compute the fault detectability indices ρ_i, i = 1, ..., q;
Compute the fault detectability matrix ϒ = CΨ;
Φ ⟵ TAΨ;
M ⟵ ϒ^†;
Ω ⟵ α(I_m − ϒM);
while k > 0 do
   Compute Ḡ*(k − 1) according to the procedure in (5.21)–(5.23);
   Obtain the optimal parametrized gain G*(k − 1) following (5.19) with Ḡ*(k − 1)
   following (5.21)–(5.23);
   Compute the state bounding zonotope ⟨c(k), H(k)⟩ by using (5.13);
   Compute the FE zonotope ⟨c_f(k), H_f(k)⟩ by using (5.15);
   Obtain the FE f̂(k) = c_f(k) with its bounds f̂_i(k) ∈ [f_i(k), f̄_i(k)], i = 1, ..., q
   with
       f̄_i(k) = c_{f_i}(k) + rs(H_f(k))_{i,i},
       f_i(k) = c_{f_i}(k) − rs(H_f(k))_{i,i},
   where c_f = [c_{f_1} ... c_{f_i} ... c_{f_q}]^⊤.
end while
```

By means of the matrix calculus in (1.14), (5.24) can be simplified as

$$W\tilde{S}(k)\bar{G}(k)^{\top} + W\tilde{S}(k)^{\top}\bar{G}(k)^{\top} - 2W\tilde{L}(k)^{\top} = 0.$$

Because $\tilde{S}(k)$ is also a symmetric matrix, we thus obtain $\bar{G}(k)$ as in (5.21). □

From the proof of Theorem 5.3, we can see the independence of $\bar{G}^*(k)$ with respect to the weighting matrix W. Thus, W can be set as an arbitrary matrix, for instance $W = I$. Besides, time-varying weighting matrix $W(k)$ will be taken into account for discussing boundedness of the proposed zonotopic FE for descriptor systems.

Remark 5.1 For the proposed zonotopic FE filter in Theorem 5.2, G that satisfies the condition $(TA - GC)\Psi = 0$ is a stabilizing gain if there exist matrices $W \in \mathbb{S}^{n_x}_{\succ 0}$, and Y

$$\begin{bmatrix} W & \star \\ WTA - W\Phi MC - Y\Omega C & W \end{bmatrix} \succ 0, \tag{5.25}$$

then the feasible solutions give $G = \Phi M - W^{-1}Y\Omega$. Note that the condition (5.25) can be found by the Lyapunov stability condition and the parametrized gain as in (5.19).

With the zonotopic FE filter defined in Theorem 5.2 and the optimal gain in Theorem 5.3, we summarize the FE algorithm in Algorithm 5.1.

5.2.4 Boundedness of Zonotopic FE

We now study the boundedness of zonotopic FE by implementing Theorem 5.2 with the designed optimal gain in Theorem 5.3. To find a sequence of time-varying weighting matrices $W(k) \in \mathbb{S}^{n_x}_{\succ 0}$, we introduce a result for discrete-time nominal descriptor systems in the following.

Proposition 5.1 *Given the descriptor system $Ex(k+1) = Ax(k)$ with a measurement output $y(k) = Cx(k)$, matrices T and N satisfying (2.42), and $\gamma \in (0, 1)$. The filter $\hat{x}(k+1) = TA\hat{x}(k) + G(k)(y(k) - C\hat{x}(k)) + Ny(k+1)$ is γ-stable (stable with a decay rate γ) if there exist matrices $G(k) \in \mathbb{R}^{n\times p}$ and $W(k) \in \mathbb{S}^{n}_{\succ 0}$, $\forall k \geq 0$ such that*

$$\begin{bmatrix} \gamma W(k) & (TA - G(k)C)^{\top} W(k+1)^{\top} \\ W(k+1)(TA - G(k)C) & W(k+1) \end{bmatrix} \succ 0. \tag{5.26}$$

Proof With matrices T and N satisfying (2.42), we reformulate the system dynamics to be $x(k+1) = TAx(k) + Ny(k+1)$. Define the state estimation error $e(k) = x(k) - \hat{x}(k)$. Therefore, we have the error dynamics

$$e(k+1) = x(k+1) - \hat{x}(k+1) = (TA - G(k)C)e(k).$$

With a sequence of matrices $W(k) \in \mathbb{S}^{n_x}_{\succ 0}$, $\forall k \geq 0$, we consider the Lyapunov candidate function as $V(k) = e(k)^{\top} W(k)e(k)$. Given $\gamma \in (0, 1)$, we have

$$\begin{aligned} \Delta V(k) &= V(k+1) - V(k) = e(k+1)^{\top} W(k+1)e(k+1) - e(k)^{\top} W(k)e(k) \\ &= e(k)^{\top} \left((TA - G(k)C)^{\top} W(k+1)(TA - G(k)C) - \gamma W(k)\right) e(k). \end{aligned}$$

For any $e(k) \neq 0$, $\Delta V(k) < 0$ implies

$$\gamma W(k) - (TA - G(k)C)^{\top} W(k+1)(TA - G(k)C) \succ 0.$$

By applying the Schur complement with $\gamma W(k) \succ 0$, we thus obtain (5.26). □

Since the zonotope reduction operator $\downarrow_{\ell,W}(\cdot)$ is used in the proposed zonotopic FE filter, we also introduce the following lemma to describe the boundedness of the use of $\downarrow_{\ell,W}(\cdot)$.

Lemma 5.1 ([13]) *Consider $H \in \mathbb{R}^{n_x\times r}$ as the generator matrix of a zonotope $\langle p, H\rangle \subset \mathbb{R}^n$, a weighting matrix $W \in \mathbb{S}^n_{\succ 0}$ with all its eigenvalues in $[\underline{\lambda}, \overline{\lambda}] \subset \mathbb{R}$.*

By means of the reduction operator $\bar{H} = \downarrow_{\ell,W}(H)$ *with* $n \leq \ell < r$, $\langle p, \bar{H} \rangle$ *is a reduced zonotope such that* $\langle c, H \rangle \subseteq \langle p, \bar{H} \rangle$. *Let* $\mu = \left(\frac{\bar{\lambda}(n+r-\ell)}{\underline{\lambda}} - 1\right)(n + r - \ell)$ *and* $\beta = 1 + \frac{\mu}{r}$. *Then, it holds*

$$\|\bar{H}\|_{F,W}^2 \leq \beta \|H\|_{F,W}^2. \tag{5.27}$$

Proof The proof of this lemma can be found in [13, Theorem 10]. □

From the structure of the proposed zonotopic FE filter in Theorem 5.2, due to that $\langle p_f(k), H_f(k) \rangle$ is a linear projection of $\langle p(k), H(k) \rangle$, $\forall k \in \mathbb{N}$, the filter dynamics is bounded by $\langle p(k), H(k) \rangle$ as defined in (5.13). Based on presented results above, we now discuss the boundedness of zonotopic FE for descriptor systems in the following theorem.

Theorem 5.4 (Boundedness of zonotopic FE) *Consider the zonotopic FE filter* $\langle p_f(k), H_f(k) \rangle$ *in (5.15) with* $\langle p(k), H(k) \rangle$ *in (5.13) and the optimal gain* $G^*(k)$ *in (5.19),* $W(k) \in \mathbb{S}_{>0}^{n_x}$, $\forall k \in \mathbb{N}$ *and* $\gamma \in (0, 1)$ *satisfying (5.26). If there exists a bounded sequence* $\psi(k)$ *such that*

$$\|TD_w\|_{F,W(k+1)}^2 + \|G(k)D_v\|_{F,W(k+1)}^2 + \|ND_v\|_{F,W(k+1)}^2 \leq \psi(k), \quad \forall k \in \mathbb{N}, \tag{5.28}$$

and when $k \to \infty$, $\bar{\psi}$ *is the upper bound of* $\psi(k)$, *then the* F_W*-radius of* $\langle c(k), H(k) \rangle$ *is bounded by*

$$\|H(k+1)\|_{F,W(k+1)}^2 \leq \bar{\gamma} \|H(k)\|_{F,W(k)}^2 + \psi(k), \quad \forall k \in \mathbb{N}, \tag{5.29}$$

with $\bar{\gamma} = \gamma\beta < 1$. *Moreover, when* $k \to \infty$, *the upper bound* $\|H(\infty)\|_{F,W(\infty)}^2$ *is given by*

$$\|H(\infty)\|_{F,W(\infty)}^2 \leq \frac{\bar{\psi}}{1 - \bar{\gamma}}. \tag{5.30}$$

Proof Considering $H(k+1)$ and the optimal gain $G^*(k)$, the F_W-radius of the zonotope $\langle p(k+1), H(k+1) \rangle$ is expressed as

$$\|H(k+1)\|_{F,W(k+1)}^2 = \left\|\left[(TA - G^*(k)C)\,\bar{H}(k),\ TD_w,\ -G^*(k)D_v,\ -ND_v\right]\right\|_{F,W(k+1)}^2.$$

Since the optimal gain $G^*(k)$ is obtained by minimizing $\|H(k+1)\|_{F,W(k+1)}^2$ with independence of $W(k+1)$, we thus have

$$\|H(k+1)\|_{F,W(k+1)}^2 \leq \left\|\left[(TA - G(k)C)\,\bar{H}(k),\ TD_w,\ -G(k)D_v,\ -ND_v\right]\right\|_{F,W(k+1)}^2,$$

for any $G(k)$ instead of $G^*(k)$ satisfying (5.26). Then, considering the boundedness in (5.28), from above inequality, we obtain a sufficient condition

$$\|H(k+1)\|_{F,W(k+1)}^2 \leq \left\|(TA - G(k)C)\,\bar{H}(k)\right\|_{F,W(k+1)}^2 + \psi(k). \tag{5.31}$$

Based on Proposition 5.1, with $W(k) \in \mathbb{S}_{\succ 0}^{n_x}$, $\forall k \in \mathbb{N}$ and $\gamma \in (0,1)$ satisfying (5.26), $(TA - G(k)C)$ is γ-stable. By applying the Schur complement to (5.26), we obtain $\gamma W(k) - (TA - G(k)C)^\top W(k+1)(TA - G(k)C) \succ 0$. Since $\bar{H}(k) \neq 0$ and by the linearity of the operator $\mathrm{tr}(\cdot)$, we have

$$\mathrm{tr}\left(\bar{H}(k)^\top (TA - G(k)C)^\top W(k+1)(TA - G(k)C)\bar{H}(k)\right) < \gamma \mathrm{tr}\left(\bar{H}(k)^\top W(k)\bar{H}(k)\right).$$

By the F_W-radius definition, we obtain $\left\|(TA - G(k)C)\,\bar{H}(k)\right\|_{F,W(k+1)}^2 < \gamma \left\|\bar{H}(k)\right\|_{F,W(k)}^2$. Therefore, with (5.31), we have

$$\|H(k+1)\|_{F,W(k+1)}^2 \leq \gamma \left\|\bar{H}(k)\right\|_{F,W(k)}^2 + \psi(k).$$

Based on the condition (5.27) in Lemma 5.1, we obtain

$$\|H(k+1)\|_{F,W(k+1)}^2 \leq \gamma\beta \|H(k)\|_{F,W(k)}^2 + \psi(k).$$

Thus, with $\bar{\gamma} = \gamma\beta$, we obtain (5.29). Considering $\gamma \in (0,1)$, $\bar{\gamma} \in (0,1)$ can also hold.

Besides, when $k \to \infty$, with the upper bound $\psi(\infty) = \bar{\psi}$, (5.29) becomes

$$\|H(\infty)\|_{F,W(\infty)}^2 \leq \bar{\gamma} \|H(\infty)\|_{F,W(\infty)}^2 + \bar{\psi},$$

which implies (5.30). □

According to Theorem 5.4, the boundedness of the state bounding zonotope $\langle p(k), H(k)\rangle$, $\forall k \in \mathbb{N}$ defined in (5.13) is provided by the boundedness condition. As a conclusion, ultimate boundedness of the proposed zonotopic FE is obtained.

5.3 Case Studies

In the following, the simulation results obtained with a numerical example and an engineering systems are shown to verify the proposed robust FE method for discrete-time descriptor systems.

5.3.1 Numerical Example

Consider a discrete-time descriptor system modeled by (5.1) with system matrices as follows:

$$E = \begin{bmatrix} 1 & 0 & 0 & 0 \\ 0 & 1 & 0 & 0 \\ 0 & 0 & 1 & 0 \\ 0 & 0 & 0 & 0 \end{bmatrix}, \quad A = \begin{bmatrix} 0.9 & 0.005 & -0.095 & 0 \\ 0.005 & 0.995 & 0.0997 & 0 \\ 0.095 & -0.0997 & 0.99 & 0 \\ 1 & 0 & 1 & 1 \end{bmatrix},$$

$$B = F = \begin{bmatrix} F_1 & F_2 \end{bmatrix} = \begin{bmatrix} 0.1 & 0 \\ 1 & 1 \\ -0.1 & 1 \\ -1 & 0 \end{bmatrix}, \quad C = \begin{bmatrix} 0 & 1 & 0 & 0 \\ 0 & 0 & 1 & 0 \\ 0 & 0 & 0 & 1 \end{bmatrix},$$

$$D_w = \begin{bmatrix} 0.3 & 0 & 0 \\ 0 & 0.3 & 0 \\ 0 & 0 & 0.3 \\ 0 & 0 & 0 \end{bmatrix}, \quad D_v = \begin{bmatrix} 0.1 & 0 & 0 \\ 0 & 0.1 & 0 \\ 0 & 0 & 0.1 \end{bmatrix}.$$

The initial state $x(0)$ is set as $x(0) = \begin{bmatrix} 0.5, \ 1, \ 0, \ -0.5 \end{bmatrix}^{\top}$ and the initial state zonotope is given by $x(0) \in \langle p(0) = x(0), 0.1 I_4 \rangle$. Besides, $w(k) \in \langle 0, I_3 \rangle$ and $v(k) \in \langle 0, I_3 \rangle$, $\forall k \in \mathbb{N}$. The input signal is set as $u(k) = \begin{bmatrix} 2\sin(k), \ 3\sin(k) \end{bmatrix}^{\top}$.

From the general solution (2.43), we choose the matrix S as

$$S = \begin{bmatrix} 1 & 0 & 0 & 0 & 0 & 0 & 0 \\ 0 & 1 & 0 & 0 & 1 & 0 & 0 \\ 0 & 0 & 1 & 0 & 0 & 1 & 0 \\ 0 & 0 & 0 & 1 & 0 & 0 & 1 \end{bmatrix},$$

and we obtain two non-empty matrices T and N satisfying the condition (2.42) as follows:

$$T = \begin{bmatrix} 1 & 0 & 0 & 0 \\ 0 & 0.5 & 0 & 0 \\ 0 & 0 & 0.5 & 0 \\ 0 & 0 & 0 & 1 \end{bmatrix}, \quad N = \begin{bmatrix} 0 & 0 & 0 \\ 0.5 & 0 & 0 \\ 0 & 0.5 & 0 \\ 0 & 0 & 1 \end{bmatrix}.$$

Since $\text{rank}(F) = \text{rank}(CF) = 2$, we have $CTF_1 \neq 0$ and $CTF_2 \neq 0$. The fault detectability indexes are $\rho_1 = 1$ and $\rho_2 = 1$ and the fault detectability matrix is $\Upsilon = C\Psi = \begin{bmatrix} 0.5 & 0.5 \\ -0.05 & 0.5 \\ -1 & 0 \end{bmatrix}$ with $\Psi = \begin{bmatrix} Tf_1 & Tf_2 \end{bmatrix} = \begin{bmatrix} 0.1 & 0 \\ 0.5 & 0.5 \\ -0.05 & 0.5 \\ -1 & 0 \end{bmatrix}$. Therefore, we obtain the matrices to obtain the optimal gain $G^*(k)$ as follows:

$$\Psi = \begin{bmatrix} 0.0973 & -0.045 \\ 0.2465 & 0.2737 \\ -0.0449 & 0.2226 \\ -0.95 & 0.5 \end{bmatrix}, \; M = \begin{bmatrix} 0.2389 & -0.2389 & -0.8686 \\ 0.8925 & 1.1075 & 0.3909 \end{bmatrix},$$

$$\Omega = \begin{bmatrix} 0.8686 & -0.8686 & 0.4777 \end{bmatrix}.$$

Therefore, the time-varying matrix $\bar{G}^*(k)$ can be obtained following (5.21)–(5.23) and we can find the optimal parameterized gain $G^*(k)$ in (5.19). Besides, as a comparison, according to Remark 2, by satisfying (5.25), we also obtain a stabilizing gain G as

$$G = \begin{bmatrix} 0.3283 & -0.4183 & 0.0878 \\ 0.4907 & 0.0566 & -0.0040 \\ -0.0279 & 0.4730 & 0.0073 \\ 0.3051 & 0.6949 & 1.0678 \end{bmatrix}.$$

Consider the actuator faults are in the following scenarios

$$f_1(k) = \begin{cases} 0 & k < 80 \\ 5 & k \geq 80 \end{cases}$$

$$f_2(k) = \begin{cases} 0 & k < 100 \\ 6\sin(0.1k) & k \geq 100 \end{cases}$$

As a result, the simulation has been carried out for $N_s = 200$ sampling steps and the robust FE results are shown in Fig. 5.2 with $G^*(k)$ and G. Note that due to $\rho_1 = 1$ and $\rho_2 = 1$, there is one-step delay in the estimation of the faults f_1 and f_2. In the figures, for allowing a better comparison, we plot the real faults delayed one sample, $f_i(k-1)$ with $i = 1, 2$. Using the proposed zonotopic FE filter, the punctual values of estimated faults are obtained altogether with the worst-case bounds of estimated faults are also found in the estimation intervals under the assumption of unknown-but-bounded disturbances and measurement noise in given zonotopes.

The actual faults in red dashed lines are bounded by estimation intervals with $G^*(k)$ and G. From the Fig. 5.2, it is obvious that the bounds obtained with G are larger than the ones obtained with $G^*(k)$. For the comparison of the performance with $G^*(k)$ and G, the MSE between the actual faults and estimation faults (centers of FE zonotopes) is computed by

$$MSE := \frac{1}{N_s} \sum_{k=1}^{N_s} \frac{1}{q} \left\| f(k) - c_f(k) \right\|,$$

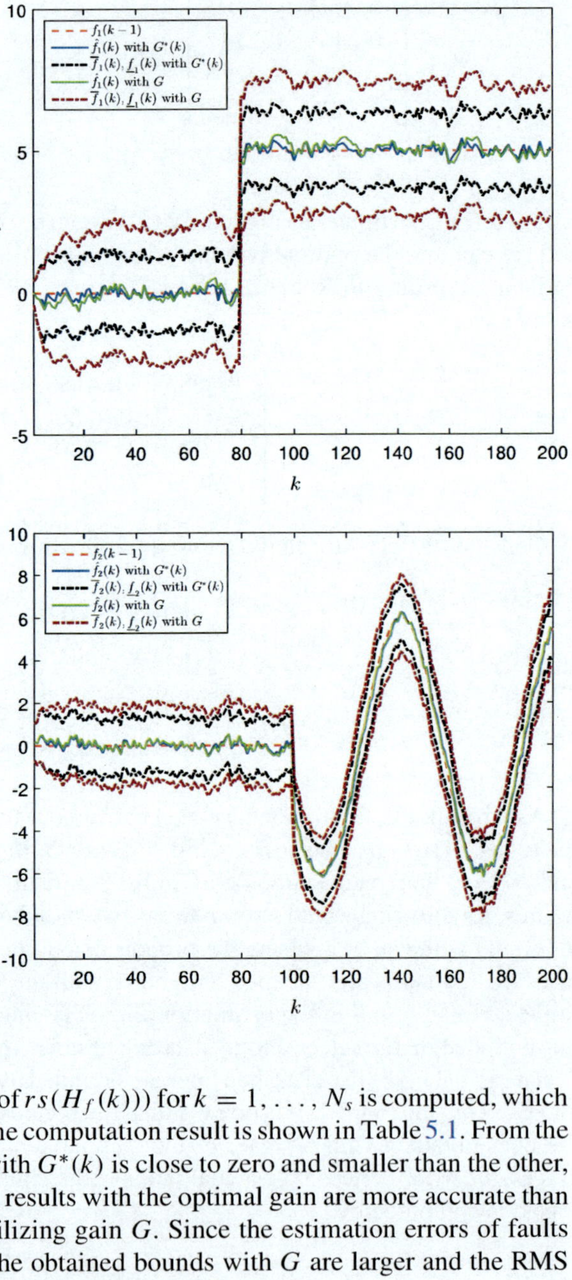

Fig. 5.2 Actuator-FE results with $G^*(k)$ and G

and the root mean squared value of $rs(H_f(k)))$ for $k = 1, \ldots, N_s$ is computed, which is denoted by $RMS(rs(H_f))$. The computation result is shown in Table 5.1. From the MSE results, the one obtained with $G^*(k)$ is close to zero and smaller than the other, which means that the estimation results with the optimal gain are more accurate than the ones obtained with the stabilizing gain G. Since the estimation errors of faults are bounded in the zonotopes, the obtained bounds with G are larger and the RMS result provides that the one with G is larger than the other.

Table 5.1 Comparison between $G^*(k)$ and G

	MSE	$RMS(rs(H_f))$
$G^*(k)$	0.0389	1.3049
G	0.0641	2.0470

5.3.2 *The Machine Infinite Bus System*

Consider a machine infinite bus system used in [14] and its linear continuous-time system with parameters described in [15] as follows:

$$\begin{aligned}
\dot{\delta}_1 &= \omega_1, \\
\dot{\delta}_2 &= \omega_2, \\
\dot{\delta}_3 &= \omega_3, \\
\dot{\omega}_4 &= \frac{1}{m_1}(p_1 - Y_{12}V_1V_2(\delta_1 - \delta_2)) - \frac{1}{m_1}(Y_{15}V_1V_2(\delta_1 - \delta_5) + c_1\omega_1), \\
\dot{\omega}_5 &= \frac{1}{m_2}(p_2 - Y_{21}V_2V_1(\delta_2 - \delta_1)) - \frac{1}{m_2}(Y_{25}V_2V_5(\delta_2 - \delta_5) + c_2\omega_2), \\
\dot{\omega}_6 &= \frac{1}{m_3}(p_3 - Y_{34}V_\infty\delta_3) - \frac{1}{m_3}(Y_{35}V_3V_5(\delta_3 - \delta_5) + c_3\omega_3), \\
0 &= P_{ch} - Y_{51}V_5V_1(\delta_5 - \delta_1) - Y_{52}V_5V_2(\delta_5 - \delta_2) - Y_{53}V_5V_3(\delta_5 - \delta_3) - Y_{54}V_5V_\infty\delta_5,
\end{aligned}$$

where δ_1, δ_2, δ_3 and δ_5 denote the phase angles of the generators, ω_1, ω_2 and ω_3 denote the speeds of the generators, p_1, p_2 and p_3 are the mechanical powers per unit that are set as $p_1 = 0.1$, $p_2 = 0.1$ and $p_3 = 0.1$, and P_{ch} is the unknown power load. From [15], the other parameters are chosen as follows: the inertia $m_1 = 0.014$, $m_2 = 0.026$ and $m_3 = 0.02$, the damping $c_1 = 0.057$, $c_2 = 0.15$ and $c_3 = 0.11$, the potential $V_1 = 1$, $V_2 = 1$, $V_3 = 1$, $V_\infty = 1$ and $V_5 = 1$, and the nominal admittance $Y_{15} = 0.5$, $Y_{25} = 1.2$, $Y_{35} = 0.8$, $Y_{45} = 1$, $Y_{35} = 0.7$ and $Y_{12} = 1$. Besides, the uncertain part of the admittance is set in the state disturbances. Let us define

$$x = \left[\delta_1, \delta_2, \delta_3, \omega_1, \omega_2, \omega_3, \delta_5\right]^\top, \quad u = \left[p_1, p_2, p_3\right]^\top.$$

We use the Euler discretization method with the sampling time $\Delta t = 0.05\,\text{s}$ to obtain the discrete-time descriptor model in the form of (5.1) with system matrices as follows:

$$E = \begin{bmatrix} 1\,0\,0\,0\,0\,0\,0 \\ 0\,1\,0\,0\,0\,0\,0 \\ 0\,0\,1\,0\,0\,0\,0 \\ 0\,0\,0\,1\,0\,0\,0 \\ 0\,0\,0\,0\,1\,0\,0 \\ 0\,0\,0\,0\,0\,1\,0 \\ 0\,0\,0\,0\,0\,0\,0 \end{bmatrix},$$

$$A = \begin{bmatrix} 1 & 0 & 0 & 0.05 & 0 & 0 & 0 \\ 0 & 1 & 0 & 0 & 0.05 & 0 & 0 \\ 0 & 0 & 1 & 0 & 0 & 0.05 & 0 \\ -5.3571 & 3.5714 & 0 & 0.7964 & 0 & 0 & 1.7857 \\ 1.9231 & -4.2308 & 0 & 0 & 0.7115 & 0 & 2.3077 \\ 0 & 0 & -3.75 & 0 & 0 & 0.725 & 2 \\ 0.025 & 0.06 & 0.04 & 0 & 0 & 0 & -0.175 \end{bmatrix},$$

$$B = F = \begin{bmatrix} F_1 \; F_2 \; F_3 \end{bmatrix} = \begin{bmatrix} 0 & 0 & 0 \\ 0 & 0 & 0 \\ 0 & 0 & 0 \\ 3.5714 & 0 & 0 \\ 0 & 1.9231 & 0 \\ 0 & 0 & 2.5 \\ 0 & 0 & 0 \end{bmatrix}, \quad C = \begin{bmatrix} 1\,0\,0\,0\,0\,0\,0 \\ 0\,1\,0\,0\,0\,0\,0 \\ 0\,0\,1\,0\,0\,0\,0 \\ 0\,0\,0\,0\,0\,0\,1 \end{bmatrix},$$

$$D_w = \begin{bmatrix} 0 & 0 & 0 & 0 \\ 0 & 0 & 0 & 0 \\ 0 & 0 & 0 & 0 \\ 0.3 & 0 & 0 & 0 \\ 0 & 0.3 & 0 & 0 \\ 0 & 0 & 0.3 & 0 \\ 0 & 0 & 0 & 0.3 \end{bmatrix}, D_v = \begin{bmatrix} 0.025 & 0 & 0 & 0 \\ 0 & 0.025 & 0 & 0 \\ 0 & 0 & 0.025 & 0 \\ 0 & 0 & 0 & 0.025 \end{bmatrix}.$$

Given the initial state $x(0) = 0$ and the initial state zonotope $x(0) \in \langle 0, 0.01 I_7 \rangle$, $w(k) \in \langle 0, I_4 \rangle$ and $v(k) \in \langle 0, I_4 \rangle$, $\forall k \in \mathbb{N}$. The input signal is set as $u(k) = \begin{bmatrix} 20, 15, 10 \end{bmatrix}^{\top}$, $\forall k \in \mathbb{N}$. From the general solution (2.43), we choose the matrix

$$S = \begin{bmatrix} 1 & 0 & 0 & 0 & 0 & 0 & 0 & 1 & 0 & 0 & 0 \\ 0 & 1 & 0 & 0 & 0 & 0 & 0 & 0 & 1 & 0 & 0 \\ 0 & 0 & 1 & 0 & 0 & 0 & 0 & 0 & 0 & 1 & 0 \\ 0 & 0 & 0 & 1 & 0 & 0 & 0 & 0 & 0 & 0 & 1 \\ 0 & 0 & 0 & 0 & 1 & 0 & 0 & 0 & 0 & 0 & 0 \\ 0 & 0 & 0 & 0 & 0 & 1 & 0 & 0 & 0 & 0 & 0 \\ 0 & 0 & 0 & 0 & 0 & 0 & 1 & 0 & 0 & 0 & 0 \end{bmatrix},$$

and we obtain two non-empty matrices T and N satisfying (2.42) and the matrix T is also non-singular as follows:

$$T = \begin{bmatrix} 0.5 & 0 & 0 & 0 & 0 & 0 & 0 \\ 0 & 0.5 & 0 & 0 & 0 & 0 & 0 \\ 0 & 0 & 0.5 & 0 & 0 & 0 & 0 \\ 0 & 0 & 0 & 1 & 0 & 0 & 0 \\ 0 & 0 & 0 & 0 & 1 & 0 & 0 \\ 0 & 0 & 0 & 0 & 0 & 1 & 0 \\ 0 & 0 & 0 & 0 & 0 & 0 & 1 \end{bmatrix}, \; N = \begin{bmatrix} 0.5 & 0 & 0 & 0 \\ 0 & 0.5 & 0 & 0 \\ 0 & 0 & 0.5 & 0 \\ 0 & 0 & 0 & 0 \\ 0 & 0 & 0 & 0 \\ 0 & 0 & 0 & 0 \\ 0 & 0 & 0 & 1 \end{bmatrix}.$$

Therefore, for the first actuator, we have $CTF_1 = 0$ and $C(TA)TF_1 \neq 0$. Hence, the fault detectability index for f_1 is $\rho_1 = 2$. Similarly, we have $\rho_2 = \rho_3 = 2$. Therefore, we have the fault detectability matrix Υ as

$$\Upsilon = C\Psi = \begin{bmatrix} 0.0893 & 0 & 0 \\ 0 & 0.0481 & 0 \\ 0 & 0 & 0.0625 \\ 0 & 0 & 0 \end{bmatrix}.$$

Therefore, we can obtain the matrices for the optimal parametrized gain $G^*(k)$ as follows:

$$\Psi = \begin{bmatrix} 0.1158 & 0 & 0 \\ 0 & 0.0582 & 0 \\ 0 & 0 & 0.0766 \\ 1.7870 & 0.1717 & 0 \\ 0.1717 & 0.7702 & 0 \\ 0 & 0 & 1.0797 \\ 0.0022 & 0.0029 & 0.0025 \end{bmatrix}, \; M = \begin{bmatrix} 11.2 & 0 & 0 & 0 \\ 0 & 20.8 & 0 & 0 \\ 0 & 0 & 16 & 0 \end{bmatrix}, \; \Omega = \begin{bmatrix} 0 & 0 & 0 & 1 \end{bmatrix}.$$

In the simulation, consider the actuator fault $f(k)$ in the following

$$f(k) = \begin{cases} 0 & k \leq 98 \\ \begin{bmatrix} 15, & 12\sin(0.1k), & 9.5\cos(0.1k) \end{bmatrix}^{\top} & k > 98 \end{cases}$$

The simulation has been carried out for $N_s = 200$ sampling time steps and the simulation results are shown in Fig. 5.3. Because of the fault detectability indices $\rho_1 = \rho_2 = \rho_3 = 2$, the fault $f(k)$ occurred at time k will be estimated in two samples. For different time-varying actuator faults, all the estimated results provide the satisfactory results including the punctual values and the worst-case bounds. By minimizing the size of the filter zonotope bounding all the uncertainties and propagated estimation errors, the obtained optimal gain $G^*(k)$ reduces the estimation

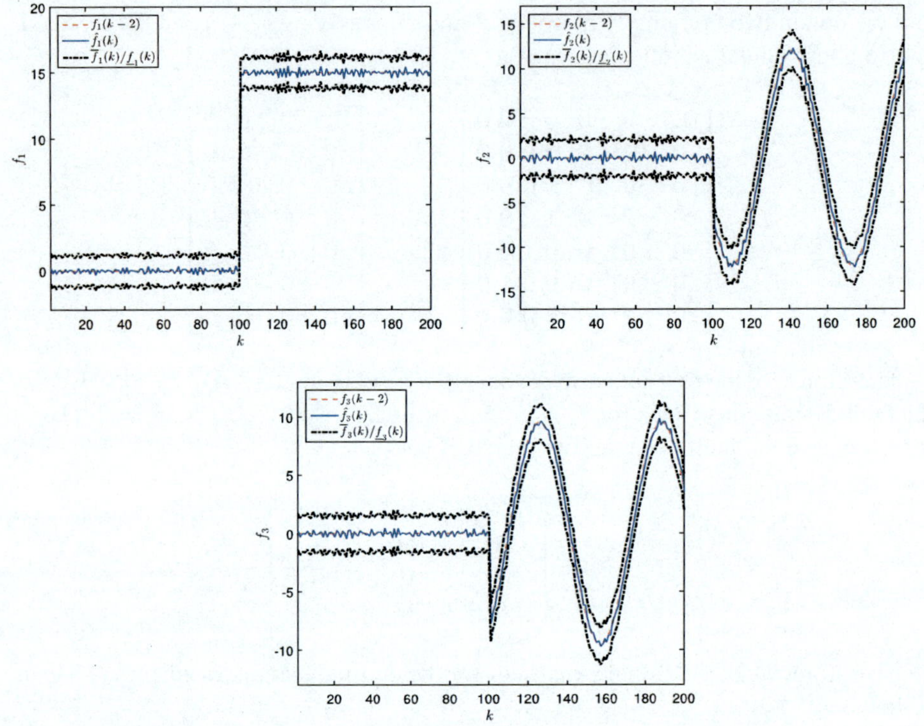

Fig. 5.3 Actuator-FE results of the machine infinite bus system

errors. Furthermore, during the propagation, the obtained FE intervals (centers of FE zonotopes and the worst-case bounds) are bounded.

5.4 Summary

This chapter has proposed a zonotopic FE filter for discrete-time descriptor systems. The system disturbances and measurement noise are bounded in given zonotopes. To achieve robustness against system uncertainties and identification of occurred actuator faults, the optimal gain is formulated in a parameterized form and following the set-based framework in Sect. 2.1.2, the optimal Kalman gain is computed. Besides, boundedness of the proposed zonotopic FE is discussed. The proposed zonotopic FE filter with the optimal FE gain is proved to be ultimately bounded. The proposed method is tested in two examples. We have compared the results with a stabilizing gain, where the robustness is not considered. The results with the optimal gain are shown to be more accurate based on the mean squared error results. As future research, the proposed FE method could be linked with set-based FI. Besides, the condition for estimating sensor faults deserves to be investigated.

References

1. Blanke M, Kinnaert M, Lunze J, Staroswiecki M (2016) Diagnosis and fault-tolerant control. Springer, Berlin Heidelberg
2. Ding S (2013) Model-based fault diagnosis techniques. Springer, London, UK
3. Wan Y, Keviczky T (2019) Real-time nonlinear moving horizon observer with pre-estimation for aircraft sensor fault detection and estimation. Int J Robust Nonlinear Control 29(16):5394–5411
4. Wan Y, Keviczky T, Verhaegen M (2018) Fault estimation filter design with guaranteed stability using Markov parameters. IEEE Trans Autom Control 63(4):1132–1139
5. Lan J, Patton R (2016) A new strategy for integration of fault estimation within fault-tolerant control. Automatica 69:48–59
6. Rotondo D, Witczak M, Puig V, Nejjari F, Pazera M (2016) Robust unknown input observer for state and fault estimation in discrete-time Takagi-Sugeno systems. Int J Syst Sci 47(14):3409–3424
7. Wang Z, Rodrigues M, Theilliol D, Shen Y (2015) Fault estimation filter design for discrete-time descriptor systems. IET Control Theory Appl 9(10):1587–1594
8. Zhang K, Jiang B, Shi P, Xu J (2015) Fault estimation observer design for discrete-time systems in finite-frequency domain. Int J Robust Nonlinear Control 25(9):1379–1398
9. Wang Y, Puig V, Cembrano G (2018) Robust fault estimation based on zonotopic Kalman observer for discrete-time descriptor systems. Int J Robust Nonlinear Control 28(16):5071–5086
10. Wang Y, Wang Z, Puig V, Cembrano G (2017) Zonotopic fault estimation filter design for discrete-time descriptor systems. In: 20th IFAC world congress, Toulouse, France, pp 5211–5216
11. Keller J (1999) Fault isolation filter design for linear stochastic systems. Automatica 35(10):1701–1706
12. Liu B, Si J (1997) Fault isolation filter design for linear time-invariant systems. IEEE Trans Autom Control 42(5):704–707
13. Combastel C (2015) Zonotopes and Kalman observers: gain optimality under distinct uncertainty paradigms and robust convergence. Automatica 55:265–273
14. Koenig D (2005) Unknown input proportional multiple-integral observer design for linear descriptor systems: application to state and fault estimation. IEEE Trans Autom Control 50(2):212–217
15. Wang Z, Shi P, Lim C (2017) Robust fault estimation observer in the finite frequency domain for descriptor systems. Int J Control 1590–1599

Chapter 6
Set-Invariance Characterizations and Active Mode Detection for Descriptor Systems

This chapter presents a general framework of set-invariance characterizations for discrete-time descriptor systems, and its application to active mode detection [1, 2] following the research line shown in Fig. 6.1. Among alternative techniques for the computation of invariant sets [3–9], we use ultimate boundedness of trajectories to obtain set-invariance characterizations for the systems subject to unknown-but-bounded disturbances [10]. The proposed computation of invariant sets relies on partitioning the state space of descriptor systems considering both causal and non-causal parts. For causal systems, we apply the ultimate boundedness method to obtain an RPI set and approximation of an mRPI set. In particular, for the case of non-causal systems, the states are split into causal and anti-causal. When the invariance property is considered asymptotically for both causal and anti-causal parts, the standard mRPI notation will be applied. For the computational result, the mRPI approximations will be discussed. On the other hand, considering the finite-time trajectories of the anti-causal states, a new notation of invariant sets, namely RNI, is introduced. For the mode detection problem, we present two strategies that use positive set invariance. The proposed solution is the design of additive active detection inputs for RPI set separations [11–13]. These input signals are obtained from the solution of two mixed-integer optimization problems. We propose two active mode detection algorithms for online monitoring of the current operating mode. Besides, the proposed active mode detection mechanism is not only limited for being used in descriptor systems. The designed active detection inputs and algorithms can also be used for standard dynamical systems.

Y. Wang, *Advances in State Estimation, Diagnosis and Control of Complex Systems*, Springer Theses, https://doi.org/10.1007/978-3-030-52440-1_6

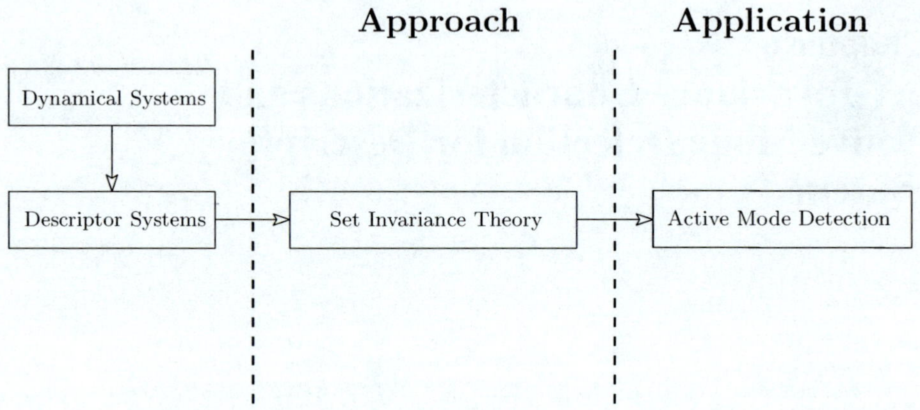

Fig. 6.1 Active mode detection based on set invariance theory

6.1 Set-Invariance Characterizations for Descriptor Systems

We now formulate explicit expressions of several RI sets and approximations of minimal RI sets for discrete-time descriptor systems in both causal and non-causal cases. Furthermore, the convergence time for each RI set is provided.

6.1.1 RPI Sets of Admissible Descriptor Systems

For an admissible descriptor system (1.1), the set analysis will be performed using the dynamics decomposition form. From Lemma 1.3, there exists a transformation (Q, P) leading to (1.3) and (1.4)–(1.6). We consider a partition of the matrix P as $P = [P_1,\ P_2]$ with $P_1 \in \mathbb{R}^{n\times r}$ and $P_2 \in \mathbb{R}^{n\times(n-r)}$. The structure of the mRPI set of the admissible descriptor system (1.1) is characterized in the following theorem.

Theorem 6.1 (mRPI set of admissible descriptor systems) *Consider an admissible descriptor system (1.1) with the dynamics decomposition form in (1.3) and $w(k) \in \mathcal{W}$, $\forall k \in \mathbb{N}$. The mRPI set Ω^c is given by*

$$\Omega^c = P_1\Phi_1 \oplus P_2\Phi_2, \tag{6.1}$$

where

$$\Phi_1 = \bigoplus_{i=0}^{\infty} \tilde{A}_1^i \tilde{B}_{w1}\mathcal{W}, \tag{6.2a}$$

$$\Phi_2 = \left(-A_4^{-1}A_3\Phi_1\right) \oplus \left(-A_4^{-1}B_{w2}\mathcal{W}\right), \tag{6.2b}$$

with $\tilde{A}_1 = A_1 - A_2 A_4^{-1} A_3$ *and* $\tilde{B}_{w1} = B_{w1} - A_2 A_4^{-1} B_{w2}$.

Proof With the transformation (Q, P), the descriptor system (1.1) is equivalent to a dynamical system including two subsystems as in (1.5). On the one hand, from (1.5) we have

$$\tilde{x}_1(k+1) = \tilde{A}_1 \tilde{x}_1(k) + \tilde{B}_{w1} w(k). \tag{6.3}$$

The stability of (1.1) implies that the matrix $\tilde{A}_1$ is Schur. Then, the characterization of the mRPI set of $\tilde{x}_1$ can be obtained as in (6.2a) using the standard LTI notions [14]. On the other hand, from (1.3) we obtain

$$\tilde{x}_2(k) = -A_4^{-1} A_3 \tilde{x}_1(k) - A_4^{-1} B_{w2} w(k), \tag{6.4}$$

Since (6.4) is an algebraic equation, we obtain the mRPI set Φ_2 by a linear projection image of the set Φ_1 in (6.2a), which leads to (6.2b).

By definition in (1.6) and using the Minkowski addition of the sets obtained via the linear mapping defined by the matrices P_1 and P_2, we determine the mRPI set Ω^c for the admissible descriptor system (1.1) as in (6.1). □

To approximate the mRPI set in (6.1), we use the ultimate bounds for dynamical systems in the following lemma which represents also a starting point for an iterative approximation of Ω^c.

Lemma 6.1 ([15]) *Consider the system (1.1) with $E = I$ and a Schur matrix $A \in \mathbb{R}^{n \times n}$, the Jordan decomposition form of $A = V \Lambda V^{-1}$ with $\Lambda = \operatorname{diag}(\lambda_1, \ldots, \lambda_n)$ and the compact disturbance set as in (1.13). The set*

$$\Phi(\varepsilon) = \left\{ x \in \mathbb{R}^n : \left| V^{-1} x \right| \le v + \varepsilon \right\}, \tag{6.5}$$

where $v = (I - |\Lambda|)^{-1} \left| V^{-1} B_w \right| \overline{w}$ and $\varepsilon \in \mathbb{R}^n$ is a vector with arbitrary small and positive components, is RPI and attractive for all the trajectories.

Corollary 6.1 *Consider the mRPI set Ω^c as in (6.1) and the Jordan decomposition $\tilde{A}_1 = V_1 \Lambda_1 V_1^{-1}$. An RPI approximation of Ω^c is given by*

$$\Omega_0^c = P_1 \hat{\Phi}_{1,0} \oplus P_2 \hat{\Phi}_{2,0}, \tag{6.6}$$

where

$$\hat{\Phi}_{1,0} = \left\{ x \in \mathbb{R}^r : \left| V_1^{-1} x \right| \le \tilde{v}_0 + \tilde{\varepsilon} \right\}, \tag{6.7a}$$

$$\hat{\Phi}_{2,0} = \left(-A_4^{-1} A_3 \hat{\Phi}_{1,0} \right) \oplus \left(-A_4^{-1} B_{w2} \mathcal{W} \right), \tag{6.7b}$$

with $\tilde{v}_0 = (I - |\Lambda_1|)^{-1} \left| V_1^{-1} \tilde{B}_{w1} \right| \overline{w}$ and $\tilde{\varepsilon} \in \mathbb{R}^r$ is a vector with arbitrary small and positive components. Moreover, any set $\Omega_i^c = P_1 \hat{\Phi}_{1,i} \oplus P_2 \hat{\Phi}_{2,i}$, $i \in \mathbb{N}$ where

$$\hat{\Phi}_{1,i} = \tilde{A}_1 \Phi_{1,i-1} \oplus \tilde{B}_{w1} \mathcal{W},$$
$$\hat{\Phi}_{2,i} = \left(-A_4^{-1} A_3 \hat{\Phi}_{1,i}\right) \oplus \left(-A_4^{-1} B_{w2} \mathcal{W}\right),$$

is also an RPI approximation of Ω^c and satisfies $\Omega_i^c \supseteq \Omega_{i+1}^c \supseteq \Omega^c$, $i \in \mathbb{N}$ with $\epsilon > 0$ satisfying $d_H\left(\Omega_i^c, \Omega^c\right) < \epsilon$.

Proof By using Lemma 6.1, the RPI set Φ_1 can be approximated by ultimate bounds as $\hat{\Phi}_{1,0}$ in (6.7a). Therefore, the mRPI set $\hat{\Phi}_{2,0}$ for $\tilde{x}_2$ can be obtained through a linear mapping as in (6.7b).

Using [10, Algorithm 1], an iterative positively invariant approximation of the mRPI set $\Omega_i^c \supseteq \Omega_{i+1}^c$ for $i \in \mathbb{N}$ can be obtained by applying the forward i-step propagation. Finally, with a constant $\epsilon > 0$, we have $d_H\left(\Omega_i^c, \Omega^c\right) < \epsilon$ for a finite index i by exploiting the convergence of the sequence Ω_i^c to Ω^c. □

Based on the above results, we present a practical condition for the compatibility check of any initial state $x(0)$ in the following corollary.

Corollary 6.2 *Consider an initial state $x(0)$ for the admissible descriptor system (1.1) in (1.4)–(1.6). If*

$$\tilde{x}_2(0) \notin \left(-A_4^{-1} A_3 \zeta^* \Phi_1\right) \oplus \left(-A_4^{-1} B_{w2} \mathcal{W}\right), \tag{6.8}$$

where $x(0) = P_1 \tilde{x}_1(0) + P_2 \tilde{x}_2(0)$ and

$$\zeta^* = \min \{\zeta \in \mathbb{R} : \tilde{x}_1(0) \in \zeta \Phi_1\},$$

then $x(0)$ is not a compatible initial state for (1.1) and it is independent of any disturbance realization $w(0) \in \mathcal{W}$.

Proof The set in (6.8) is not an RPI set but it also represents a constraint for the descriptor part of states whenever this constraint is violated. As a consequence, it leads to algebraic equations cannot be satisfied. □

Remark 6.1 By Definition 1.11 and its characterization in Theorem 6.1, the consistency in terms of initial state $x(0)$ with the descriptor model (1.1) can be tested. In presence of the disturbance $w(0) \in \mathcal{W}$, $x(0)$ may not be a compatible initial state. This shows that $x(0)$ should be understood as an implicit function of $w(0)$, i.e. $x(w(0))$, by means of the solution of algebraic equations.

To complete the study of admissible descriptor systems, the computation result of the convergence time for discrete-time admissible descriptor systems is provided. This is equivalent to an upper bound for the total number of steps necessary for the system trajectories to reach the set Ω^c from a given initial state.

Lemma 6.2 (Convergence time [16]) *Consider the system (1.1) with $E = I$ and a Schur matrix $A \in \mathbb{R}^{n \times n}$, the Jordan decomposition form of $A = V \Lambda V^{-1}$ with $\Lambda =$ diag $(\lambda_1, \ldots, \lambda_n)$. Let $\xi(k) = V^{-1} x(k)$ with $\xi(k) = [\xi_1(k), \ldots, \xi_n(k)]^\top$ and*

$$\xi(k+1) = \Lambda\xi(k) + V^{-1}B_w w(k),$$

and the initial condition $\xi(0) = \xi^* = \left[\xi_1^*, \ldots, \xi_n^*\right]^\top \in \mathbb{R}^n$. *Consider the RPI set* $\Phi\ (\varepsilon)$ *in* (6.5) *and define the vector* $v^* = \left[v_1^*, \ldots, v_n^*\right]^\top$ *with*

$$v^* = \arg\min_{\bar{v}} \left|\xi^* - \bar{v}\right| \quad \textit{subject to} \quad |\bar{v}| \leq v,$$

where the minimum is computed element-wise. Then, the system trajectory $x(k)$ *with the initial state* $x(0) = V\xi(0)$ *belongs to* $\Phi\ (\varepsilon)$, $\forall k \geq T_c$, *where*

$$T_c = \max\left(\ell_1, \ldots, \ell_n\right),$$

with

$$\ell_i = \begin{cases} 0, & \text{if } \ \xi_i^* = v_i^*, \\ \max\left(0, \log_{|\lambda_i|}\left(\frac{\varepsilon_i}{|\xi_i^* - v_i^*|}\right)\right), & \text{otherwise,} \end{cases}$$

for $i = 1, \ldots, n$.

Based on Lemma 6.2, from any compatible initial state $x(0)$, the convergence time of the admissible descriptor system (1.1) is given in the following theorem.

Theorem 6.2 (Convergence time of admissible descriptor systems) *Consider an admissible descriptor system (1.1),* $w(k) \in \mathcal{W}$, $\forall k \in \mathbb{N}$ *and the set* $\Omega_0^c \supseteq \Omega^c$ *in Corollary 6.1. For a compatible initial state* $x(0)$*, the system trajectory* $x(k)$ *belongs to* Ω_0^c*, that is,* $\tilde{x}_1(k)$ *defined in (1.6) belongs to* $\hat{\Phi}_{1,0}$*, for* $k \geq T_{c_a}$*, where* T_{c_a} *is the convergence time corresponding to (1.1) and depends on* $x(0)$ *and* $\tilde{\varepsilon}$.

Proof Based on Lemma 1.3, $\tilde{x}_2(k)$ has no dynamics and is a linear mapping of $\tilde{x}_1(k)$ and $w(k)$. By directly applying the result in Lemma 6.2 to $\tilde{x}_1(k)$ with its dynamics $\tilde{x}_1(k+1) = \tilde{A}_1\tilde{x}_1(k) + \tilde{B}_{w1}w(k)$, we can obtain the convergence time T_{c_a}. □

6.1.2 RPI Sets of Non-causal Descriptor Systems

In case that the descriptor system (1.1) is regular and stable but not causal, there might exist a unique solution at each time [17]. We now consider a non-causal and stable descriptor system (1.1) and use the Kronecker canonical form in (1.7) for the RPI characterization.

From Lemma 1.4, a non-causal descriptor system (1.1) can be transformed in (1.7) with a nilpotent matrix $\bar{N}$ satisfying $\bar{N} \neq 0$. As introduced in [17, Chap. 8], for a regular matrix pair (E, A), there exists a suitable transformation $(\bar{Q}, \bar{P})$ with $\bar{P} = \left[\bar{P}_1,\ \bar{P}_2\right]$, $\bar{P}_1 \in \mathbb{R}^{n\times p}$, $\bar{P}_2 \in \mathbb{R}^{n\times(n-p)}$ yielding to (1.7).

For the transformed system in the Kronecker form, we use the following partitioning form

$$\bar{x}(k) = \begin{bmatrix} \bar{x}_1(k) \\ \bar{x}_2(k) \end{bmatrix} = \bar{P}^{-1}x(k),\ \bar{Q}B_w = \begin{bmatrix} \bar{B}_{w1} \\ \bar{B}_{w2} \end{bmatrix}, \tag{6.9}$$

with $\bar{x}_1(k) \in \mathbb{R}^p$, $\bar{x}_2(k) \in \mathbb{R}^{(n-p)}$.

Based on the Kronecker canonical form in Lemma 1.4, we have that

$$\bar{x}_1(k+1) = \bar{A}\bar{x}_1(k) + \bar{B}_{w1}w(k), \tag{6.10a}$$

$$\bar{N}\bar{x}_2(k+1) = \bar{x}_2(k) + \bar{B}_{w2}w(k). \tag{6.10b}$$

The structure in (6.10) highlights the fact that the non-causal descriptor system (1.1) is stable if and only if the matrix $\bar{A}$ is Schur. We now formulate the mRPI set of discrete-time non-causal descriptor systems.

Theorem 6.3 (mRPI set of non-causal descriptor systems) *Consider a non-causal descriptor system (1.1) with the Kronecker canonical form in (1.7) and $w(k) \in \mathcal{W}$, $\forall k \in \mathbb{N}$. The mRPI set Ω^n is given by*

$$\Omega^n = \bar{P}_1\Theta_1 \oplus \bar{P}_2\Theta_2, \tag{6.11}$$

with

$$\Theta_1 = \bigoplus_{i=0}^{\infty} \bar{A}^i \bar{B}_{w1}\mathcal{W}, \tag{6.12a}$$

$$\Theta_2 = \bigoplus_{i=0}^{n-p-1} \left(-\bar{N}^i \bar{B}_{w2}\mathcal{W}\right). \tag{6.12b}$$

Proof The non-causal descriptor system can be decomposed in two subsystems, where (6.10a) is an ordinary difference equation. Hence, the mRPI set of $\bar{x}_1$ can be constructed as in (6.12a). On the other hand, from (6.10b), the anti-causal state $\bar{x}_2(k)$ can be propagated as follows:

$$\bar{x}_2(k) = \bar{N}\bar{x}_2(k+1) - \bar{B}_{w2}w(k),$$

$$\bar{x}_2(k+1) = \bar{N}\bar{x}_2(k+2) - \bar{B}_{w2}w(k+1),$$

and after the $(n-p)$-step iterations, this inequality becomes

$$\bar{x}_2(k) = \bar{N}^{(n-p)}\bar{x}_2(k+n-p) - \sum_{i=0}^{n-p-1} \bar{N}^i \bar{B}_{w2}w(k+i). \tag{6.13}$$

Since $\bar{N}$ is a nilpotent matrix with $\bar{N}^{n-p} = 0$, we know that for $k > n - p$, $\bar{N}^k = 0$. Therefore, (6.13) becomes $\bar{x}_2(k) = -\sum_{i=0}^{n-p-1} \bar{N}^i \bar{B}_{w2} w(k+i)$. With $w(k) \in \mathcal{W}$, $\forall k \in \mathbb{N}$, the set for $\bar{x}_2$ can be computed by $\Theta_2 = \bigoplus_{i=0}^{n-p-1} (-\bar{N}^i \bar{B}_{w2}\mathcal{W}) = (-\bar{B}_{w2}\mathcal{W}) \oplus (-\bar{N}\bar{B}_{w2}\mathcal{W}) \oplus \cdots \oplus (-\bar{N}^{n-p-1}\bar{B}_{w2}\mathcal{W})$. Finally, we derive the mRPI set Ω^n for the non-causal descriptor system (1.1) by the linear mapping as in (6.11). □

Remark 6.2 Theorem 6.3 builds on the assumption that the time domain of solution to the system (1.1) is $\mathbb{N}$. The existence of this infinite-time trajectory leads to a positive invariance property although the system is not causal. Theorem 6.3 should be reconsidered in case that the trajectories are defined only for a finite-time window.

Corollary 6.3 *Consider the mRPI set Ω^n as in (6.11) and the Jordan decomposition $\bar{A} = \bar{V}_1 \bar{\Lambda}_1 \bar{V}_1^{-1}$. An RPI approximation of Ω^n is given by*

$$\Omega_0^n = \bar{P}_1 \hat{\Theta}_{1,0} \oplus \bar{P}_2 \Theta_2, \tag{6.14}$$

with

$$\hat{\Theta}_{1,0} = \left\{x \in \mathbb{R}^{n_{\bar{x}_1}} : \left|\bar{V}_1^{-1} x\right| \leq \bar{v}_0 + \bar{\varepsilon}\right\}, \tag{6.15a}$$

$$\Theta_2 = \bar{B}_{w2}\mathcal{W} \oplus \bar{N}\bar{B}_{w2}\mathcal{W} \oplus \cdots \oplus \bar{N}^{s-1}\bar{B}_{w2}\mathcal{W}, \tag{6.15b}$$

where $\bar{v}_0 = \left(I - \left|\bar{\Lambda}_1\right|\right)^{-1} \left|\bar{V}_1^{-1}\bar{B}_{w1}\right| \overline{w}$ and $\bar{\varepsilon} \in \mathbb{R}^p$ is a vector with arbitrary small and positive components. Moreover, any set $\Omega_i^n = \bar{P}_1 \hat{\Theta}_{1,i} \oplus \bar{P}_2 \Theta_2$, $i \in \mathbb{N}$ where

$$\hat{\Theta}_{1,i} = \tilde{A}_1 \Phi_{1,i-1} \oplus \tilde{B}_{w1}\mathcal{W},$$

is also an RPI approximation of Ω^n and $\Omega_i^n \supseteq \Omega_{i+1}^n \supseteq \Omega^n$, $i \in \mathbb{N}$ with $\epsilon > 0$ satisfying $d_H\left(\Omega_i^n, \Omega^n\right) < \epsilon$.

Proof From Lemma 6.1, the mRPI set Θ_1 related to $\bar{x}_1$ can be approximated by an RPI set $\Theta_1 \subseteq \hat{\Theta}_{1,0}$ based on (6.15a). Similar to the proof of Corollary 6.1, we thus obtain Ω_i^n by the iterative forward mapping all by preserving the positive invariance. □

For a non-causal descriptor system (1.1), we also present a practical condition for the compatibility check of any initial state $x(0)$ in the following corollary.

Corollary 6.4 *Consider an initial state $x(0)$ of a non-causal descriptor system (1.1). If $\bar{x}_2(0) \notin \Theta_2$ where $x(0) = \bar{P}_1\bar{x}_1(0) + \bar{P}_2\bar{x}_2(0)$, then $x(0)$ is a compatible initial state for (1.1) irrespective of any disturbance realization $w(0) \in \mathcal{W}$.*

Proof Similar to Corollary 6.2, thus the proof is omitted. □

For any compatible initial state $x(0)$ of a non-causal descriptor system (1.1), the computation result of the convergence time is presented as follows.

Theorem 6.4 (Convergence time of non-causal descriptor systems) *Consider a non-causal descriptor system (1.1) affected by disturbances $w(k) \in \mathcal{W}$, $\forall k \in \mathbb{N}$ and let the set $\Omega_0^n \supseteq \Omega^n$. For a compatible initial state $x(0)$, the system trajectory $x(k)$ converges to Ω_0^n in T_{C_n} iterations, that is, $\bar{x}_1(k)$ defined in (6.9) belongs to $\hat{\Theta}_{1,0}$, for $k \geq T^{C_n}$, where T^{C_n} is the convergence time corresponding to (1.1) and depends on $x(0)$ and $\bar{\varepsilon}$.*

Proof In terms of the mRPI set Ω_0^n, again based on Lemma 6.2, the convergence time T^{C_n} of the non-causal descriptor system (1.1) is determined by the partitioned state $\bar{x}_1(k)$ with its dynamics described in (6.10b). □

6.1.3 RPI Sets for Finite-Time Trajectories of Non-causal Descriptor Systems

As an extension for a non-causal descriptor system (1.1), we now focus on trajectories defined only on a finite-time window, that is $x(k)$, $k \in \mathbb{Z}_{[0,L]}$ with $L > 0$. The dynamics of a non-causal descriptor system (1.1) obey the equivalent subsystems in (6.10) but the set-invariance characterization need to be relaxed in order to consider the finite number of dynamical constraints as well as the structural particularities (algebraic equations) related to anti-causality.

The difficulties are related to a combination of causal and anti-causal dynamics in (6.10a) and (6.10b). For (6.10a), the positive invariance will be the appropriate concept while for (6.10b), the negative invariance offers the suitable framework in a predefined finite-time window L.

Theorem 6.5 (L-step RNI set of non-causal descriptor systems)*Consider the anti-causal subsystem (6.10b). A set Υ is L-step RNI if*

$$\Upsilon \supseteq \bar{N}\Upsilon \oplus \{-\bar{B}_{w2}\mathcal{W}\} \supseteq \cdots \supseteq \bar{N}^L\Upsilon \bigoplus_{i=0}^{L-1} \{-\bar{N}^i\bar{B}_{w2}\mathcal{W}\}. \tag{6.16}$$

Proof From (6.10b), we have $\bar{x}_2(k) = \bar{N}\bar{x}_2(k+1) - \bar{B}_{w2}w(k)$. For a finite time window $L > 0$, $\bar{x}_2(L) \in \Upsilon$. By the backward propagations of $\bar{x}_2(k+L) \in \Upsilon$ for any $k \in \mathbb{Z}_{[-L,0]}$, we can derive (6.16). □

Corollary 6.5 *Given L_1- and L_2-step RNI sets Υ_1 and Υ_2 with $L_1 \geq L_2 \geq n - p$ satisfying $\Upsilon_1 \supseteq \Upsilon_2$, then for any $l \geq 0$, it holds*

$$\bar{N}^l\Upsilon_1 \bigoplus_{i=0}^{l-1} \{-\bar{N}^i\bar{B}_{w2}\mathcal{W}\} \supseteq \bar{N}^l\Upsilon_2 \bigoplus_{i=0}^{l-1} \{-\bar{N}^i\bar{B}_{w2}\mathcal{W}\}. \tag{6.17}$$

Proof The relationship (6.16) holds for $l = 0$ as $\Upsilon_1 \supseteq \Upsilon_2$. Suppose $\bar{N}^l \Upsilon_1 \bigoplus_{i=0}^{l-1} \{-\bar{N}^i \bar{B}_{w2}\mathcal{W}\} \supseteq \bar{N}^l \Upsilon_2 \bigoplus_{i=0}^{l-1} \{-\bar{N}^i \bar{B}_{w2}\mathcal{W}\}$ holds for some $l \geq 0$. Then, by pre-multiplying with $\bar{N}$ and Minkowski summing the set $\{-\bar{B}_{w2}\mathcal{W}\}$ on both sides, we obtain $\bar{N}^{l+1}\Upsilon_1 \bigoplus_{i=0}^{l} \{-\bar{N}^i \bar{B}_{w2}\mathcal{W}\} \supseteq \bar{N}^{l+1}\Upsilon_2 \bigoplus_{i=0}^{l} \{-\bar{N}^i \bar{B}_{w2}\mathcal{W}\}$. The proof is completed by induction. □

Remark 6.3 The set Θ_2 in (6.12b) is L-step RNI with respect to (6.10b), $\forall L > 0$.

Remark 6.4 Consider the set Θ_2 as in (6.12b). An L-step RNI set with respect to (6.10b) can be constructed iteratively starting from $\Upsilon_0 = \Theta_2$ and for $i \in \mathbb{Z}_{[1,L]}$, the recursive construction is given by

$$\Upsilon_i = \left\{x \in \mathcal{X}_2 : \exists w \in \mathcal{W}, \bar{N}x - \bar{B}_{w2}w \in \Upsilon_{i-1}\right\}, \tag{6.18}$$

and $\mathcal{X}_2 \subseteq \mathbb{R}^{(n-p)}$ is a pre-defined set of state constraints for $\bar{x}_2$.

Theorem 6.6 (L-step RI set of non-causal descriptor systems) *Consider a non-causal descriptor system (1.1) in with the Kronecker form in (1.7). The set*

$$\Omega = \bar{P}_1\Theta_1 \oplus \bar{P}_2\Upsilon, \tag{6.19}$$

guarantees that $x(k) \in \Omega$, $\forall k \in \mathbb{Z}_{[0,L]}$ if $\bar{x}_1(0) \in \Theta_1$ and $\bar{x}_2(L) \in \Upsilon$.

Proof From (6.12a), the set Θ_1 is RPI for the dynamics of $\bar{x}_1(k)$. If $\bar{x}_1(0) \in \Theta_1$, then it follows $\bar{x}_1(k) \in \Theta_1$, $\forall k \in \mathbb{Z}_{[0,L]}$. Meanwhile, the set Υ is L-step RNI for $\bar{x}_2(k)$ as discussed in Theorem 6.5. If $\bar{x}_2(L) \in \Upsilon$, then it follows $\bar{x}_2(k) \in \Upsilon$, $\forall k \in \mathbb{Z}_{[0,L]}$. Thus, we obtain Ω by a linear mapping of Θ_1 and Υ as in (6.19). □

Proposition 6.1 *Consider a non-causal descriptor system (1.1) in the restricted equivalent form (1.7) and define a finite-time trajectory $x(k)$ for $k \in \mathbb{Z}_{[0,L]}$ with $L > 0$. If $x(0) \in \Omega_0$ for $L > s = n - p$ with $\bar{N}^s = 0$ and $\bar{N}^{s-1} \neq 0$, then $x(k) \in \Omega_0$ for $k \in \mathbb{Z}_{[0,L-s]}$ and $x(k) \in \Omega_{k-(L-s)}$ for $k \in \mathbb{Z}_{[L-s,L]}$, where $\Omega_i = \bar{P}_1\Theta_1 \oplus \bar{P}_2\Upsilon_i$ with Υ_i in (6.18).*

Proof For $k \in \mathbb{Z}_{[0,L-s]}$, from (6.13), $x(k)$ is contained in the RI set $\Omega_0 = \Theta_2$ as defined in (6.19). On the other hand, for $k \in \mathbb{Z}_{[L-s,L]}$, the anti-causal component is contained in Υ_i, which can be propagated by using (6.18) leading to the confinement of the finite time trajectories for $L - s < k < L$. □

Example Consider the closed-loop dynamical Leontief model described from [18] in the form of (1.1), where

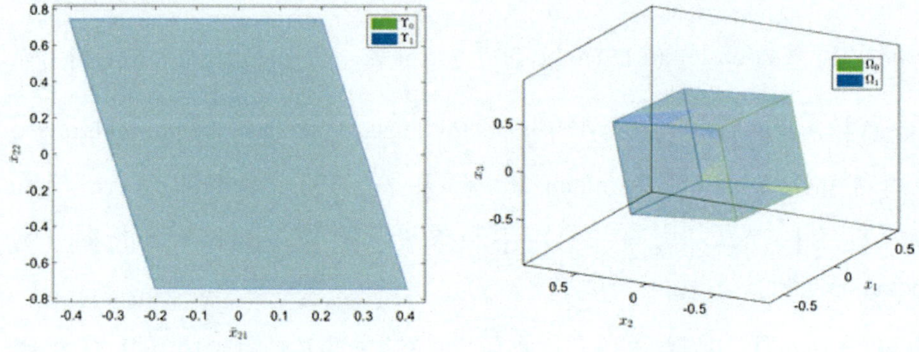

(a) RNI sets of $\bar{x}_2$ (b) RI sets

Fig. 6.2 RNI and RI sets of the dynamical Leontief model

$$E = \begin{bmatrix} 1 & 0.5 & 0.75 \\ 0.25 & 0 & 0.5 \\ 0 & 0 & 0 \end{bmatrix}, \; A = \begin{bmatrix} 1.1328 & 0.1427 & -0.3413 \\ -0.1172 & 0.6427 & -0.1913 \\ 0.1328 & 0.1427 & -0.0913 \end{bmatrix},$$

$$B_w = \begin{bmatrix} -0.3828 & -0.1427 & -0.4087 \\ -0.3828 & -0.1427 & -0.4087 \\ -0.3828 & -0.1427 & -0.4087 \end{bmatrix},$$

and $w(k) \in \mathcal{W}, \forall k \in \mathbb{N}$ with $\bar{w} = [0.2,\ 0.3,\ 0.1]^\top$. By applying the computational method in [19], we obtain a transformation $(\bar{Q}, \bar{P})$ with

$$\bar{Q} = \begin{bmatrix} -0.5524 & -0.7530 & 3.8890 \\ 0.5393 & -1.1540 & -0.0193 \\ 0 & 0 & 4.6456 \end{bmatrix}, \; \bar{P} = \begin{bmatrix} -0.2576 & 0.6414 & 0.5385 \\ -0.3391 & -0.8020 & 0.5596 \\ -0.9048 & -0.3206 & -0.6998 \end{bmatrix},$$

yielding the Kronecker form in (1.7) for this Leontief model with $\bar{A} = 0.0284$ and $\bar{N} = \begin{bmatrix} 0 & 0.4067 \\ 0 & 0 \end{bmatrix}$. With $n - p = 2$, by means of (6.18), we obtain the RNI sets Υ_0 and Υ_1 for $\bar{x}_2 = [\bar{x}_{21},\ \bar{x}_{22}]^\top \in \mathbb{R}^2$ as shown in Fig. 6.2a. By means of (6.19), we obtain the RI sets Ω_0 and Ω_1 of this dynamical Leontief model as shown in Fig. 6.2b. Note that at each figure, both computed sets are overlapped.

6.2 Active Mode Detection for Multi-mode Descriptor Systems

We now propose an active mode detection mechanism based on the RPI set characterizations for systems with multiple modes of operation and no switch between different modes. The objective is the identification of the current operating mode in a finite time with any initial state $x(0)$. This operating mode will be detected from a (finite) predefined set of modes of operation. The algorithmic procedures are able to detect the current operating mode based on the offline design of active detection inputs and the online monitoring. We formulate two mixed-integer optimization problems to find suitable active detection inputs for guaranteed set separations in a finite time window.

6.2.1 Problem Formulation in Active Mode Detection

Consider a family of discrete-time descriptor systems corresponding to multiple modes of operation as

$$E^{\sigma}x(k+1) = A^{\sigma}x(k) + B^{\sigma}u(k) + B_w^{\sigma}w(k), \tag{6.20}$$

where $E^{\sigma} \in \mathbb{R}^{n\times n}$ with $\operatorname{rank}(E^{\sigma}) \leq n$, $A^{\sigma} \in \mathbb{R}^{n\times n}$, $B^{\sigma} \in \mathbb{R}^{n\times m}$, $B_w^{\sigma} \in \mathbb{R}^{n\times q}$, and $\sigma \in \Sigma_d = \{1, \ldots, d\}$ denotes the constant mode index and $u(k) \in \mathbb{R}^m$ denotes an additive input vector at time instant k. It is assumed that the descriptor system (6.20) is regular and stable for any $\sigma \in \Sigma_d$, then it follows that matrices $(E^{\sigma} - A^{\sigma})$ are non-singular.

To simplify the notation for analysis, based on the Kronecker canonical form in Lemma 1.4, let us denote the partitioning form

$$x = \left[x_1^{\top},\ x_2^{\top}\right]^{\top}, \tag{6.21}$$

where $x_1 \in \mathbb{R}^p$ is the dynamical part corresponding to the dynamics (6.10a) and $x_2 \in \mathbb{R}^{(n-p)}$ is the algebraic part corresponding to the algebraic equation (6.10b). Based on this notation, we also denote $B^{\sigma} = \left[B_1^{\sigma\top},\ B_2^{\sigma\top}\right]^{\top}$ and $B_w^{\sigma} = \left[B_{w_1}^{\sigma\ \top},\ B_{w_2}^{\sigma\ \top}\right]^{\top}$.

The objective of the mode detection is to decide which mode $\sigma \in \Sigma_d$ is active in (6.20) by monitoring the current state $x(k)$ and without prior knowledge on $w(k) \in \mathcal{W}$. The initial state $x(0)$ is assumed to be known and we make use of the RPI sets of (6.20) of each mode $\sigma \in \Sigma_d$ as $\tilde{\mathcal{P}}^{\sigma}$ when $u \equiv 0$. For a state $x(k)$ of (6.20), $\forall k \in \mathbb{N}$, the system (6.20) in the mode $i \in \Sigma_d$ can be performed by $x(k) = \bar{x}^i(k) + \tilde{x}^i(k)$ with the nominal and perturbed dynamics

$$E^i\bar{x}^i(k+1) = A^i\bar{x}^i(k), \tag{6.22a}$$

$$E^i\tilde{x}^i(k+1) = A^i\tilde{x}^i(k) + B_w^i w(k), \tag{6.22b}$$

where $\bar{x}^i \in \mathbb{R}^n$ and $\tilde{x}^i \in \mathbb{R}^n$.

The basic *passive* mode detection mechanism ($u \equiv 0$) can be summarized as follows:

Proposition 6.2 *Consider the compatible initial state* $x(0) = \bar{x}_i(0) + \tilde{x}^i(0)$ *satisfying*

$$x(0) - \bar{x}^i(0) \in \tilde{\mathcal{P}}^i,$$

and let the set of viable modes be initialized as $\Sigma(0) = \Sigma_d$. *Given the state measured at time* k, *if* $x(k) \notin \left\{\bar{x}^i(k) \oplus \tilde{\mathcal{P}}^i\right\}$, *then the mode* i *is not the current operating mode, that is,*

$$\Sigma(k) = \Sigma(k) \setminus \{i\}.$$

Proof The error dynamics $\tilde{x}^i(k) = x(k) - \bar{x}^i(k)$ satisfy (6.22b) and the initialization ensures $\tilde{x}^i(k) \in \tilde{\mathcal{P}}^i$. If the system (6.20) is operating in mode i, then the positive invariance of $\tilde{\mathcal{P}}^i$ is guaranteed with respect to (6.22b). Whenever $x(k) \notin \left\{\bar{x}^i(k) \oplus \tilde{\mathcal{P}}^i\right\}$, the positive invariance is violated and the mode i cannot represent the current operating mode. □

Let us also denote the transformation $\left(\bar{Q}^i, \bar{P}^i\right)$ for the descriptor system (6.20) at mode $i \in \Sigma_d$ such that $\bar{Q}^i E^i \bar{P}^i$ and $\bar{Q}^i A^i \bar{P}^i$ satisfy the Kronecker canonical form in (1.7). From the RPI set characterizations in Sect. 6.1, the RPI set $\tilde{\mathcal{P}}^i$ composed by $\tilde{\mathcal{P}}^i = \bar{P}_1^i \Phi_1^i \oplus \bar{P}_2^i \Phi_2^i$ with $\bar{P}^i = \left[\bar{P}_1^i, \ \bar{P}_2^i\right]$.

Theorem 6.7 *A state* $x(k) = \left[x_1(k)^\top, \ x_2(k)^\top\right]^\top$ *in the form of* (6.21) *is compatible with respect to the descriptor system* (6.20) *in an operating mode* $i \in \Sigma_d$ *only if* $x_2(k)$ *satisfies*

$$x_2(k) \in \bar{P}_2^i \Phi_2^i. \tag{6.23}$$

Proof Based on the Kronecker canonical form in (1.7), with the transformation $\left(\bar{Q}^i, \bar{P}^i\right)$ in mode $i \in \Sigma_d$, for a compatible state $x(k)$, the corresponding algebraic equation (6.10b) should be satisfied. Thus, the condition (6.23) could be used for checking the operating mode $i \in \Sigma_d$. □

Based on the above theorem, we state the following corollary without proof.

Corollary 6.6 *For an initial state* $x(0) = \left[x_1(0)^\top, \ x_2(0)^\top\right]^\top$ *in the form of* (6.21), *if* $x_2(0) \notin \bar{P}_2^i \Phi_2^i$, *then the initial operating mode set* $\Sigma_d(0) = \Sigma_d \setminus \{i\}$.

Remark 6.5 Assuming $\Sigma_d(k) = \Sigma_d(k-1) \setminus \{i\}, \forall i \in \Sigma_d$ such that $x(k) - \bar{x}^i(k) \notin \tilde{\mathcal{P}}^i$, then Card $(\Sigma_d(k))$ is monotonically decreasing as time k increases. However, one cannot guarantee Card $(\Sigma_d(k)) \to 1$.

Example Consider three modes of operation in (6.20).[1] As shown in Fig. 6.3, from an initial state $x(0)$, the mode shown in blue sets is detected after several steps. As time k increases, the modes in red and green sets are discarded. Note that the

[1]Numerical values are provided later in the numerical example chapter.

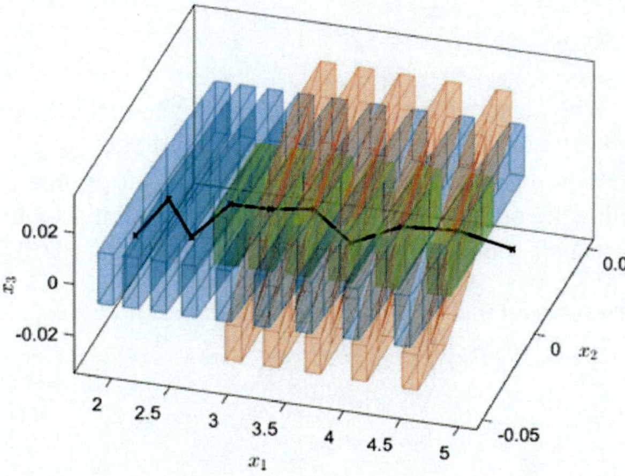

Fig. 6.3 A *passive* mode detection example

system state trajectory $x(k)$ may always stay in the intersection of three sets during propagation. In this case, we cannot discard any mode.

This *passive* mode detection does not guarantee the mode identifiability regardless of the initial conditions. Indeed, $\bigcap_{\sigma \in \Sigma_d} \tilde{\mathcal{P}}^\sigma \neq \varnothing$ and thus there exists at least a realization $w(k)$, $\forall k \in \mathbb{N}$, which does not allow to decrease the Cardinality of $\Sigma_d(k)$ and eventually identify the current mode of operation. The *active* mode detection is intended to enhance the monitoring process by the injection of an excitation signal.

6.2.2 Design of Active Detection Input

In the following, we would like to design two types of active detection inputs: (i) constant detection input; (ii) a sequence of variable detection inputs. For any two different modes $i, j \in \Sigma_d$, the active detection input denoted by $u(k)$ is designed to guarantee $\mathcal{P}^i(k) \cap \mathcal{P}^j(k) = \varnothing$ for some $k \in \mathbb{N}$, where $\mathcal{P}^i(k)$ and $\mathcal{P}^j(k)$ denote the tube of trajectories parameterized by $u(k)$. From (6.20), the system (6.20) in modes i and j can be formulated as

$$E^i x(k+1) = A^i x(k) + B^i u(k) + B_w^i w(k), \tag{6.24a}$$

$$E^j x(k+1) = A^j x(k) + B^j u(k) + B_w^j w(k). \tag{6.24b}$$

Recall that for $u(k) = 0$ in (6.24), it follows $\mathcal{P}^i(k) = \tilde{\mathcal{P}}^i$ and $\mathcal{P}^j(k) = \tilde{\mathcal{P}}^j$.

Similar to (6.22), assuming the system (6.20) in mode $i \in \Sigma_d$, we split $x(k) = \bar{x}^i(k) + \tilde{x}^i(k)$ with

$$E^i \bar{x}^i(k+1) = A^i \bar{x}^i(k) + B^i u(k), \tag{6.25a}$$

$$E^i \tilde{x}^i(k+1) = A^i \tilde{x}^i(k) + B^i_w w(k). \tag{6.25b}$$

With an active detection input $u(k)$, $\forall k \in \mathbb{N}$, the state $x(0)$ has to be decomposed as $x(0) = \bar{x}^i(0) + \tilde{x}^i(0)$ (for instance in mode $i \in \Sigma_d$) to satisfy the algebraic equations in the descriptor model (6.20). Based on this observation, we introduce the following proposition to check whether the initial state $x(0)$ is compatible by testing the satisfaction of algebraic equations in (6.20) for different modes.

Proposition 6.3 *Given the set of modes Σ_d. For any $i \in \Sigma_d$ such that* $\mathrm{rank}(E^i) < n$, *if $B^i_2 \neq 0$, then $\exists u(0)$ such that*

$$x(0) \notin \tilde{\mathcal{P}}^i. \tag{6.26}$$

Proof From (6.25b), we know $x(0) = \bar{x}^i(0) + \tilde{x}^i(0)$ and $\tilde{x}^i(0) \in \tilde{\mathcal{P}}^i$. Based on the nominal descriptor dynamics (6.25a), $\bar{x}^i(0)$ is also constrained by $u(0)$ at time $k = 0$. If $B^i_2 \neq 0$, then $\bar{x}^i(0) \neq 0$. Considering the boundedness of $\tilde{\mathcal{P}}^i$ and the fact that $x(0) = \bar{x}^i(0) + \tilde{x}^i(0)$, there exists $u(0)$ acting on $\bar{x}^i(0)$ that satisfies (6.26). □

The result in Proposition 6.3 shows that descriptor systems have structural advantages in view of mode detection, that is, the algebraic equations in a descriptor systems must hold. When an additional detection input signal is applied, by checking (6.26), some modes can be discarded.

6.2.2.1 Constant Active Detection Input

We first present the procedure to design a constant active detection input $\bar{u} \neq 0$ that can be applied to the system (6.20) with a finite detection time N_T as

$$u(k) = \begin{cases} \bar{u}, & \text{if } k \leq N_T - 1, \\ 0, & \text{otherwise.} \end{cases} \tag{6.27}$$

With this constant input $\bar{u}$, (6.25) becomes

$$E^i \bar{x}^i(k+1) = A^i \bar{x}^i(k) + B^i \bar{u}, \tag{6.28a}$$

$$E^i \tilde{x}^i(k+1) = A^i \tilde{x}^i(k) + B^i_w w(k). \tag{6.28b}$$

Recall $x(0) = \left[x_1(0)^\top, x_2(0)^\top\right]^\top$. The initial condition is given by $\bar{x}^i_1(0) = x_2(0)$ and $\bar{x}^i_2(0)$ satisfies (6.28a) with $\bar{u}$.

By definition of the RPI set, we denote $\tilde{x}^i(k+1) \in \tilde{\mathcal{P}}^i, \forall \tilde{x}^i(k) \in \tilde{\mathcal{P}}^i, \forall w(k) \in \mathcal{W}$, $\forall k \in \mathbb{N}$. The system trajectory in mode i belongs to the parameterized RPI set, that is,

$$x(k) \in \mathcal{P}^i(k) = \left\{\bar{x}^i(k) \oplus \tilde{\mathcal{P}}^i\right\}, \tag{6.29}$$

with $\bar{x}^i(k)$ obtained from (6.28a) and $\forall w(k) \in \mathcal{W}$, $\forall k \in \mathbb{N}$.

From the nominal dynamics (6.28a), the stability is guaranteed when the system evolves towards the equilibrium point

$$\bar{x}^i_\infty = \left(E^i - A^i\right)^{-1} B^i \bar{u}. \tag{6.30}$$

In the following theorem, we present the set separation condition for the design of $\bar{u}$.

Theorem 6.8 *For any two modes $i, j \in \Sigma_d$, the sets*

$$\mathcal{P}^i_\infty = \left\{\bar{x}^i_\infty \oplus \tilde{\mathcal{P}}^i\right\},\ \mathcal{P}^j_\infty = \left\{\bar{x}^j_\infty \oplus \tilde{\mathcal{P}}^j\right\} \tag{6.31}$$

satisfy $\mathcal{P}^i_\infty \cap \mathcal{P}^j_\infty = \varnothing$ if and only if there exists an active detection input $\bar{u}$ such that

$$\left((E^i - A^i)^{-1}B^i - (E^j - A^j)^{-1}B^j\right)\bar{u} \notin \mathcal{S}^{ij}. \tag{6.32}$$

Proof From (6.31), $\mathcal{P}^i_\infty \cap \mathcal{P}^j_\infty = \varnothing$ is equivalent to

$$\left\{\bar{x}^i_\infty \oplus \tilde{\mathcal{P}}^i\right\} \cap \left\{\bar{x}^j_\infty \oplus \tilde{\mathcal{P}}^j\right\} = \varnothing. \tag{6.33}$$

By adding $-\bar{x}^\infty_j$ to the above both sets in (6.33), we obtain

$$\left\{\bar{x}^i_\infty \oplus (-\bar{x}^j_\infty) \oplus \tilde{\mathcal{P}}^i\right\} \cap \left\{\bar{x}^j_\infty \oplus (-\bar{x}^j_\infty) \oplus \tilde{\mathcal{P}}^j\right\} = \varnothing,$$

which can be simplified as

$$\left\{\left(\bar{x}^i_\infty - \bar{x}^j_\infty\right) \oplus \tilde{\mathcal{P}}^i\right\} \cap \tilde{\mathcal{P}}^j = \varnothing, \tag{6.34}$$

leading to (6.32) based on (6.30). □

Let us denote the half-space representation of the set $\mathcal{S}_{ij}$ as

$$\mathcal{S}_{ij} = \left\{x \in \mathbb{R}^n : H_{ij}x \leq b_{ij}\right\},\ \forall i, j \in \Sigma_d,$$

where $H_{ij} \in \mathbb{R}^{p_{ij} \times n}$, $b_{ij} \in \mathbb{R}^{p_{ij}}$, and p_{ij} is the total number of the linear constraints corresponding to $\mathcal{S}_{ij}$.

Based on the set separation condition in (6.32), the constant active detection input $u \in [u_{\min}, u_{\max}]$ can be obtained by solving offline the following mixed-integer optimization problem.

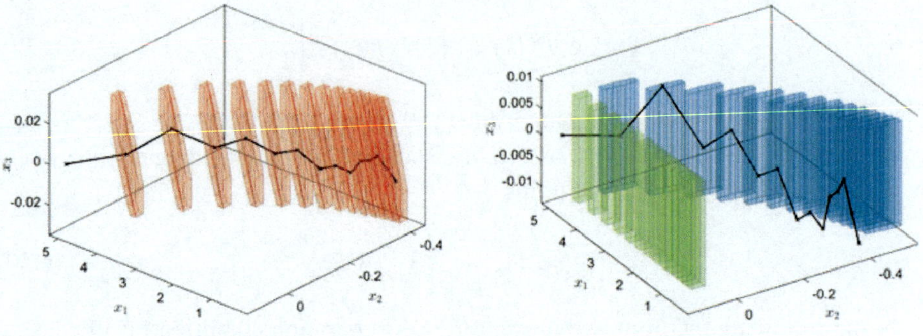

(a) The operating mode (b) The other modes

Fig. 6.4 Propagated RPI sets with a constant active detection input $\bar{u}$

Problem 6.1 (Constant active detection input)

$$\underset{u}{\text{minimize}} \ u^2, \tag{6.35a}$$

subject to

$$\bar{x}^i_\infty = \left(E^i - A^i\right)^{-1} B^i u, \tag{6.35b}$$

$$\bar{x}^j_\infty = \left(E^j - A^j\right)^{-1} B^j u, \tag{6.35c}$$

$$u_{\min} \leq u \leq u_{\max}, \tag{6.35d}$$

$$H_{ij}\left(\bar{x}^i_\infty - \bar{x}^j_\infty\right) \geq b_{ij} - M_r \Delta_{ij} + \epsilon_r, \tag{6.35e}$$

$$\Delta_{ij} = \left\{\delta_1, \ldots, \delta_{p_{ij}}\right\} \in \{0, 1\}, \tag{6.35f}$$

$$\sum_{l=1}^{p_{ij}} \delta_l = p_{ij} - 1, \forall i, j \in \Sigma_d, i \neq j \tag{6.35g}$$

with an arbitrary large positive scalar M_r and an arbitrary small positive scalar ϵ_r.

The optimal solution of Problem 6.1 defines the constant active detection input $\bar{u} = u$.

Example Consider the same three modes of operation in (6.20). By solving Problem 6.1, a constant active detection input $\bar{u}$ can be obtained. A example of the propagated RPI sets with $\bar{u}$ is shown in Fig. 6.4 for causal and non-causal descriptor systems. From an initial state $x(0)$, the system state trajectory only stays in the RPI sets of the red mode during propagation which sometimes stays outside the others (blue and green modes).

With the constant detection input obtained from solving Problem 6.1, the guaranteed mode detection result is presented in the following theorem.

Theorem 6.9 *If $\bar{u}$ is a feasible solution of Problem 6.1, then for any initial state $x(0)$, there exists a finite time $N_T(x(0))$ such that the detection* $\text{Card}(\Sigma_d(k)) = 1$ *is achieved in $k \leq N_T(x(0))$. Moreover, the convergence time from $x(0)$ to the set $\mathcal{P}^i_\infty$ denoted as T^i_c can be computed explicitly for any $i \in \Sigma_d$. Then, the upper bound for the detection time is*

$$N_T(x(0)) = \max_i T^i_c. \tag{6.36}$$

Proof By the design of $\bar{u}$, it is guaranteed $\mathcal{P}^i_\infty \cap \mathcal{P}^j_\infty = \varnothing$ for any two modes $i, j \in \Sigma_d$. For a given initial state $x(0)$ compatible with the mode i in (6.28), one has $x(T^i_c) \in \mathcal{P}^i_\infty$ independent of the operating mode

$$x\big(N_T(x(0))\big) \in \mathcal{P}^i_\infty. \tag{6.37}$$

But $\mathcal{P}^i_\infty \cap \mathcal{P}^j_\infty = \varnothing$ for all $i \neq j$ and (6.37) only holds for the current operating mode. □

6.2.2.2 Variable Active Detection Inputs

The previous result shows that for any initial state $x(0)$, the mode detection can be achieved in a finite time. However, this finite time or the energy of the active detection input can be further optimized with variable signals [20]. Assume that the system (6.20) is in mode i. Considering a horizon N_t, we would like to design offline for a given initial state $x(0)$, a variable active detection input sequence $u^*(l)$, $l = 0, \ldots, N_t - 1$ such that the operating mode of (6.20) will be detected in no more than N_t time steps. The applied active detection input for (6.20) will be

$$u(k) = \begin{cases} u^*(k), & \text{if } k \leq N_t - 1, \\ 0, & \text{otherwise.} \end{cases} \tag{6.38}$$

According to the discussion above, for any mode $i \in \Sigma_d$, the state $x(k)$ can be split to be $x(k) = \bar{x}^i(k) + \tilde{x}^i(k)$, where $\bar{x}^i(k)$ and $\tilde{x}^i(k)$ are propagated based on (6.25).

By definition of the RPI set, we know $\tilde{x}^i(k + l + 1) \in \tilde{\mathcal{P}}^i$ for $l = 0, \ldots, N_t - 1$, $\forall w(l) \in \mathcal{W}$. From (6.25), when the system (6.20), the following condition should be satisfied:

$$x(k + l + 1) \in \mathcal{P}^i(k + l + 1) = \left\{\bar{x}^i(k + l + 1) \oplus \tilde{\mathcal{P}}^i\right\},$$

for $l = 0, \ldots, N_t - 1$.

For any two modes $i, j \in \Sigma_d$, the variable active detection input sequence $u^*(l)$ for $l = 0, \ldots, N_t - 1$ along the horizon of N_t can be designed to guarantee that at least one of the following condition:

$$\left\{\bar{x}^i(l) \oplus \tilde{\mathcal{P}}^i\right\} \cap \left\{\bar{x}^j(l) \oplus \tilde{\mathcal{P}}^j\right\} = \varnothing, \tag{6.39}$$

holds for $l = 0, \ldots, N_t - 1$.

We now propose the following offline mixed-integer optimization problem to design variable active detection inputs satisfying (6.39).

Problem 6.2 (Variable active detection inputs)

$$\underset{u(0),\ldots,u(N_t-1)}{\text{minimize}} \quad \sum_{l=0}^{N_t-1} u(l)^2, \tag{6.40a}$$

subject to

$$E^i\bar{x}^i(l+1) = A^i\bar{x}^i(l) + B^iu(l), \tag{6.40b}$$

$$E^j\bar{x}^j(l+1) = A^j\bar{x}^j(l) + B^ju(l), \tag{6.40c}$$

$$u_{\min} \le u(l) \le u_{\max}, \tag{6.40d}$$

$$u(N_t) = 0, \tag{6.40e}$$

$$\bar{x}_1^i(0) = \bar{x}_1^j(0) = x_1(0), \tag{6.40f}$$

$$H_{ij}\left(\bar{x}^i(l+1) - \bar{x}^j(l+1)\right) \ge b_{ij} - M_r\Delta_{ij}(l) + \epsilon_r, \tag{6.40g}$$

$$\Delta_{ij}(l) = \left\{\delta_1(l), \ldots, \delta_{p_{ij}}(l)\right\} \in \{0, 1\}^{p_{ij}}, \tag{6.40h}$$

$$\sum_{l=0}^{N_t-1}\sum_{n_p=1}^{p_{ij}} \delta_{n_p}(l) \le N_t p_{ij} - 1, \tag{6.40i}$$

for $l = 0, \ldots, N_t - 1$ and $\forall i, j \in \Sigma_d, i \neq j$, with an arbitrary large positive scalar M_r and an arbitrary small positive scalar ϵ_r.

The optimal solution of Problem 6.2 defines the variable active detection inputs $u^*(l)$ for $l = 0, \ldots, N_t - 1$.

The variable detection input sequence obtained from solving Problem 6.2 guarantees that the mode detection result is also presented in the following theorem.

Theorem 6.10 *Given an initial state $x(0)$. If Problem 6.2 is feasible for a horizon N_t, then the detection is guaranteed* $\text{Card}(\Sigma_d(k)) = 1$ *for $k \le N_t$.*

Proof Similar to the proof of Theorem 6.9, the feasible solution ensures that the set separation $\mathcal{P}^i(k) \cap \mathcal{P}^j(k) = \varnothing$ for some $0 \le k \le N_t$. Thus, the monotonic decrease of $\Sigma_d(k)$ is guaranteed, and consequently the convergence in finite time to $\text{Card}(\Sigma_d(k)) = 1$. □

Algorithm 6.1 Active mode detection with constant detection input

(Offline procedure) For any $\Sigma \subseteq \Sigma_d$ with $\text{Card}(\Sigma) \geq 2$, compute $\bar{u}_\Sigma$ as the solution of Problem 6.1;
(Online procedure) Input an initial state $x(0)$;
Compute the compatible state $\bar{x}^i(0)$ with $u(0) = \bar{u}_{\Sigma_d(0)}$ and $x(0)$;
$k = 0$;
while $\text{Card}(\Sigma_d(k)) > 1$ **do**
    for $i \in \Sigma_d(k)$ **do**
        if $x(k) \notin \left\{\bar{x}^i(k) \oplus \tilde{\mathcal{P}}^i\right\}$ **then**
            $\Sigma_d(k) = \Sigma_d(k) \setminus \{i\}$;
        end if
    end for
    $u(k) = \bar{u}_{\Sigma_d(k)}$;
    Update the nominal state $\bar{x}^i(k+1)$ by (6.28a);
    $k = k + 1$;
end while
Obtain $\text{Card}(\Sigma_d(k)) = 1$ and the operating mode is detected.

Corollary 6.7 *Given* $x(0)$ *and assuming Problem 6.1 is feasible. Then, a feasible solution for Problem 6.2 exists with* $N_t \leq N_T(x(0))$.

Proof The feasibility of Problem 6.1 ensures a mode detection in $N_T(x(0))$ steps. The sequence $u(0) = u(1) = \cdots = u(N_T(x(0))) = \bar{u}$ represents a feasible solution of Problem 6.2 with $N_t = N_T(x(0))$. Thus, the proof is complete and the optimal solution of Problem 6.2 can only improve the detection time $N_t \leq N_T(x(0))$. □

6.2.3 *Active Mode Detection Algorithms*

Based on the above results, we next propose two algorithms for active mode detection. The first algorithm exploits the separation based on the constant input signal (Problem 6.1) and achieves the mode detection by updating online the active input according to the monitoring of the compatible modes. Overall, this leads to a piecewise constant signal and a detection time upper-bounded by $N_T(x(0))$.

The second algorithm builds with a variable detection input sequence, which guarantees the active mode detection in N_t time steps. This algorithm can be enhanced by recomputing the active detection input sequence $u^*(l)$ for $l = 0, \ldots, N_t - 1$ after each update of the set Σ_d.

Algorithm 6.2 Active mode detection with variable detection inputs

1: *(Offline procedure)* Given an initial state $x(0)$, solve Problem 6.2 and obtain the active detection input sequence $u^*(l)$ for $l = 0, \ldots, N_t - 1$;
2: *(Online procedure)* Initialize $\Sigma_d(0) = \Sigma_d$;
3: Compute the compatible state $\bar{x}^i(0)$ with $u(0) = u^*(0)$ and $x(0)$;
4: $k = 0$;
5: **while** $\text{Card}(\Sigma_d(k)) > 1$ **do**
6:   **for** $i \in \Sigma_d(k)$ **do**
7:     **if** $x(k) \notin \{\bar{x}^i(k) \oplus \tilde{\mathcal{P}}^i\}$ **then**
8:       $\Sigma_d(k) = \Sigma_d(k) \setminus \{i\}$;
9:     **end if**
10:   **end for**
11:   $u(k) = u^*(k)$;
12:   Update the nominal state $\bar{x}^i(k+1)$ by (6.25a);
13:   $k = k + 1$;
14: **end while**
15: Obtain $\text{Card}(\Sigma_d(k)) = 1$ and the operating mode is detected.

6.3 Numerical Example

Given the descriptor system (6.20) with three modes ($d = 3$) where system matrices are given by

$$E^1 = \begin{bmatrix} 1 & 0 & 0 \\ 0 & 1 & 0 \\ 0 & 0 & 1 \end{bmatrix}, \; A^1 = \begin{bmatrix} 0.8558 & -0.1692 & 0.2212 \\ 0.0186 & 0.7203 & -0.0929 \\ 0.0292 & -0.0467 & 0.7540 \end{bmatrix},$$

$$E^2 = \begin{bmatrix} 1.5 & 0.8 & 0 \\ 0 & 1.6 & 0 \\ 0.7 & 0.8 & 0 \end{bmatrix}, \; A^2 = \begin{bmatrix} 1.275 & 0.64 & 2 \\ 0 & 1.28 & 2.5 \\ 0.595 & 0.64 & 1 \end{bmatrix},$$

$$E^3 = \begin{bmatrix} 0.5 & 0 & 1.8 \\ 0 & 0 & -1.2 \\ 0 & 0 & 1.5 \end{bmatrix}, \; A^3 = \begin{bmatrix} 0.32 & 1.8 & -2 \\ 0 & -1.2 & 1.5 \\ 0 & 1.5 & 0.8 \end{bmatrix},$$

from which mode 1 is a standard dynamical system, mode 2 is an admissible descriptor system, and mode 3 is a non-causal descriptor system. Besides,

$$B^1 = \begin{bmatrix} 1 \\ 1 \\ 0.5 \end{bmatrix}, \; B^2 = \begin{bmatrix} 3.3 \\ 2.85 \\ 2 \end{bmatrix}, \; B^3 = \begin{bmatrix} 1.3 \\ -0.45 \\ 1.9 \end{bmatrix},$$

$$B_w^1 = \begin{bmatrix} 1 \\ 1 \\ 1 \end{bmatrix}, \; B_w^2 = \begin{bmatrix} 6.3 \\ 6.6 \\ 3.5 \end{bmatrix}, \; B_w^3 = \begin{bmatrix} -3.7 \\ 3.3 \\ 3.9 \end{bmatrix},$$

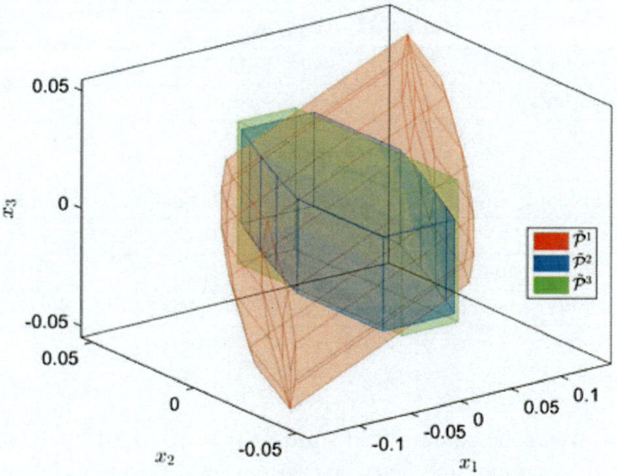

Fig. 6.5 mRPI sets of three modes

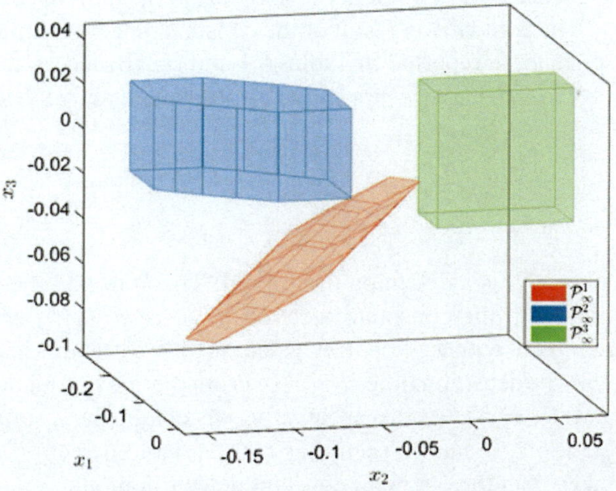

Fig. 6.6 Separated RPI sets of three modes

and $w(k) \in \mathcal{W}$, $\forall k \in \mathbb{N}$, where $\mathcal{W}$ is defined in (1.13) with $\bar{w} = 0.01$. By means of results in Sect. 6.1, the mRPI sets of three modes $\tilde{\mathcal{P}}^\sigma$ for $\sigma \in \Sigma_3 = \{1, 2, 3\}$ can be obtained as shown in Fig. 6.5 with $u \equiv 0$. Since the coordinate origin is the equilibrium point of three modes, these three mRPI sets $\tilde{\mathcal{P}}^\sigma$ for $\sigma \in \Sigma_3$ overlap.

By solving Problem 6.1 with three modes in this example, we can obtain a constant input $\bar{u}_{123} = -0.0202$. The separated RPI sets $\mathcal{P}^\sigma_\infty$ for $\sigma \in \Sigma_3$ are shown in Fig. 6.6. Besides, from an initial state $x(0) = [0, 0, 0]^\top$, using the results in Lemma 6.2, Theorems 6.2 and 6.4, the convergence time corresponding to three modes can be computed as $T_c^1 = 22$, $T_c^2 = 33$ and $T_c^3 = 10$. Hence, based on Theorem 6.9, the upper bound for the detection time is $N_T(x(0)) = \max_i T_c^i = 33$. Furthermore, for any two modes $i, j \in \Sigma_3$, by solving Problem 6.1, constant detection inputs can be obtained: $\bar{u}_{12} = 0.0202$, $\bar{u}_{23} = 0.0139$ and $\bar{u}_{13} = 0.0167$. These constant detection inputs will be used in the simulation by applying Algorithm 6.1.

Table 6.1 Computation result with constant active detection input

	U with $\bar{u}$	Detection time $N_T(x(0))$
Upper bound	0.1353	33
Online simulations	7.9686×10^{-4}	3

Table 6.2 Computation result with variable active detection input

	U with $u^*(k)$	Detection time N_t
Upper bound	0.0047	10
Online simulations	1.5948×10^{-4}	1

With the same initial state $x(0) = [0, 0, 0]^\top$, by solving Problem 6.2 with $N_t = 10$, we obtain the active detection input sequence $u^*(l)$ for $l = 0, \ldots, 9$. This input sequence will be used in the simulation by applying Algorithm 6.2.

The comparison result of the constant detection input and variable detection input sequence is reported in Tables 6.1 and 6.2. Based on the objective of Problem 6.1 or Problem 6.2, a measure of the effort used in active detection inputs is given by

$$U := \sum_{k=0}^{N_h} u(k)^2,$$

where N_h is a detection time. From Theorem 6.9, the upper-bound of the detection time with the constant detection input is $N_T(x(0)) = \max_i T_c^i = 33$. From Theorem 6.10, with a given initial state $x(0)$, we obtain a feasible solution to Problem 6.2 with the detection time $N_t = 10$. From the above equation with $N_h = N_T(x(0)) = 33$ and $N_h = N_t = 10$ respectively, we obtain the computation results of energy. In Table 6.2, it can be seen that with variable detection input sequence, less effort is taken than the one with constant detection input.

The online simulations of applying Algorithms 6.1 and 6.2 have been carried out under the same initial conditions: $x(0)$ and $w(k) \in \mathcal{W}$ within the simulation window. The results are shown in Fig. 6.7. By applying Algorithm 6.1, the operating mode can be detected at time $k = 3$ and the system (6.20) is in mode 2 since the state trajectory only stays in the blue set at time $k = 3$. By applying Algorithm 6.2, the current operating mode can be detected at time $k = 1$ since the state trajectory only stays in the blue set, which is faster than the previous case. Moreover, from Tables 6.1 and 6.2, the effort associated with constant detection input is larger than the one with variable detection input sequence.

Fig. 6.7 *Active* mode detection results

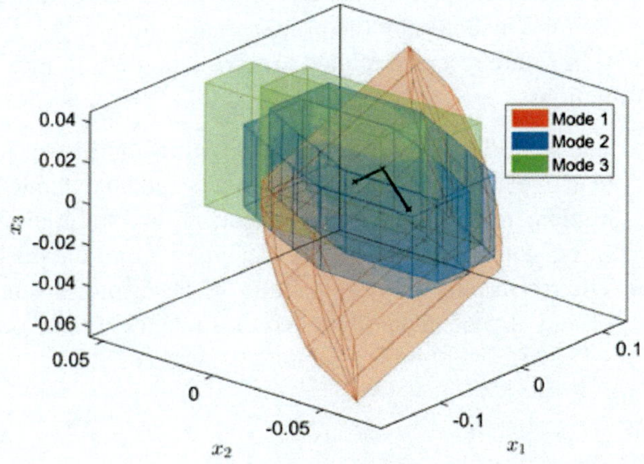

(a) by Algorithm 6.1

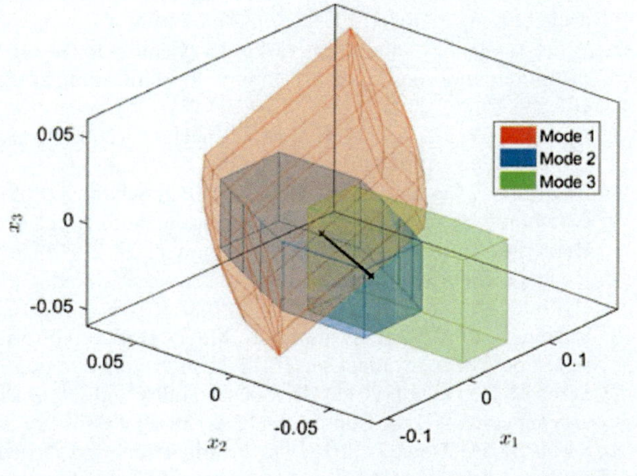

(b) by Algorithm 6.2

6.4 Summary

This chapter has proposed robust invariant set characterizations of discrete-time descriptor systems in both causal and non-causal cases. Two restricted equivalent forms of descriptor systems have been revisited. Based on these forms, the explicit results on robust invariant set characterizations are provided. Besides, we have also proposed an active mode detection mechanism based on set invariance for discrete-time descriptor system with multiple modes of operation. To separate RPI sets of descriptor systems, we have proposed two methods to design active detection inputs, from which we present two active mode detection algorithms. Finally, through a numerical example, the results show that the operating mode can be detected in a finite time by applying the proposed algorithms.

The future research related to this chapter is summarized in the following potential directions:

- The *unstable* descriptor systems with a stabilizing feedback could be considered in active mode detection. The RI sets can be extended to be controlled RI sets. For implementing active mode detection, the stabilizing feedback control input and active detection input can be designed simultaneously.
- The constraints on system states of descriptor systems can be considered, which should also be taken into account in the design of active mode detection.

References

1. Wang Y, Olaru S, Valmorbida G, Puig V, Cembrano G (2017) Robust invariant sets and active mode detection for discrete-time uncertain descriptor systems. In: 56th IEEE conference on decision and control (IEEE-CDC), Melbourne, Australia, pp 5648–5653
2. Wang Y, Olaru S, Valmorbida G, Puig V, Cembrano G (2019) Set-invariance characterizations of discrete-time descriptor systems with application to active mode detection. Automatica 107:255–263
3. Alamo T, Cepeda A, Fiacchini M, Camacho E (2009) Convex invariant sets for discrete-time Lur'e systems. Automatica 45(4):1066–1071
4. Blanchini F, Casagrande D, Miani S (2010) Modal and transition dwell time computation in switching systems: a set-theoretic approach. Automatica 46(9):1477–1482
5. Heidari R, Braslavsky J, Seron M, Haimovich H (2016) Ultimate bound minimisation by state feedback in discrete-time switched linear systems under arbitrary switching. Nonlinear Anal Hybrid Syst 21:84–102
6. Raković S, Kerrigan E, Kouramas K, Mayne D (2005) Invariant approximations of the minimal robust positively invariant set. IEEE Trans Autom Control 50(3):406–410
7. Seron M, De Doná J (2015) On robust stability and set invariance of switched linear parameter varying systems. Int J Control 88(12):2588–2597
8. Seron M, De Doná J (2016) On invariant sets and closed-loop boundedness of Lure-type nonlinear systems by LPV-embedding. Int J Robust Nonlinear Control 26(5):1092–1111
9. Stoican F, Oară C, Hovd M (2015) RPI approximations of the mRPI set characterizing linear dynamics with zonotopic disturbances. In: Developments in model-based optimization and control: distributed control and industrial applications. Springer, pp 361–377
10. Olaru S, De Doná J, Seron M, Stoican F (2010) Positive invariant sets for fault tolerant multisensor control schemes. Int J Control 83(12):2622–2640
11. Kofman E, Seron M, Haimovich H (2008) Control design with guaranteed ultimate bound for perturbed systems. Automatica 44(7):1815–1821
12. Marseglia G, Raimondo D (2017) Active fault diagnosis: a multi-parametric approach. Automatica 79:223–230
13. Stoican F, Olaru S, Seron M, De Doná J (2012) Reference governor design for tracking problems with fault detection guarantees. J Process Control 22(5):829–836
14. Kolmanovsky I, Gilbert E (1998) Theory and computation of disturbance invariant sets for discrete-time linear systems. Math Problems Eng 4(4):317–367
15. Kofman E, Haimovich H, Seron M (2007) A systematic method to obtain ultimate bounds for perturbed systems. Int J Control 80(2):167–178

16. Seron M, De Doná J, Olaru S (2012) Fault tolerant control allowing sensor healthy-to-faulty and faulty-to-healthy transitions. IEEE Trans Autom Control 57(7):1657–1669
17. Dai L (1989) Singular control systems. Springer, Berlin Heidelberg, Germany
18. Zhang L, Lam J, Zhang Q (1999) Lyapunov and Riccati equations of discrete-time descriptor systems. IEEE Trans Autom Control 44(11):2134–2139
19. Gerdin M (2004) Computation of a canonical form for linear differential-algebraic equations. Technical report, Linköping University
20. Blanchini F, Casagrande D, Giordano G, Miani S, Olaru S, Reppa V (2017) Active fault isolation: a duality-based approach via convex programming. SIAM J Control Optim 55(3):1619–1640

Part III
Control

Chapter 7
Economic Model Predictive Control Strategies Based on a Periodicity Constraint

Periodic behavior appears in some specific systems, such as WDNs [1–3] and electrical networks [4]. One specific example stems from the periodic behavior of customer demands in WDNs. A WDN generally consists of a large number of hydraulic elements, such as storage tanks, pressurized pipelines, pumping stations (including several parallel pumps) and valves. EMPC is suitable for optimizing the economic performance of operations in WDNs, as shown in [5–7], but these methods do not take specific advantage of the periodic nature of the consumer demands and energy costs. Taking into account the daily water demand patterns and periodic electricity prices, periodic operations can also be considered in the EMPC design.

In this chapter, a novel EMPC framework for periodic operation is first proposed. We formulate an EMPC optimization problem without setting a terminal state [8, 9]. Hence, it does not need to know a periodic steady trajectory as a priori knowledge. Therefore, the economic cost function is optimized with a periodicity constraint considering all the periodic trajectories including the current state along the prediction horizon. Unlike the conventional MPC optimization formulation, the current state is set as a shifted position and not necessarily being the first prediction state. In order to investigate the closed-loop convergence, an optimal periodic steady trajectory can be obtained by the proposed finite-horizon optimization problem that is called the EMPC planner. Recursive feasibility and the closed-loop convergence to the optimal periodic steady trajectory are discussed and an optimality certificate is provided based on the Karush-Kuhn-Tucker (KKT) optimality conditions. Furthermore, the proposed EMPC framework is extended into the robust case [10]. The tube-based approach is used to achieve robust constraint satisfaction as well as recursive feasibility in the presence of disturbances. The mismatches between the nominal model and the closed-loop system with perturbations are limited using a local control law. Also under convexity assumption, robust stability of the closed-loop system is analyzed using KKT optimality conditions and an optimality certificate is provided to check if the closed-loop trajectories reach a neighborhood of optimal nominal periodic steady trajectories.

Y. Wang, *Advances in State Estimation, Diagnosis and Control of Complex Systems*, Springer Theses, https://doi.org/10.1007/978-3-030-52440-1_7

7.1 EMPC Based on a Periodicity Constraint

Consider the class of discrete-time LTI systems

$$x(k+1) = Ax(k) + Bu(k), \tag{7.1}$$

where $x \in \mathbb{R}^{n_x}$ and $u \in \mathbb{R}^{n_u}$ denote the system state vector and the control input vector, respectively. Moreover, $A \in \mathbb{R}^{n_x \times n_x}$ and $B \in \mathbb{R}^{n_x \times n_u}$ are system matrices.

For (7.1), system states and control inputs are limited by the following constraints:

$$x(k) \in \mathcal{X}, \ u(k) \in \mathcal{U}, \ \forall k \in \mathbb{N}, \tag{7.2}$$

where $\mathcal{X}$ and $\mathcal{U}$ are strictly convex sets of states and inputs.

The economic performance of the system (7.1) is measured by a time-varying economic cost function

$$\ell\left(x(k), u(k), p_i\right), \ i = \operatorname{mod}(k, T) \tag{7.3}$$

where $T \in \mathbb{Z}_+$ is a period index and p_i is a time-varying exogenous signal usually indicating the unit prices, which is stored in a known sequence p as

$$p = \{p_i\}, \ i = 1, \ldots, T,$$

and exhibiting a periodic behavior is implemented using the modulo operator $\operatorname{mod}(k, T)$. It is worth mentioning that $\ell\left(x(k), u(k), p_i\right)$ is not necessarily a quadratic function that depends on a sequence of references for tracking. The main control objective is to minimize the closed-loop economic cost measured by $\ell\left(x(k), u(k), p_i\right)$ that is a strictly convex function, $\forall k \in \mathbb{N}$ and the periodicity of this economic stage cost function is given by $\ell(x(k), u(k), p_i) = \ell(x(k+T), u(k+T), p_i)$ with $i = \operatorname{mod}(k, T)$.

In this chapter, we propose an EMPC formulation that by guaranteeing the closed-loop system convergence to a periodic steady trajectory minimizes the economic cost while satisfying all the constraints. A procedure to certify that the reached trajectory is optimal with respect to the optimal economic cost. In addition, the proposed controller does not lose feasibility even in the presence of sudden changes in the economic cost.

In principle, MPC controllers are based on solving a finite horizon optimization problem. If a steady state trajectory is known, a terminal constraint is included forcing the predictions to reach this steady trajectory at the end of the MPC prediction horizon. While several controllers proposed in the literature are based on a standard terminal region/constraint approach, we assume that a steady state trajectory is unknown in the EMPC design. We propose a different approach in which the MPC controller seeks to minimize the economic cost function over a single period that includes the current state. Besides, we propose an optimization problem to find an optimal periodic steady trajectory that will be used for the analysis of the closed-loop convergence.

The proposed controller guarantees recursive feasibility and hence the closed-loop convergence even in the presence of sudden changes in the economic cost function, because the constraints of the optimization problem are independent of this cost function. Note that standard approaches that depend on terminal constraints often lead to optimization problems that have to be modified if the economic cost function changes, which in general lead to a more complex control scheme and even to possible loss of feasibility issues [11, 12].

7.1.1 EMPC Planner

We first present a finite-horizon optimization problem, the so-called *planner*, to find the optimal periodic steady trajectory that will be used for the analysis. Because of the periodic nature discussed above, it can be proved that the infinite horizon problem is equivalent to the following finite horizon optimization problem in which a single period is taken into account [1]. This optimization problem yields the same solution if the time frame to be considered is any period.

$$\underset{\substack{x(0),\ldots,x_T,\\ u(0),\ldots,u(T-1)}}{\text{minimize}} \; J_T(x,u,p) = \sum_{i=0}^{T-1} \ell\left(x(i), u(i), p_i\right), \tag{7.4a}$$

subject to

$$x(i+1) = Ax(i) + Bu(i), \tag{7.4b}$$
$$x(i) \in \mathcal{X}, \tag{7.4c}$$
$$u(i) \in \mathcal{U}, \tag{7.4d}$$
$$x(0) = x(T). \tag{7.4e}$$

Remark 7.1 Note that in formulation above, the time step $i = 0$ is chosen as the first step of one period. If a different initial step is chosen, the functions would be different but would lead to an equivalent problem. This choice will affect the proposed EMPC optimization problem as it will be based on solving a finite horizon optimization problem in a period that starts at some multiple of T, that is, at the same time step used to define the planner.

7.1.2 EMPC Controller

The EMPC strategy is proposed by implementing the following optimization problem. Considering the periodicity, the current state $x(k)$ at time step $k \in \mathbb{N}$ is inserted into the shifted position.

$$\underset{\substack{x(0),\ldots,x(T),\\u(0),\ldots,u(T-1)}}{\text{minimize}} \; J_T(x, u, p), \tag{7.5a}$$

subject to

$$x(i+1) = Ax(i) + Bu(i), \tag{7.5b}$$

$$x(i) \in \mathcal{X}, \tag{7.5c}$$

$$u(i) \in \mathcal{U}, \tag{7.5d}$$

$$x(0) = x(T), \tag{7.5e}$$

$$x(j) = x(k), \quad j = \text{mod}\,(k, T)\,. \tag{7.5f}$$

Due to the periodic system behavior, the optimization problem (7.5) is always initialized from time step $i = 0$. At each time step, this optimization problem is solved with a fixed prediction horizon of T. Note that the current state $x(k)$ is not always set as the first state prediction.

Let $u^*(i)$, $i = 0, \ldots, T-1$ be a set of optimal solutions of the optimization problem (7.5) with the initialization of $x(k)$. According to the receding horizon strategy, the optimal control action $u(k)$ applied to the closed-loop system at time step k is chosen by

$$u(k) = u^*(j), \; j = \text{mod}\,(k, T)\,. \tag{7.6}$$

Remark 7.2 In the formulations of the optimization problems (7.4) and (7.5), the subscript k, $\forall k \in \mathbb{N}$ corresponds to a time instant while the index i with $i = 0, 1, \ldots, T-1$ refers to a prediction step in the optimization problem.

We now provide the following two remarks regarding the properties of the proposed EMPC controller.

Remark 7.3 Note that the constraints of the optimization problem (7.5) do not depend on the economic cost function, so recursive feasibility is guaranteed even in the presence of a sudden change.

Remark 7.4 The optimization problem (7.5) is feasible if there exists a feasible periodic trajectory over a length of T that includes the current state $x(k)$. This implies that the domain of attraction, that is, the feasibility region of (7.5) is in general very large, as it is not constrained to reach a specific target in a fixed time as in standard MPC formulations with terminal regions.

7.1.3 The Closed-Loop Properties with the EMPC Controller

We now discuss the closed-loop properties of the system (7.1) with the EMPC controller implemented by (7.5). Recursive feasibility and convergence analysis are standard notions in MPC designs [12]. In the following, we summarize and prove these closed-loop properties of the proposed controller. In particular, under a certain assumption, the closed-loop system trajectory converges to the optimal periodic steady trajectory obtained by the planner (7.4).

Theorem 7.1 *The system (7.1) in closed-loop with the EMPC implemented by the optimization problem (7.5) is stable and converges to a periodic steady trajectory. This trajectory is equal to the optimal trajectory obtained from the optimization problem (7.4), if there exists a time step $M > 0$ such that for any $k \geq M$, the dual variables corresponding to the equality constraints (7.5f) in KKT optimality conditions are zero.*

Proof We first discuss the recursive feasibility of the closed-loop control system.

(Recursive feasibility) If the optimization problem (7.5) is feasible at time step $k \in \mathbb{N}$, then it is also feasible at time step $k+1$. Let us denote $x(j)$ and $u(j)$ for $j = \text{mod}(k, T)$ be feasible solutions of the optimization problem (7.5) at time step $k \in \mathbb{N}$. All the constraints (7.5b)–(7.5f) are satisfied at time step $k \in \mathbb{N}$. Thanks to the formulation in (7.5), as discussed in Remark 7.2, constraints (7.5b)–(7.5e) are also satisfied at time step $k+1$ since they do not depend on the time step $k+1$.

From (7.5f), we have

$$x(j) = x(k+1), \quad j = \text{mod}\,(k+1, T)\,,$$

which is equivalent to

$$x(j+1) = x(k+1), \quad j = \text{mod}\,(k, T)\,. \tag{7.7}$$

If the constraint (7.7) holds, then the optimization problem (7.5) is feasible at time step $k+1$. Recall the feasible solutions $x(j)$ and $u(j)$ for $j = \text{mod}(k, T)$ and we know $x(j) = x(k)$ and $u(j) = u(k)$ hold. From (7.5b), we can derive

$$x(j+1) = Ax(j) + Bu(j),$$

and with the control action $u(k)$ chosen in (7.6) and the system (7.1), the constraint (7.7) is satisfied. Thus, we can conclude that the optimization problem (7.5) is also feasible at time step $k+1$.

Since the optimization problem (7.5) is recursively feasible, we denote its optimal MPC cost as $V(k, x, p)$ at time step k. By optimality [13], we know that

$$V(k+1, x, p) \leq V(k, x, p), \tag{7.8}$$

which implies the cost of the optimization problem (7.5) is a non-increasing sequence and therefore the closed-loop system is stable.

Next, we discuss about the convergence of the closed-loop control system.

(Convergence analysis) For the analysis below, we first reformulate the optimization problems (7.4) and (7.5) into standard convex formulations. Let us define the vector

$$z = \left[x(0)^\top, \ldots, x(T)^\top, u(0)^\top, \ldots, u(T-1)^\top\right]^\top, \tag{7.9}$$

and the cost function $J_T(x, u, p)$ becomes $J_T(z, p)$. Then, we rewrite the optimization problem (7.4) to be in a standard convex form as follows

$$\underset{z}{\text{minimize}}\ J_T(z, p), \tag{7.10a}$$

subject to

$$h_r(z) \leq 0, \quad r = 1, \ldots, m, \tag{7.10b}$$
$$g_i(z) = 0, \quad i = 1, \ldots, n. \tag{7.10c}$$

where the functions h_r for $r = 1, \ldots, m$ correspond to the constraints (7.4c)–(7.4d), and the functions g_i for $i = 1, \ldots, n$ represent the prediction model (7.4b) and the periodicity constraint (7.4e). Besides, let us denote the optimal solution of the optimization problem (7.10) (the optimization problem (7.4)) as z^p, that is, the optimal periodic steady trajectory.

By the convexity assumption, it follows that there exist dual variables

$$\lambda^p = \begin{bmatrix} \lambda_1^p \\ \vdots \\ \lambda_m^p \end{bmatrix}, \quad \mu^p = \begin{bmatrix} \mu_1^p \\ \vdots \\ \mu_n^p \end{bmatrix},$$

such that the following KKT optimality conditions of (7.10) hold

$$\nabla J_T\left(z^p, p\right) + \sum_{r=1}^{m} \lambda_r^p \nabla h_r\left(z^p\right) + \sum_{i=1}^{n} \mu_i^p \nabla g_i\left(z^p\right) = 0, \tag{7.11a}$$
$$h_r\left(z^p\right) \leq 0, \quad r = 1, \ldots, m, \tag{7.11b}$$
$$g_i\left(z^p\right) = 0, \quad i = 1, \ldots, n, \tag{7.11c}$$
$$\lambda_r^p \geq 0, \quad r = 1, \ldots, m, \tag{7.11d}$$
$$\lambda_r^p h_r\left(z^p\right) = 0, \quad r = 1, \ldots, m. \tag{7.11e}$$

Similarly, with the vector z defined in (7.9), the optimization problem (7.5) can be reformulated to a standard convex form as

$$\underset{z}{\text{minimize}}\ J_T\left(z, p\right), \tag{7.12a}$$

subject to

$$h_r\left(z\right) \le 0, \quad r = 1, \ldots, m, \tag{7.12b}$$
$$g_i\left(z\right) = 0, \quad i = 1, \ldots, n, \tag{7.12c}$$
$$Q_j z = x(k), \quad j = \operatorname{mod}\left(k, T\right), \tag{7.12d}$$

where Q_j with $j = \operatorname{mod}\left(k, T\right)$ is defined based on (7.5f). Besides, we denote the optimal solution of the optimization problem (7.12) at time k as $z(k)$. Then, there exist dual variables

$$\lambda(k) = \begin{bmatrix} \lambda_1(k) \\ \vdots \\ \lambda_m(k) \end{bmatrix}, \mu(k) = \begin{bmatrix} \mu_1(k) \\ \vdots \\ \mu_n(k) \end{bmatrix}, \nu(k) = \begin{bmatrix} \nu_1(k) \\ \vdots \\ \nu_{n_x}(k) \end{bmatrix},$$

such that the following KKT optimality conditions of (7.12) hold

$$\nabla J_T\left(z(k), p\right) + \sum_{r=1}^{m} \lambda_r(k) \nabla h_r\left(z(k)\right) + \sum_{i=1}^{n} \mu_i(k) \nabla g_i\left(z(k)\right) + \sum_{l=1}^{n_x} \nu_l(k) Q^l_{\operatorname{mod}(k,T)} = 0, \tag{7.13a}$$

$$h_r\left(z(k)\right) \le 0, \quad r = 1, \ldots, m, \tag{7.13b}$$
$$g_i\left(z(k)\right) = 0, \quad i = 1, \ldots, n, \tag{7.13c}$$
$$Q_j z(k) = x(k), \quad j = \operatorname{mod}(k, T), \tag{7.13d}$$
$$\lambda_r(k) \ge 0, \quad r = 1, \ldots, m, \tag{7.13e}$$
$$\lambda_r(k) h_r\left(z(k)\right) = 0, \quad r = 1, \ldots, m, \tag{7.13f}$$

where Q^l_j is the l-th row of Q_j transposed.

As discussed above, the cost of the optimization problem (7.5) is a non-increasing sequence. Taking into account that by assumption that the cost function $J_T\left(z, p\right)$ is strictly convex, with recalling $V(k, x, p)$ as the cost at time step k, it is not possible that there exist two consecutive time steps k and $k+1$ such that the costs $V\left(k, x, p\right) = V\left(k+1, x, p\right)$ with $z(k) \neq z(k+1)$ due to optimality. Hence, if $V\left(k+1, x, p\right) = V\left(k, x, p\right)$, $\forall k \ge M$, the system (7.1) in closed-loop reaches a periodic steady trajectory, that is, $z(M) = z(M+1) = \cdots$.

Without loss of generality, we assume that $\operatorname{mod}\left(M, T\right) = 0$. Let us denote $z^s = z(M)$ as this periodic steady trajectory. The solution z^s is also feasible for the optimization problem (7.4). On the one hand, the closed-loop solution $z(k)$ should be equal to the optimal solution z^s, that is $z^s = z(k)$, $\forall k \ge M$. On the other hand, z^s is an optimal solution of the optimization problem (7.5) such that z^s, $\forall k \geq M$ satisfies the KKT conditions in (7.13). If the dual variables in $\nu(k)$ are zero, then (7.13d) can be disabled. As a result, z^s also satisfies the KKT conditions in (7.11). Hence, we have

$$z^s = z^p,$$

which means that the closed-loop trajectory $z(k)$ converges to the optimal periodic steady trajectory z^p, $\forall k \geq M$. □

Remark 7.5 For a periodic steady trajectory z^s, all the constraints of (7.12) must be satisfied for all $k = 1, \ldots, T$ with $z(k) = z^s$ and $x(k) = Q_j z^s$ with $j = \text{mod}(k, T)$. The solution provided by the planner (7.10) satisfies this condition by definition with dual variables of constraint (7.12d) equal to zero. Although it is rarely found because the number of constraints is larger than the number of variables, other trajectories may also satisfy the condition. In this case, the closed-loop system may converge to a periodic trajectory different from the planner.

7.1.4 Example

To better illustrate Remark 7.5, we present the following example in which the closed-loop system converges to a periodic trajectory different from the planner.

Consider the following system subject to additive known signal

$$x(k+1) = Ax(k) + Bu(k) + B_d d(k),$$

with the following system matrices

$$A = \begin{bmatrix} 0.5 & 0.5 \\ 1 & 0.25 \end{bmatrix}, \quad B = \begin{bmatrix} 1 \\ 1 \end{bmatrix}, \quad B_d = \begin{bmatrix} 1 \\ 0 \end{bmatrix},$$

where d is a periodic known disturbance signal with a period $T = 3$. The values of these periodic signals are given by $d(k) = d_i$ with $i = \text{mod}(k, T)$, where $d_1 = -0.1$, $d_2 = -0.2$ and $d_3 = -0.1$ for $i = 1, 2, 3$. This system is controlled by the proposed EMPC. In this example, consider the formulations in (7.10) and (7.12) and the quadratic cost function $J_T(z) = \frac{1}{2} z^\top H z + f^\top z$ with

$$H = \text{diag}\left([1\ 1\ 10\ 1\ 1\ 20\ 1\ 1\ 10]\right),$$
$$f = \left[0.1\ 0.1\ 0.1\ 0.1\ 0.1\ 0.1\ 0.1\ 0.1\ 0.1\right]^\top,$$

where diag$(\cdot)$ returns a diagonal matrix with diagonal elements defined by its argument. The input must belong to the set $\mathcal{U} = \{u \in \mathbb{R} : -0.1 \leq u \leq 0.1\}$ and no constraints are considered for the states.

Fig. 7.1 Closed-loop state trajectory of the example

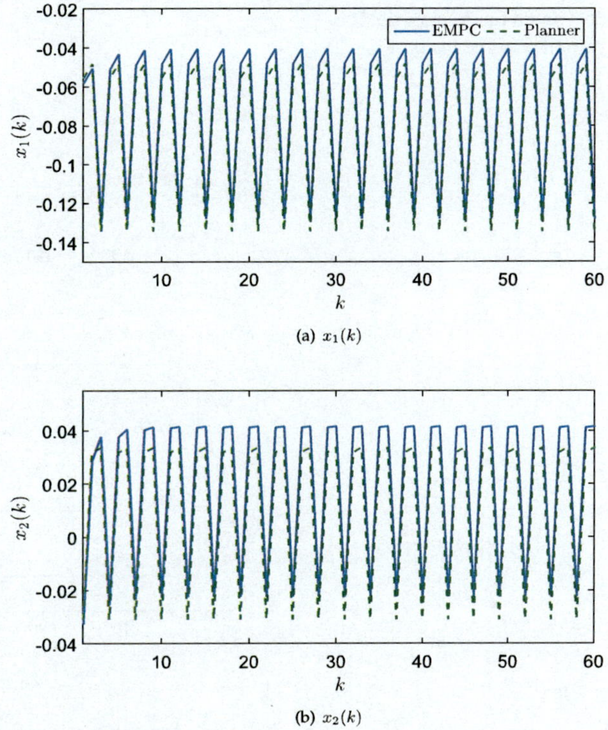

The simulation with this example has been carried out for 60 sampling steps. As shown in Fig. 7.1, the closed-loop trajectories of both entries $x_1(k)$ and $x_{2(k)}$ of the state states converge to a periodic trajectory that are different from the one corresponding to the planner. Figure 7.2a shows that the cost of the MPC optimization problem is a non-increasing sequence that reaches a constant value when the system converges to the periodic trajectory. In this case, because this trajectory is different from the optimal one computed by the planner, its corresponding cost is higher that the planner. Besides, we show a measurement (defined by the 2-norm) of the dual variable $\nu(k)$ in Fig. 7.2b. From this figure, we can see that dual variables corresponding to (7.12d) are not zero at any time since the closed-loop trajectory cannot reach the planner.

7.2 REMPC Based on a Periodicity Constraint

Consider the system (7.1) subject to additive disturbances

$$x(k+1) = Ax(k) + Bu(k) + w(k), \tag{7.14}$$

where $w \in \mathbb{R}^{n_x}$ denotes the disturbance vector.

Fig. 7.2 Closed-loop MPC cost and a measure of dual variables of the example

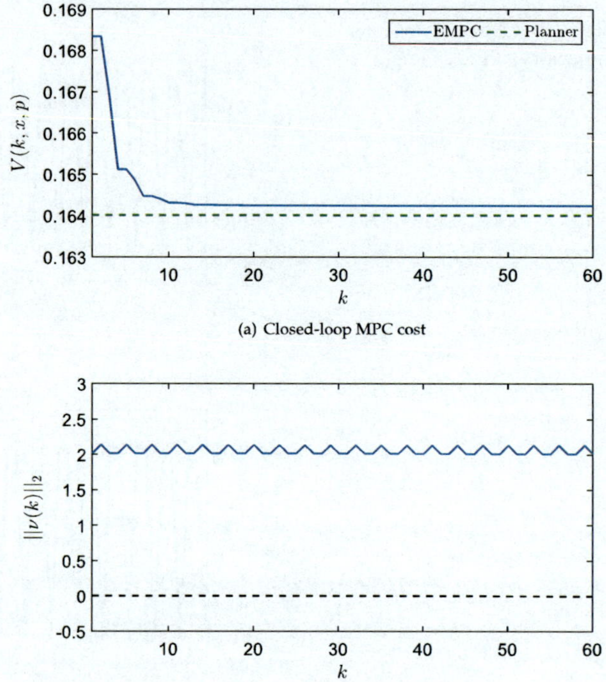

(a) Closed-loop MPC cost

(b) A measure of $\nu(k)$

Remark 7.6 From application point of view, the disturbance vector $w(k)$ may include two parts as

$$w(k) = B_d d(k) + \tilde{w}(k), \tag{7.15}$$

where $d \in \mathbb{R}^{n_d}$ denotes the vector of deterministic disturbances that is considered to be known also following a periodic behavior, that is $d(k) = d(k+T)$ with a period $T \in \mathbb{Z}_+$ (see, e.g. [2, 14, 15]). $\tilde{w}(k)$ is the unknown disturbances vector. $B_d \in \mathbb{R}^{n_x \times n_d}$ is a distribution matrix.

For notation simplicity, we consider that in general $w(k)$ is unknown and the following assumption is made.

Assumption 7.1 The disturbance vector $w(k)$ is assumed to be unknown but bounded by a convex set $\mathcal{W}$, that is

$$w(k) \in \mathcal{W}, \quad \forall k \in \mathbb{N}. \tag{7.16}$$

The state and control input vectors, $x(k)$ and $u(k)$ are required to satisfy the constraints in (7.2). Denote the nominal state and input vectors as $\bar{x} \in \mathbb{R}^n$ and $\bar{u} \in \mathbb{R}^m$, which follow (7.1). Let us recall it as

$$\bar{x}(k+1) = A\bar{x}(k) + B\bar{u}(k). \tag{7.17}$$

In principle, the nominal system (7.17) could be used as the prediction model in an MPC design. However, due to the existence of $w(k) \in \mathcal{W}$, $\forall k \in \mathbb{N}$, the predicted states have a mismatch with the real states of the system (7.14). Hence, an REMPC controller is required to guarantee recursive feasibility and robust constraint satisfaction in closed-loop.

With the strictly convex economic cost function (7.3), the control objective is to minimize the closed-loop economic cost $\sum_{k=0}^{\infty} \ell(x(k), u(k), p_i)$ of system (7.14) in the presence of disturbances. The mismatch between the closed-loop perturbed states and the open-loop nominal predicted states is corrected using a local control law.

We consider that state x of system (7.14) is fully measured and the pair (A, B) is controllable. Following the so-called tube based approach [16], to guarantee recursive feasibility, we will use a robustly stabilizing local control gain $K \in \mathbb{R}^{n_x \times n_u}$ such that $(A + BK)$ is Schur stable to tighten the sets $\mathcal{X}$ and $\mathcal{U}$. Based on Definition 1.12 and Lemma 6.1, let an RPI set $\mathcal{Z}$ be a polytopic form as

$$\mathcal{Z} := \left\{ x \in \mathbb{R}^n : H^z x \le b^z, H^z \in \mathbb{R}^{n_x \times n}, b^z \in \mathbb{R}^{n_x} \right\}.$$

Based on the robust tube-based technique, we refine the sets $\mathcal{X}$ and $\mathcal{U}$ to be $\mathcal{X}^r$ and $\mathcal{U}^r$, where

$$\mathcal{X}^r = \mathcal{X} \ominus \mathcal{Z}, \tag{7.18a}$$

$$\mathcal{U}^r = \mathcal{U} \ominus K\mathcal{Z}. \tag{7.18b}$$

Assumption 7.2 The sets $\mathcal{X}^r$ and $\mathcal{U}^r$ are assumed to be not empty.

7.2.1 REMPC Planner

We now extend EMPC planner in (7.4) to the robust case that provides the best possible periodic trajectory with respect to the economic cost taking into account the set of tightened constraints that will be used in the MPC formulation. The resulting trajectory provides the optimal nominal periodic steady trajectory. The control objective of the proposed robust MPC controller is to drive the closed-loop system to a neighborhood of the optimal nominal periodic steady trajectory while robustly satisfying all the constraints.

$$\underset{\substack{\bar{x}(0),\ldots,\bar{x}(T),\\ \bar{u}(0),\ldots,\bar{u}(T-1)}}{\text{minimize}} \; J_T(\bar{x}, \bar{u}, p) = \sum_{i=0}^{T-1} \ell(\bar{x}(i), \bar{u}(i), p_i), \tag{7.19a}$$

subject to

$$\bar{x}(i+1) = A\bar{x}(i) + B\bar{u}(i), \tag{7.19b}$$
$$\bar{x}(i) \in \mathcal{X} \ominus \mathcal{Z}, \tag{7.19c}$$
$$\bar{u}(i) \in \mathcal{U} \ominus K\mathcal{Z}, \tag{7.19d}$$
$$\bar{x}(0) = \bar{x}(T). \tag{7.19e}$$

By solving the optimization problem (7.19) offline, we can obtain the optimal solution denoted as $\bar{x}^*(0), \ldots, \bar{x}^*(T)$ and $\bar{u}^*(0), \ldots, \bar{u}^*(T-1)$.

7.2.2 REMPC Controller

In addition, the controller formulation is based on a tube-based approach [10]. This implies two facts: (i) the constraints are tightened using an RPI set; (ii) the periodic trajectory will not meet through the current state k at prediction time $j = \text{mod}(k, T)$, but instead the difference has to be included in the aforementioned RPI set. These ingredients, together with the controller equations aiming to reduce this difference between the real state and the predicted state, will provide recursive robust constraint satisfaction. The closed-loop properties of the proposed REMPC controller will be demonstrated in the following. In general, the REMPC controller is formulated by the following optimization problem:

$$\underset{\substack{\bar{x}(0),\ldots,\bar{x}(T),\\ \bar{u}(0),\ldots,\bar{u}(T-1)}}{\text{minimize}} \; J_T(\bar{x}, \bar{u}, p), \tag{7.20a}$$

subject to

$$\bar{x}(i+1) = A\bar{x}(i) + B\bar{u}(i), \tag{7.20b}$$
$$\bar{x}(i) \in \mathcal{X} \ominus \mathcal{Z}, \tag{7.20c}$$
$$\bar{u}(i) \in \mathcal{U} \ominus K\mathcal{Z}, \tag{7.20d}$$
$$\bar{x}(0) = \bar{x}(T), \tag{7.20e}$$
$$x(k) - \bar{x}(j) \in \mathcal{Z}, \quad j = \text{mod}(k, T). \tag{7.20f}$$

From the optimal solutions of (7.20), with the local control gain $K \in \mathbb{R}^{n_x \times n_u}$, the control action at time instant k is chosen as

$$u(k) = \bar{u}(j) + K(x(k) - \bar{x}(j)), \; j = \text{mod}(k, T). \tag{7.21}$$

Using the formulation in (7.21), the mismatch between the predicted state $\bar{x}(j)$ for $j = \text{mod}(k, T)$ and the closed-loop state $x(k)$ is attenuated by the local control gain K. In this case, thanks to constraint (7.20f), the closed-loop state trajectory $x(k)$ can always stay in a neighborhood of $\bar{x}(j)$, that is the tube defined by the RPI set $\mathcal{Z}$.

Besides, a periodic operation with the proposed REMPC can be achieved using the periodicity constraint defined in (7.20e).

Remark 7.7 In the optimization problems (7.19) and (7.20), the index $i = 0, \ldots, T-1$ is a prediction step along the MPC prediction horizon while the index $k \in \mathbb{N}$ is a time instant for the simulation loop.

7.2.3 *The Closed-Loop Properties with the REMPC Controller*

We now study the properties of the system (7.14) in closed-loop with the robust economic MPC controller implemented by (7.20), which are summarized in the following theorem.

Theorem 7.2 *Consider the system (7.14) with the robust economic MPC controller implemented by (7.20), the following closed-loop properties hold:*

(a) If the optimization problem (7.20) is feasible from an initial state $x(0)$, then the closed-loop system satisfies all the constraints for all possible disturbances satisfying Assumption 7.1 and the optimal MPC cost $V(k, x, p)$, $\forall k \in \mathbb{N}$ is a non-increasing sequence.

(b) If there exists a time step $M > 0$ such that for any $k \geq M$, all the variables in the dual vector corresponding to the constraint (7.5f) are zero in the KKT optimality conditions, then the closed-loop system has reached a neighborhood (enclosed by the RPI set $\mathcal{Z}$) of the optimal nominal periodic steady trajectory $\bar{x}^*(j)$ with $j = \mathrm{mod}(k, T)$ obtained from the planner (7.19).

Proof We first prove the closed-loop property expressed in the statement (a). In the following, we discuss recursive feasibility and robust constraint satisfaction of the closed-loop system. From these result, the closed-loop convergence is provided.

(*Recursive feasibility*) Let $\bar{x}(j)$ and $\bar{u}(j)$ be feasible solutions of the optimization problem (7.20) at time instant k. We now prove that the optimization problem (7.20) is also feasible at time $k+1$. From the REMPC formulation in (7.20), the constraints (7.20b)–(7.20e) do not depend on the time instant k so $\bar{x}(j)$ and $\bar{u}(j)$ satisfy them by definition. The only constraint that depends on the time instant k is (7.20f). From (7.20b), we have that

$$\bar{x}(j+1) = A\bar{x}(j) + B\bar{u}(j), \quad j = \mathrm{mod}\,(k, T)\,.$$

Taking into account (7.14) and the control action $u(k)$ chosen in (7.21), we can obtain

$$\begin{aligned}x(k+1) &= Ax(k) + B\Big(\bar{u}(j) + K\big(x(k) - \bar{x}(j)\big)\Big) + w(k)\\ &= A\big(x(k) - \bar{x}(j) + \bar{x}(j)\big) + B\Big(\bar{u}(j) + K\big(x(k) - \bar{x}(j)\big)\Big) + w(k)\\ &= A\bar{x}(j) + B\bar{u}(j) + A\big(x(k) - \bar{x}(j)\big) + BK\big(x(k) - \bar{x}(j)\big) + w(k).\end{aligned}$$

Therefore, by subtracting above two equations, we have

$$x(k+1) - \bar{x}(j+1) = (A + BK)\big(x(k) - \bar{x}(j)\big) + w(k).$$

Considering the constraint (7.20f), we obtain

$$x(k+1) - \bar{x}(j+1) \in (A + BK)\,\mathcal{Z} \oplus \mathcal{W} \subseteq \mathcal{Z},$$

for any $w(k) \in \mathcal{W}$. Hence, the constraint (7.20f) holds at time $k+1$ and the optimization problem (7.20) is also feasible at time $k+1$.

(*Robust constraint satisfaction*) With the feasible solution $\bar{x}(j)$ and $\bar{u}(j)$ at the time instant time k, we know $\bar{x}(j) \in \mathcal{X} \ominus \mathcal{Z}$ and $\bar{u}(j) \in \mathcal{U} \ominus K\mathcal{Z}$ for $j = \text{mod}(k, T)$. Taking into account that constraint (7.20f) holds, the control action $u(k)$ at time instant k is chosen in (7.21), which implies

$$u(k) \in \mathcal{U} \ominus K\mathcal{Z} \oplus K\mathcal{Z} \subseteq \mathcal{U}.$$

From constraint (7.20f), we also have

$$x(k) \in \bar{x}(j) \oplus \mathcal{Z} \in \mathcal{X} \ominus \mathcal{Z} \oplus \mathcal{Z} \subseteq \mathcal{X}.$$

We have proved that the optimization problem (7.20) is recursively feasible with an initial condition $x(0)$ and the constraints in (7.2) are satisfied. Since the optimal solution of the previous time step is always feasible, by optimality, we can know the optimal MPC cost $V(k, x, p)$ is a non-increasing sequence along the time step k, that is

$$V(k+1, x, p) \leq V(k, x, p), \quad \forall k \in \mathbb{N}. \tag{7.22}$$

We next prove the closed-loop property in the statement (b). For notation simplicity, we denote a periodic trajectory including states and control inputs over the MPC prediction horizon as the vector $z \in \mathbb{R}^{n_x + n_u}$, where

$$z = \left[\bar{x}(0)^\top \cdots \bar{x}(T)^\top \; \bar{u}(0)^\top \cdots \bar{u}(T-1)^\top\right]^\top, \tag{7.23}$$

and therefore the economic cost function $J_T(\bar{x}, \bar{u}, p)$ becomes $J_T(z, p)$. For the planner (7.19), the optimal cost can be denoted as $J_T(z^*, p)$. Taking into account that the optimization problems (7.19) and (7.20) are strictly convex, we reformulate them in the following convex forms. The optimization problem (7.19) is equivalent to

$$\underset{z}{\text{minimize}}\ J_T\left(z, p\right), \tag{7.24a}$$

subject to

$$h_r\left(z\right) \leq 0, \quad r = 1, \ldots, m, \tag{7.24b}$$
$$g_i\left(z\right) = 0, \quad i = 1, \ldots, n, \tag{7.24c}$$

where (7.24b)–(7.24c) are linear constraints. Specifically, (7.24b) corresponds to the refined constraints on states and inputs in (7.18), and (7.24c) corresponds to the nominal prediction model.

Similarly, the optimization problem (7.20) is equivalent to

$$\underset{z}{\text{minimize}}\ J_T\left(z, p\right), \tag{7.25a}$$

subject to

$$h_r\left(z\right) \leq 0, \quad r = 1, \ldots, m, \tag{7.25b}$$
$$g_i\left(z\right) = 0, \quad i = 1, \ldots, n, \tag{7.25c}$$
$$H_j^z x(k) - H_j^z Q^z(\sigma) z - b_j^z \leq 0, \quad j = 1, \ldots, n_x, \ \sigma = \text{mod}(k, T), \tag{7.25d}$$

where H_j^z and b_j^z denote the j-th row of H^z and b^z, and $Q^z(\sigma)z = \bar{x}(\sigma)$ with $\sigma = \text{mod}(k, T)$.

For the optimization problem (7.25) at the time instant k, we denote

$$z(k) = \arg\min_{z} V\left(k, x, p\right). \tag{7.26}$$

According to [13, Sect. 5.5.3], we can obtain the KKT optimality conditions of (7.25) as follows:

$$\nabla J_T\left(z(k), p\right) + \sum_{r=1}^{m} \lambda_r(k) \nabla h_r\left(z(k)\right) + \sum_{i=1}^{n} \mu_i(k) \nabla g_i\left(z(k)\right) + \sum_{j=1}^{n_x} \nu_j(k) H_j^z Q^z(\sigma) = 0, \tag{7.27a}$$

$$h_r\left(z(k)\right) \leq 0, \quad r = 1, \ldots, m, \tag{7.27b}$$
$$g_i\left(z(k)\right) = 0, \quad i = 1, \ldots, n, \tag{7.27c}$$
$$H_j^z x(k) - H_j^z Q^z(\sigma) z(k) - b_j^z \leq 0, \quad j = 1, \ldots, n_x, \ \sigma = \text{mod}(k, T), \tag{7.27d}$$
$$\lambda_r(k) \geq 0, \quad r = 1, \ldots, m, \tag{7.27e}$$
$$\lambda_r(k) h_r\left(z(k)\right) = 0, \quad r = 1, \ldots, m, \tag{7.27f}$$
$$\nu_j(k) \geq 0, \quad j = 1, \ldots, n_x, \tag{7.27g}$$

$$\nu_j(k)\left(H_j^z x(k) - H_j^z Q^z(\sigma)z(k) - b_j^z\right) = 0,\ \ j = 1,\ldots,n_x,\ \sigma = \mathrm{mod}(k,T), \tag{7.27h}$$

where $\lambda_r(k)$, $\mu_i(k)$ and $\nu_j(k)$ are dual variables. Denote the following vectors

$$\lambda(k) = \begin{bmatrix} \lambda_1(k) \\ \vdots \\ \lambda_m(k) \end{bmatrix}, \mu(k) = \begin{bmatrix} \mu_1(k) \\ \vdots \\ \mu_n(k) \end{bmatrix}, \nu(k) = \begin{bmatrix} \nu_1(k) \\ \vdots \\ \nu_{n_x}(k) \end{bmatrix}. \tag{7.28}$$

In terms of the REMPC planner in (7.19), the equivalent convex form can be written in a similar form as (7.25) excluding the constraint (7.25d).

Recall the optimal nominal periodic steady trajectory as z^*, where the variable assignment for z is defined in (7.23). Therefore, there exists a set of dual vectors λ^* and μ^* this optimal solution z^* also satisfies the KKT optimality conditions:

$$\nabla J_T\left(z^*, p\right) + \sum_{r=1}^{m} \lambda_r^* \nabla h_r\left(z^*\right) + \sum_{i=1}^{n} \mu_i^* \nabla g_i\left(z^*\right) = 0, \tag{7.29a}$$

$$h_r\left(z^*\right) \le 0,\ \ r = 1,\ldots,m, \tag{7.29b}$$

$$g_i\left(z^*\right) = 0,\ \ i = 1,\ldots,n, \tag{7.29c}$$

$$\lambda_r^* \ge 0,\ \ r = 1,\ldots,m, \tag{7.29d}$$

$$\lambda_r^* h_r\left(z^*\right) = 0,\ \ r = 1,\ldots,m. \tag{7.29e}$$

(*Convergence*) We have proved that the optimal MPC cost $V(k, x, p)$, $\forall k \in \mathbb{N}$ is a non-increasing sequence. Without loss of generality, we also consider that this optimal MPC cost is lower bounded by the optimal MPC cost corresponding to the planner (7.19). This implies that as the time step $k \to +\infty$, the optimal MPC cost can reach a constant value. In this case, there exists a time instant M such that for any $k \ge M$, $V(k+1, x, p) = V(k, x, p)$ holds, that is, we have reached a constant cost. Because the economic cost function $J_T(z, p)$ is strictly convex, it follows $z(k+1) = z(k) = z^s$. It means that after M time steps, we can obtain a steady periodic trajectory z^s.

(*Optimality Certificate*) Since z^s is a feasible solution of the optimization problem (7.25), there exist dual vectors λ^s, μ^s and ν^s such that the KKT optimality conditions (7.27) hold. Recall z^* and $J_T(z^*, p)$ as the optimal planner trajectories and the economic planner cost obtained by solving the optimization problem (7.25). If the dual vector $\nu^s = 0$, then λ^s and μ^s satisfy the KKT conditions (7.29) of the planner, which implies $z^s = z^*$ and $V(k, x, p) = J_T(z^*, p)$.

The condition $\nu_j^s = 0$ is called the *optimality certificate*. If this certificate is satisfied, then from the trajectory $z^s = z^*$, we denote $\bar{x}^*(\sigma) = Q^z(\sigma)z^*$, $\sigma = \mathrm{mod}(k, T)$ corresponding to states. From constraint (7.5f), we obtain $x(k) - \bar{x}^*(\sigma) \in \mathcal{Z}$ for any $k \ge M$, which means the closed-loop system can reach a neighborhood (enclosed by the RPI set $\mathcal{Z}$) of the periodic nominal steady trajectory that is obtained by the planner.

Summing up, the proposed controller guarantees robust constraint satisfaction, recursive feasibility and a non-increasing optimal cost of the optimization problem (7.20), which guarantees the convergence to a neighborhood of the optimal nominal periodic steady trajectory when the optimality certificate is satisfied. Based on the constraint (7.5f), as the time step $k \to +\infty$, the deviation of the closed-loop system trajectory from the nominal steady trajectory is bounded in the RPI set $\mathcal{Z}$. □

Remark 7.8 From Theorem 7.2, we have provided an optimality certificate, that is all the variables in the dual vector corresponding to the inequality constraint (7.25d) are zero in the KKT optimality conditions, which can be verified online to known if the closed-loop convergence is optimal, that is it reaches a neighborhood of the optimal periodic steady trajectory after M time steps.

Remark 7.9 For some certain systems, due to tight constraints leading to low degree of freedom, the optimality certificate may be not satisfied. Then, from the recursive feasibility, the corresponding KKT optimality conditions (7.27) can be still satisfied. In this case, he optimization problem (7.25) is also possible to reach a steady solution z^s with $|V(k+1, x, p) - V(k, x, p)| \leq \epsilon$, $\forall k \geq \bar{M}$ with an arbitrary small scalar ϵ. However, from the KKT optimality conditions (7.29), z^s is a suboptimal solution. Thus, we can conclude that $z^s \neq z^*$ and $V_T(k, x, p) > J_T(z^*, p)$.

In this REMPC design, the tube-based technique is used. As an example shown in Fig. 7.3, the optimal nominal periodic steady trajectory obtained by the planner (7.19) is plotted in red dashed line, the tubes defined by the RPI set $\mathcal{Z}$ are plotted in blue boundaries, and a closed-loop trajectory of system (7.14) with the proposed REMPC (7.20) is plotted in the blue line. Hence, we can conclude that once the closed-loop trajectory is close to the optimal nominal periodic steady trajectory, the optimal solution does not change because the state is in a tube and the input applied in (7.21) guarantees that it will not go outside the tube because it is defined as an RPI set.

7.2.4 Example: The Mass Model

The mass model with a spring and a damper taken from [14] is shown in Fig. 7.4. Consider a discrete-time model of this mass model in the form as in (7.14) with the following system matrices

$$A = \begin{bmatrix} 0.9952 & 0.0950 \\ -0.0950 & 0.9002 \end{bmatrix}, \quad B = \begin{bmatrix} 0.0048 \\ 0.0950 \end{bmatrix}, \quad B_d = B,$$

and $w(k) := B_d d(k) + \tilde{w}(k)$, where the displacement and the velocity of the mass model are chosen as state variables in x, d is a periodic known signal with a period

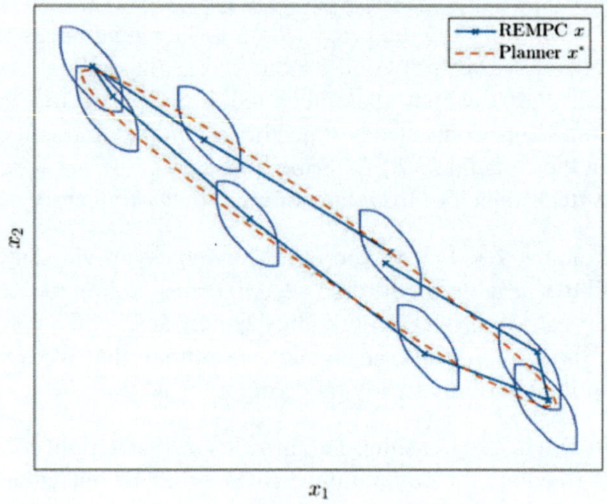

Fig. 7.3 Closed-loop trajectory and optimal nominal steady periodic trajectory with tubes

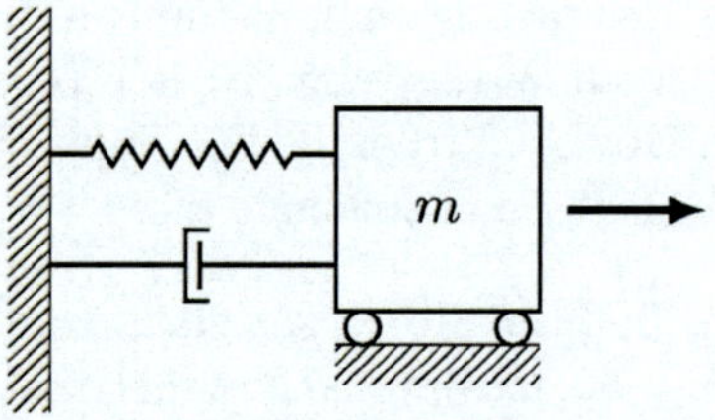

Fig. 7.4 Mass model with a spring and a damper

$T = 10$ that is given by a sequence $d(k) = d_i$ with $i = \mathrm{mod}(k, T)$. The disturbance $\tilde{w}(k) \in \mathcal{W}$, $\forall k \in \mathcal{W}$, where the set $\mathcal{W}$ is given by $\mathcal{W} = \{w \in \mathbb{R}^2 \mid |w| \leq [0.005 \;\; 0.01]^\top\}$. The constraints on states and inputs are given by the following convex sets

$$\mathcal{X} = \{x \in \mathbb{R}^2 \mid |x| \leq [1.5 \;\; 0.75]^\top\},$$
$$\mathcal{U} = \{u \in \mathbb{R} \mid |u| \leq 8\}.$$

The local control law $K \in \mathbb{R}^{1\times 2}$ is computed using the LQR method with weighting matrices $Q = \begin{bmatrix} 0.1 & 0 \\ 0 & 0.1 \end{bmatrix}$ and $R = 0.01$ obtaining

$$K = [-1.8635 \quad -2.5172].$$

The initial state is $x(0) = [-0.0890 \;\; 0.3570]^\top$. As defined in [14], the economic cost function is chosen to be $\ell(\bar{x}, \bar{u}, p) = 10(\bar{x}_2(i) - p_i)^2 + (\bar{u}(i))^2$ with a periodic signal p. In order to test the proposed controller with sudden changes. These sudden changes are given by choosing different values of periodic signals d and p. In the simulation, the following two scenarios are considered.

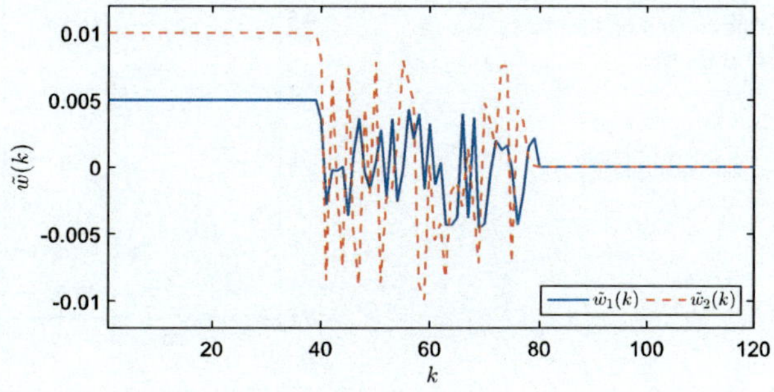

Fig. 7.5 Sampled bounded disturbances

- **Scenario 1**: For $k < 50$,

$$d_i = 5\cos\left(\frac{2\pi i}{T}\right), \quad i = 0, \ldots, T-1,$$
$$p_i = 0.1\sin\left(\frac{2\pi i}{T}\right), \quad i = 0, \ldots, T-1.$$

- **Scenario 2**: For $k \geq 50$,

$$d_i = 0.1\sin\left(\frac{2\pi i}{T}\right) + 0.5, \quad i = 0, \ldots, T-1,$$
$$p_i = 1.2\cos\left(\frac{2\pi i}{T}\right), \quad i = 0, \ldots, T-1.$$

The optimization problems (7.19) and (7.20) are solved using the YALMIP toolbox [17] and the MOSEK solver [18] in the MATLAB environment. For the previous scenarios, the planner has been applied. Then, two optimal nominal periodic steady trajectories and two different optimal MPC costs can be obtained.

The closed-loop simulation has been carried out for 120 sampling time steps with a sudden change defined in the previous two scenarios. As shown in Fig. 7.5, the unknown disturbance $\tilde{w}(k)$ is defined as follows:

$$\tilde{w}(k) = \begin{cases} \bar{w}, & k < 40, \\ \tilde{w} \in \mathcal{W}, & 40 \leq k < 80, \\ 0, & k \geq 80. \end{cases}$$

The closed-loop results of state and control input trajectories are shown in Figs. 7.6 and 7.7. For $k < 50$ (Scenario 1), starting from the feasible initial state $x(0)$, the closed-loop state and input trajectories converge to a neighborhood of the optimal nominal periodic trajectories obtained by the Planner 1. At the time step $k = 50$, there is a sudden change of the periodic signals d and p as defined in Scenario 2. For $k \geq 50$ (Scenario 2), the closed-loop system is also feasible and the closed-

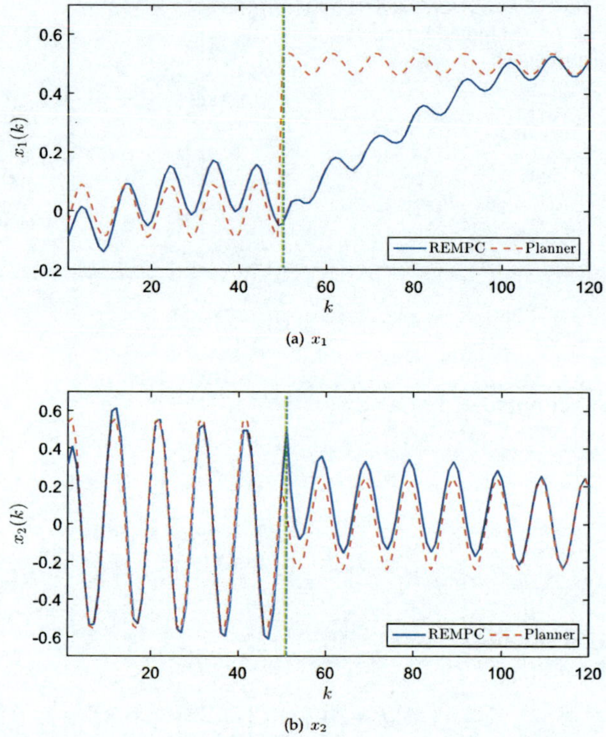

Fig. 7.6 Closed-loop state trajectories of the mass model

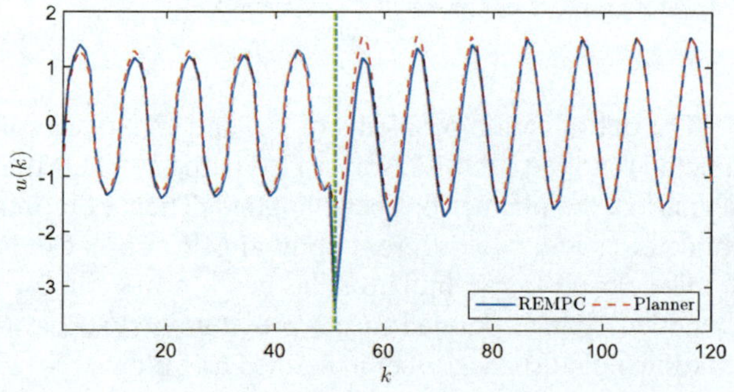

Fig. 7.7 Closed-loop input trajectory of the mass model

loop state and input trajectories converge to a neighborhood of the optimal nominal periodic trajectories obtained by the Planner 2. From these results, it proves that the closed-loop system is always feasible from an initial state even with a sudden change.

Since we have discussed that the recursive feasibility mainly relies on the equality constraint (7.5f), this constraint should be satisfied with the closed-loop state $x(k)$, $\forall k \in \mathbb{N}$. As shown in Fig. 7.8, the mismatch between the closed-loop state and the optimal nominal state should be always inside the RPI set $\mathcal{Z}$. Hence, this result also proves that the closed-loop system can be always recursively feasible in the presence of unknown-but-bounded additive disturbances.

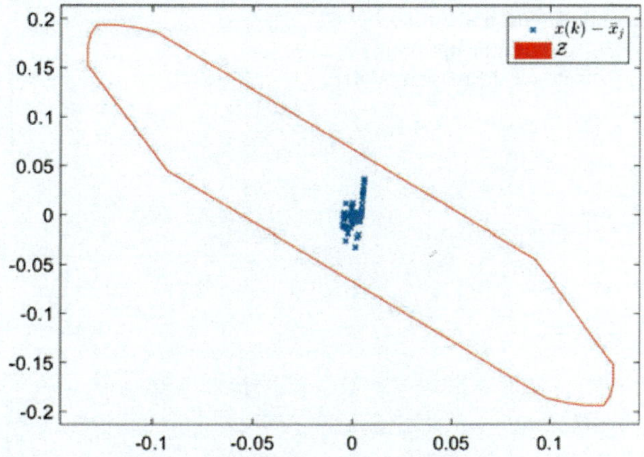

Fig. 7.8 Validation of the recursive feasibility for the mass model

Taking into account three different selections of bounded additive disturbances, for $k < 40$, the closed-loop state and input trajectories are periodic based on the periodicity constraints and meanwhile approaching to the optimal nominal periodic steady trajectories obtained by the Planner 1. For $40 \leq k < 80$, the closed-loop trajectories are close to the optimal nominal periodic steady trajectories and with the sudden change, the optimal nominal periodic steady trajectories are switched to the ones obtained by the Planner 2. Besides, the closed-loop trajectories in Figs. 7.6b and 7.7 stay close to the optimal nominal periodic steady trajectories in the tube (defined by the RPI set $\mathcal{Z}$). For $k \geq 80$, since $w(k) = 0$, the closed-loop state and input trajectories are able to match the optimal nominal periodic steady trajectories of the Planner 2 after a transient time.

Moreover, from the offline computation results of the planners, two optimal MPC costs are also shown in Fig. 7.9a. Since the optimality certificate is verified online, the closed-loop optimal MPC cost can converge to the optimal one for each scenario with a sudden change in the closed-loop cost between two scenarios.

As discussed in Theorem 7.2, the optimality certificate is given by checking whether all the variables in the dual vector $\nu(k)$ corresponding to (7.5f) are zero. From the online closed-loop simulation, these dual variables can be extracted together with the optimal solution of (7.20) at each time step. To verify the optimality certificate, the 2-norm of $\nu(k)$ as $\|\nu(k)\|_2$ is shown in Fig. 7.9b. For these scenarios considered, two steady situations are expected to be observed. Despite sudden changes in the controller design parameters, after a transient time, the 2-norm of $\nu(k)$ converges to zero. Also as shown in Figs. 7.6 and 7.7, the closed-loop trajectories are able to reach a neighborhood of the optimal nominal periodic steady trajectories obtained by each planner.

Fig. 7.9 Closed-loop economic cost and online verification of the optimality certificate for the mass model

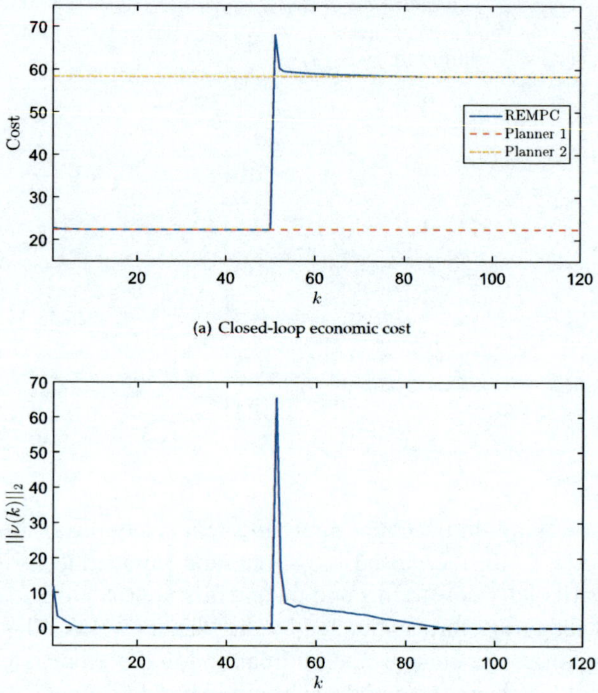

(a) Closed-loop economic cost

(b) Optimality certificate

7.3 Summary

This chapter has presented an EMPC framework based on a periodicity constraint for both nominal and uncertain linear systems. We have proved that with the proposed both EMPC controllers, the closed-loop system is able to converge to a (neighborhood of) periodic trajectory and all the system constraints can be satisfied even in the presence of disturbances and sudden changes in the economic cost function. Besides, an optimality certificate of the proposed EMPC has been given. We have proved that if this optimality certificate is satisfied, the closed-loop trajectories can reach (a neighborhood of) the optimal periodic steady trajectory obtained by the planner. In the robust case, the neighborhood region is defined by the considered RPI set. Besides, in some particular cases, due to constraints of the corresponding optimization problem are set too hard, the closed-loop trajectory may be trapped in another periodic trajectory and the optimality certificate cannot be satisfied.

Some future directions for this EMPC framework could be

- Extension to nonlinear systems;
- Enforcing the convergence to the optimal periodic steady trajectory. In this case, additional constraint or penalty cost function might be used;
- Application to real case studies.

References

1. Limon D, Pereira M, Muñoz de la Peña D, Alamo T, Grosso J (2014) Single-layer economic model predictive control for periodic operation. J Process Control 24(8):1207–1224
2. Wang Y, Puig V, Cembrano G (2017) Non-linear economic model predictive control of water distribution networks. J Process Control 56:23–34
3. Wang Y, Salvador J, Muñoz de la Peña D, Puig V, Cembrano G (2017) Periodic nonlinear economic model predictive control with changing horizon for water distribution networks. In: 20th IFAC world congress, Toulouse, France, pp 6588–6593
4. Pereira M, Limon D, Muñoz de la Peña D, Valverde L, Alamo T (2015) Periodic economic control of a nonisolated microgrid. IEEE Trans Ind Electron 62(8):5247–5255
5. Cembrano G, Wells G, Quevedo J, Perez R, Argelaguet R (2000) Optimal control of a water distribution network in a supervisory control system. Control Eng Pract 8(10):1177–1188
6. Ocampo-Martinez C, Puig V, Cembrano G, Quevedo J (2013) Application of MPC strategies to the management of complex networks of the urban water cycle. IEEE Control Syst 33(1):15–41
7. Puig V, Ocampo-Martinez C, Pérez R, Cembrano G, Quevedo J, Escobet T (2017) Real-time monitoring and operational control of drinking-water systems. Springer
8. Wang Y, Muñoz de la Peña D, Puig V, Cembrano G (2018) A novel formulation of economic model predictive control for periodic operations. In: European control conference (ECC), Limassol, Cyprus, pp 1015–1020
9. Wang Y, Salvador J, Muñoz de la Peña D, Puig V, Cembrano G (2018) Economic model predictive control based on a periodicity constraint. J Process Control 68:226–239
10. Wang Y, Muñoz de la Peña D, Puig V, Cembrano G (2019) Robust economic model predictive control based on a periodicity constraint. Int J Robust Nonlinear Control 29(11):3296–3310
11. Limon D, Pereira M, Muñoz de la Peña D, Alamo T, Jones C, Zeilinger M (2016) MPC for tracking periodic references. IEEE Trans Autom Control 61(4):1123–1128
12. Mayne D (2014) Model predictive control: recent developments and future promise. Automatica 50(12):2967–2986
13. Boyd S, Vandenberghe L (2004) Convex optimization. Cambridge University Press
14. Broomhead T, Manzie C, Shekhar R, Hield P (2015) Robust periodic economic MPC for linear systems. Automatica 60:30–37
15. Wang Y, Puig V, Cembrano G (2016) Economic MPC with periodic terminal constraints of nonlinear differential-algebraic-equation systems: application to drinking water networks. In: European control conference (ECC), Aalborg, Denmark, pp 1013–1018
16. Mayne D, Seron M, Raković S (2005) Robust model predictive control of constrained linear systems with bounded disturbances. Automatica 41(2):219–224
17. Löfberg J (2004) YALMIP: a toolbox for modeling and optimization in MATLAB
18. MOSEK ApS (2015) The MOSEK optimization toolbox for MATLAB manual. Version 7.1 (Revision 28)

Chapter 8
Applications of Economic Model Predictive Control Strategies for Complex Systems

This chapter presents three application results of EMPC strategies for realistic water distribution networks and power systems. The control-oriented model of all these systems is built in a descriptor form. The importance of this chapter is to demonstrate the proposed EMPC strategies in real case studies. Meanwhile, some additional difficulties encountered from these applications appear. To address these, a two-layer control strategy and a nonlinear constraint relaxation approach are presented. Specially, this chapter includes three parts:

- The first part presents a two-layer NEMPC of WDNs with a real simulation platform [1–3]. The upper layer includes a real-time NEMPC controller to provide an optimal flow set-point while the lower layer is based on a pumping scheduling approach to translate this optimal set-point into an ON/OFF sequence. The detailed WDN is simulated in a realistic simulator, namely EPANET;
- The second part proposes an iterative algorithm of nonlinear constraint relaxation [4, 5], which is used to implement with the EMPC controller designed in Sect. 7.1. According to the descriptor model of WDNs, nonlinearities only appear in algebraic equations and thus the relaxation approach is used to obtain a set of linear constraints for bounding these algebraic equations;
- The third part presents an extension of the REMPC controller designed in Sect. 7.2 for the descriptor model of smart micro-grids [6]. Since both differential/difference and algebraic equations are affected by disturbances, in particular, algebraic equations should be satisfied at any time with unknown disturbances. In order to guarantee recursive feasibility, the tube-based approach presented in Sect. 7.2 is improved using change of variable.

Y. Wang, *Advances in State Estimation, Diagnosis and Control of Complex Systems*, Springer Theses, https://doi.org/10.1007/978-3-030-52440-1_8

8.1 A Two-Layer NEMPC of WDNs

We first present a two-layer control scheme that combines an NEMPC strategy in the upper layer, and a pump scheduling approach in the lower layer. The NEMPC strategy is implemented by using a nonlinear programming technique and the pump scheduling approach is realized by solving a local optimization problem. The proposed two-layer control strategy is validated using a hydraulic simulator that emulates the real WDN behavior. The D-Town water network, a well known benchmark, is used as the case study. The closed-loop simulation is implemented using a simulation platform with a virtual-reality hydraulic simulator that emulates the online operation.

8.1.1 *Control-Oriented Modeling WDNs*

We first introduce the control-oriented mathematical modeling methodology of the WDN including the flow and hydraulic head relations for the different network components. As result of the application of this methodology to a particular WDN, a set of dynamic and static relationships that lead to a descriptor system in discrete-time ready to be used in the implementation of the MPC is obtained. A WDN can be decomposed by a set of constitutive elements: *reservoirs/tanks, control valves, pumping stations, nodes* and *water demand sectors*, each being characterized by means of flow-head relations [7–9].

8.1.1.1 Tanks

Water tanks supply and provide the entire WDN with the storage capacity of drinking water to consumers guaranteeing adequate water pressure service. The mass balance expression relating the stored volume ϖ in the m-th tank can be written as the discrete-time difference equation which describes the tank dynamical evolution as

$$\varpi_m(k+1) = \varpi_m(k) + \Delta t \left(\sum_i q_{i,m}^{\text{in}}(k) - \sum_j q_{m,j}^{\text{out}}(k) \right), \tag{8.1}$$

where $q_{i,m}^{\text{in}}(k)$ denotes the inflows from the i-th element to the m-th tank and $q_{m,j}^{\text{out}}(k)$ denotes the outflows from the m-th tank to the j-th element. Δt is the sampling time and k is the discrete-time instant. The physical limitation related to the storage volume in the m-th tank is described as

$$\underline{\varpi_m} \leq \varpi_m(k) \leq \overline{\varpi_m}, \quad \forall k \in \mathbb{N}, \tag{8.2}$$

where $\underline{\varpi_m}$ and $\overline{\varpi_m}$ denote the minimum and maximum admissible storage capacity, respectively.

The head model in WDN is typically written in terms of the hydraulic head that relates the energy in an incompressible fluid to the height of an equivalent static column of that fluid. Note that the head is usually expressed in units of height.

Using this concept, the head related to the m-th tank with respect to the volume of storage water inside can be determined as follows:

$$h_m(k) = \frac{\varpi_m(k)}{\mathrm{S}_m} + E_m, \quad \forall k \in \mathbb{N}, \tag{8.3}$$

where S_m is the cross-sectional area of the m-th tank and E_m corresponds the m-th tank elevation.

8.1.1.2 Pumping Stations

Pumps located in pumping stations of a WDN can be of several types: *fixed-speed* pumps, *variable-speed* pumps and *variable throttle* pumps [7] depending on how they are controlled. We will consider fixed-speed pumps that are the most used in WDN because of the simplicity of operation, i.e., they are operated in an ON/OFF manner. However, such simplicity introduces an additional problem when implementing an MPC strategy since the ON/OFF operation would involve including discrete variables in the optimization problem.

Pump flows are regarded as the manipulated variables. Therefore, the flow limitations for pumps can be regarded as input constraints, which can be expressed as

$$\underline{q_{u_n}} \leq q_{u_n}(k) \leq \overline{q_{u_n}}, \quad \forall k \in \mathbb{N}, \tag{8.4}$$

where q_{u_n} represents the manipulated flow of the n-th pump (or valve), $\underline{q_{u_n}}$ and $\overline{q_{u_n}}$ represent the minimum and maximum flow capacity of the n-th pump, respectively. These limitations vary with the pressure according to hydraulic flow/head curve of the pump.

The hydraulic characteristic of a pump is formulated by a nonlinear function related to the flow and head variables. Therefore, for a pump, the hydraulic characteristics are bounded by the following constraints

$$\Delta h_p(k) = h_d(k) - h_s(k) \geq 0, \quad \forall k \in \mathbb{N}, \tag{8.5a}$$

$$h_d(k) \in \left[\underline{h_d}, \overline{h_d}\right], \tag{8.5b}$$

$$h_s(k) \in \left[\underline{h_s}, \overline{h_s}\right], \tag{8.5c}$$

where $h_d(k)$ and $h_s(k)$ denote the *suction head* and the *delivery head* at time step k, respectively, with the physical limitation of $h_d(k) \geq h_s(k)$. Moreover, $\underline{h_d}$ and $\underline{h_s}$

denote the minimum values of the suction and delivery heads. $\overline{h_d}$ and $\overline{h_s}$ denote the maximum values of the suction and delivery heads.

8.1.1.3 Valves

In terms of the type of valves, there is a variety of options, such as pressure modulating, non-return, pressure reducing, flow variable control, head control and so on [7]. For simplicity, valves considered are of the flow-control type.

It is worth mentioning that unlike pumps, the characteristic of valves is difficult to model, because different degrees of opening of the valve produce different characteristic curves (head-flow relationships). Due to this, it is not possible to include these curves in the control-oriented model of a WDN. Then, from control point of view, the pressure (head) variables are left to be free decision variables within considered bounds in the closed-loop optimization. Thus, the valve model considers only the following constraints

$$\Delta h_v(k) = h_{us}(k) - h_{ds}(k) \geq 0, \quad \forall k \in \mathbb{N}, \tag{8.6a}$$

$$h_{us}(k) \in \left[\underline{h_{us}}, \overline{h_{us}}\right], \tag{8.6b}$$

$$h_{ds}(k) \in \left[\underline{h_{ds}}, \overline{h_{ds}}\right], \tag{8.6c}$$

where $h_{us}(k)$ and $h_{ds}(k)$ denote the heads at the nodes around the valve in the upstream and downstream at time step k, respectively. $\underline{h_{us}}$ and $\underline{h_{ds}}$ denote the minimum values of the upstream and downstream heads. $\overline{h_{us}}$ and $\overline{h_{ds}}$ denote the maximum values of the upstream and downstream heads.

8.1.1.4 Nodes

Water flow through each node of the network must fulfill the mass balance relations. The expression of the mass conservation in these nodes can be written as

$$\sum_i q_{i,l}^{\text{in}}(k) = \sum_j q_{l,j}^{\text{out}}(k), \quad \forall k \in \mathbb{N}, \tag{8.7}$$

where $q_{i,l}^{\text{in}}$ represents the non-manipulated inflow through l-th node from the i-th element and $q_{l,j}^{\text{out}}$ represents the non-manipulated outflow through l-th node to the j-th element.

8.1.1.5 Water Demand Sectors

A demand sector represents water demand of the network users of a certain physical area. At a certain time step k, the consumed water in the r-th demand sector can be expressed as $d_r(k)$. Since the optimal control strategy is considered as a predictive one, the short-term demand forecasts are able to obtain by using a suitable demand forecasting algorithm, such as [10, 11].

8.1.1.6 Pipes

Pipes convey water from one place in the network to another. Water inside pressurized pipes flows from the higher hydraulic head to that at lower head. Therefore, the head-flow relationship for a pipe can be described as

$$q_{i,j}(k) = \Phi_{i,j}\left(h_i(k) - h_j(k)\right), \tag{8.8}$$

where $\Phi_{i,j}$ is a nonlinear relationship, usually described by an empirical equation, for instance, the *Hazen-Williams* formula. Hence, the head drop through a pipe $\Delta h_d(k)$, $\forall k \in \mathbb{N}$ can be calculated as

$$\Delta h_d(k) = h_i(k) - h_j(k) = R_{i,j} q_{i,j}(k)\left|q_{i,j}(k)\right|^{0.852}, \tag{8.9}$$

with

$$R_{i,j} := \frac{10.67 L_{i,j}}{C_{i,j}^{1.852} D_{i,j}^{4.87}},$$

where $L_{i,j}$, $D_{i,j}$ and $C_{i,j}$ denote the pipe length, diameter and roughness coefficient, respectively.

Basically, pipes can be classified based on the flow sense into unidirectional and bidirectional. Therefore, $\Delta h_d(k)$, $\forall k \in \mathbb{N}$ in unidirectional pipe is always positive with its selected direction while in bidirectional pipe $\Delta h_d(k)$, $\forall k \in \mathbb{N}$ could be varying between positive and negative since the direction of the flow can be reversed.

8.1.2 The Upper Layer: NEMPC

8.1.2.1 Control-Oriented Model of WDNs

Considering the modeling methodology of each component in WDNs presented above, the control-oriented model of WDNs can be formulated as

Table 8.1 Variable assignments in the control-oriented model of the WDN

Type of variable	Related symbols	Description
Difference states: x	h_m	Hydraulic heads at the storage nodes (i.e. storage tanks)
Algebraic states: z	h_d, h_s, h_i, h_j	Hydraulic heads at the non-storage nodes
Control inputs: u	q_{u_n}	Manipulated flows through actuators (pumps and valves)
Non-control inputs: v	$q_{i,j}$	Non-manipulated flows through interconnected pipes
System disturbances: d	d_r	Water demands

$$x(k+1) = Ax(k) + B_u u(k) + B_v v(k) + B_d d(k), \tag{8.10a}$$
$$0 = E_u u(k) + E_v v(k) + E_d d(k), \tag{8.10b}$$
$$0 = P_x x(k) + P_z z(k) + \psi(v(k)). \tag{8.10c}$$

where $x \in \mathbb{R}^{n_x}$ represents the vector of hydraulic heads at storage nodes (tanks) as difference states, $z \in \mathbb{R}^{n_z}$ represents the vector of hydraulic heads at non-storage nodes as algebraic states, $u \in \mathbb{R}^{n_u}$ denotes the vector of the manipulated flows through actuators (pumps and valves) as control inputs, $v \in \mathbb{R}^{n_v}$ denotes the vector of non-manipulated flows through interconnected pipes and $d \in \mathbb{R}^{n_d}$ corresponds to the vector of water demands as system disturbances. $k \in \mathbb{N}$ denotes the time step. All the considered variables are classified as control-oriented variables in Table 8.1. Moreover, $\psi(\cdot)$ denotes the vector of nonlinear *Hazen-Williams* mapping functions.

Remark 8.1 Note that units of all the control-oriented variables need to be consistent. The unit of the head is selected as m (meter). The water flows is with unit of m^3/s (cubic-meter per second).

8.1.2.2 Cost Function Settings

According to [9], the operational goals for the management of WDNs include:

- Economic: To provide a reliable water supply with the required pressure minimizing operational costs;
- Safety: To guarantee the availability of enough water with suitable pressure in each storage tank to satisfy its underlying uncertain water demands;
- Smoothness: To operate actuators (pumps and valves) in the WDN under smooth control actions.

The main control objective is to minimize the water distribution costs that includes water acquisition costs and electrical costs especially for pumping water through the pumps. The water is delivered into the nodes with different heads (including elevations) through the distribution network implying many electrical costs on the booster pumping. Therefore, the cost function associated to this objective can be formulated as

$$\ell_1(k) := p(k)^\top u(k), \tag{8.11}$$

with $p(k) = \alpha_1 + \alpha_2(k)$, where α_1 denotes the single-column vector of static economic costs of the water depending on the selected water sources and $\alpha_2(k)$ represents the vector of the time-varying electrical costs. Considering the variable daily electrical tariff, $\alpha_2(k)$ is time-varying.

For the purpose of maintaining the water supply in spite of the variation of water demands between two consecutive MPC sampling steps, a suitable safety head for each storage tank must be maintained. Hence, the mathematical expression for this objective is formulated with a quadratic penalty as

$$\ell_2(k) := \begin{cases} \|x(k) - x^s\|_2^2, & \text{if } x(k) \leq x^s, \\ 0, & \text{otherwise}, \end{cases} \tag{8.12}$$

where x^s denotes the vector of the safety heads for all the tanks and $\|\cdot\|_2^2$ is the squared 2-norm symbol. This cost function can also be realized by means of a soft constraint with adding a slack variable $\xi(k)$, which can be reformulated as

$$\ell_2(k) := \|\xi(k)\|_2^2, \tag{8.13}$$

together with the following soft constraint:

$$x(k) \geq x^s - \xi(k). \tag{8.14}$$

The actuators in WDN mainly include pumps and valves. Thus, the flow-based control actions found by the EMPC controller is required to be smooth in order to maximize the lifespan of the actuators. In addition, the use of the smooth operations is benefit for the lower-layer regulatory performance. To achieve a sequence of smooth operations, the slew rate of the control actions between two consecutive time steps is penalized. Hence, the cost function for this part can be written as

$$\ell_3(k) := \|\Delta u(k)\|_2^2, \tag{8.15}$$

with $\Delta u(k) := u(k) - u(k-1)$.

In general, the multi-objective cost function that gathers all the control objectives for the operational management of the WDN can be summarized as

$$\sum_{j=1}^{\Gamma} \lambda_j \ell_j(k), \tag{8.16}$$

where λ_j denotes the weighting term that indicates the prioritization of control objectives and $\Gamma = 3$ is the number of the selected control objectives.

8.1.2.3 Constraint Settings

In the real components of a WDN, there are the physical limitations associated to the system variables. Therefore, these constraints should complement the mass balance principles and physical relations between flow and head introduced in (8.10). In the following, these physical constraints are described in detail.

The hard constraint on the system states x comes from the tank capacity in the WDN, which can be described as

$$\underline{x_i} \leq x_i(k) \leq \overline{x_i}, \quad \forall k \in \mathbb{N}, \quad i \in [1, m] \subset \mathbb{Z}_+, \tag{8.17}$$

where $\underline{x_i}$ and $\overline{x_i}$ represent the minimum and maximum heads with respect to capacities of the i-th tank, respectively. The tank volumetric capacity can be transformed into hydraulic head constraints by (8.3).

Taking into account the physical capacity of different actuators, the manipulated flows are under the following constraint

$$\underline{u_i} \leq u_i(k) \leq \overline{u_i}, \quad \forall k \in \mathbb{N}, \quad i \in [1, n] \subset \mathbb{Z}_+, \tag{8.18}$$

where $\underline{u_i}$ and $\overline{u_i}$ denote the minimum and maximum manipulated flows of the i-th actuator, respectively. On the other hand, the non-manipulated flows throughout the interconnected pipes can be limited between $\underline{v_i}$ and $\overline{v_i}$ as

$$\underline{v_i} \leq v_i(k) \leq \overline{v_i}, \quad \forall k \in \mathbb{N}, \quad i \in [1, n_p] \subset \mathbb{Z}_+, \tag{8.19}$$

where n_p is the number of pipes.

The heads at some certain non-storage nodes are required to be up to some minimum levels as in the case of the water demand sectors. Hence, the following inequality constraint is necessary to be considered:

$$z_i(k) \geq \underline{z_i}, \quad \forall k \in \mathbb{N}, \quad i \in [1, n_h] \subset \mathbb{Z}_+, \tag{8.20}$$

where $\underline{z_i}$ are the required heads at the water demand sectors. Moreover, n_h is the total number of the water demand sectors.

8.1.2.4 NEMPC Formulation

In general, the NEMPC strategy can be implemented by solving a finite-horizon optimization problem over a prediction horizon H_p, where the multi-objective cost function is minimized subject to the prediction model and a set of system constraints.

Thus, the optimization problem associated to the NEMPC strategy can be formulated as follows:

$$\underset{\substack{x(0),\ldots,x(H_p)\\u(0),\ldots,u(H_p-1)}}{\text{minimize}} \quad \sum_{i=0}^{H_p-1}\sum_{j=1}^{\Gamma} \lambda_j \ell_j(i), \tag{8.21a}$$

subject to

$$x(i+1) = Ax(i) + B_u u(i) + B_v v(i) + B_d d(i), \tag{8.21b}$$
$$0 = E_u u(i) + E_v v(i) + E_d d(i), \tag{8.21c}$$
$$0 = P_x x(i) + P_z z(i) + \psi\big(v(i)\big), \tag{8.21d}$$
$$\underline{x} \leq x(i) \leq \overline{x}, \tag{8.21e}$$
$$\underline{u} \leq u(i) \leq \overline{u}, \tag{8.21f}$$
$$z(i \geq \underline{z} \tag{8.21g}$$
$$x(i) \geq x^s - \xi(k+i), \tag{8.21h}$$
$$x(i) = x(k). \tag{8.21i}$$

Since the control-oriented model of the WDN includes the nonlinear relations in (8.21d), the above optimization problem naturally becomes nonlinear. Thus, the optimization problem (8.21) should be solved using a suitable nonlinear programming technique. Assuming that the optimization problem (8.21) is feasible, the sequence of control actions is

$$u^* = \left[u^*(0)^\top, \ldots, u^*(H_p-1)^\top\right]^\top. \tag{8.22}$$

And then by deploying the receding-horizon strategy, the optimal control action at time step k is the first component of the sequence of control actions denoted by

$$u(k) = u^*(0). \tag{8.23}$$

8.1.3 The Lower Layer: Pumping Scheduling Approach

In practice, the main energy consumption is used for pumping water through the pumping stations. In case of the pumps with ON/OFF operation, the flows in (8.22) become discrete values and subsequently (8.21) becomes a nonlinear mixed-integer problem. In the lower layer, we propose the following pumping scheduling approach. Denote $Q_j^* = u_j(k)$, $\forall j \in [1, n_s] \subset \mathbb{Z}_+$ with $\sum_{j=1}^{n_s} u_j(k) = u(k)$ as the optimal hourly flow set-point of the j-th pumping station obtained from the upper layer, where n_s is the total number of pumping stations in WDN. The control objectives of the lower layer can be summarized as follows:

- To provide enough water to reach the optimal water flow set-points.
- To use the minimum possible number of parallel pumps and avoid too many switches in order to maximize their working lives.

In terms of the j-th pumping station, the pumping flow of the i-th pump is affected by the factors of the suction and delivery heads. Hence, if these boundary heads are given, the actual flow $q^r_{i,j}$ through the pump is considered within an interval, which can be formulated as

$$q^r_{i,j} \in \left[q^n_{i,j} - \sigma_{i,j},\ \ q^n_{i,j} + \sigma_{i,j}\right], \tag{8.24}$$

where $q^n_{i,j}$ denotes nominal pumping flow produced through the i-th pump, and $\sigma_{i,j}$ represents the variance of the pumping flow depending on the uncertainty of the boundary heads. It is assumed that the actual flow $q^r_{i,j}$ can be measured. In some cases, only one pump cannot provide enough flows to maintain the optimal flow set-point. Hence, parallel pumps are set in each pumping station. Ideally, the optimal pumping flow Q^*_j can be satisfied when all the pumps are open in the lower layer such that the following condition holds:

$$Q^*_j \Delta t_u = \sum_{i=1}^{n_{cj}} \sum_{t=1}^{H_l} q^r_{i,j} \Delta t_l, \tag{8.25}$$

where n_{c_j} is the total number of parallel pumps in the j-th pumping station and H_l is the control horizon of the lower layer.

Consider that the parallel pumps are operated in ON/OFF way, the binary variable $\chi_{i,j}(t) \in \{0, 1\}$ at time step t is chosen, where $\chi_{i,j}(t) = 0$ describes the OFF-status and $\chi_{i,j}(t) = 1$ presents the ON-status. Therefore, the actual flow of the i-th pump can be computed by

$$q_{i,j}(t) = \chi_{i,j}(t) q^r_{i,j}, \quad \forall i \in \left[1, n_{c_j}\right] \subset \mathbb{Z}_+, \forall t \in [1, n_s] \subset \mathbb{Z}_+. \tag{8.26}$$

Furthermore, the minimum usages of required parallel pumps and switches are necessary to be taken into account. It is considered that the parallel pumps are selected in a sequence order from $i = 1$ to $i = n_{c_j}$. Therefore, the required parallel pumps for j-th pumping station can be constrained by the following condition

$$\chi_{i+1,j}(t) + \left(1 - \chi_{i,j}(t)\right) \leq 1, \quad \forall i \in \left[1, n_{c_j}\right] \subset \mathbb{Z}_+, \tag{8.27}$$

which means if i-th pump is not used, then $i + 1$-th pump is also not used. Additionally to (8.27), the minimum required parallel pumps with their selection orders can be decided by maximizing the following term

$$J_p = \sum_{i=1}^{n_{c_j}} \sum_{t=1}^{H_l} \mu_i \chi_{i,j}(t), \tag{8.28}$$

where $\mu_i > 0$ for $\forall i \in \left[1, n_{c_j}\right] \subset \mathbb{Z}_+$. Considering the pump operations in an order, the lexicographic prioritization is used to set this sequence of weights as $\mu_1 > \cdots > \mu_{n_{c_j}}$.

On the other hand, during total horizon of the lower layer, the used parallel pumps are switched once in order to get smooth control actions. The required pumps are used at the beginning and then turned off when the optimal set-point is satisfied. Therefore, this objective can be realized by

$$\chi_{i,j}(t+1) - \chi_{i,j}(t) \leq 0, \quad \forall t \in [1, n_s] \subset \mathbb{Z}_+, \tag{8.29}$$

which means that the required parallel pumps can be switched from ON-status to OFF-status only once.

The pump scheduling approach in the lower layer can be implemented by solving the following optimization problem

$$\underset{\chi_{i,j}(t)}{\text{minimize}} \ \phi_v \left\| V_j^p - V_j^* \right\|_2^2 - \phi_p \sum_{i=1}^{n_{c_j}} \sum_{t=1}^{H_l} \mu_i \chi_{i,j}(t), \tag{8.30a}$$

subject to

$$V_j^p = \sum_{i=1}^{n_{c_j}} \sum_{t=1}^{H_l} q_{i,j}(t) \Delta t_l, \tag{8.30b}$$

$$V_j^* = Q_j^* \Delta t_u, \tag{8.30c}$$

$$q_{i,j}(t) = \chi_{i,j}(t) q_{i,j}^r, \tag{8.30d}$$

$$\chi_{i+1,j}(t) + \left(1 - \chi_{i,j}(t)\right) \leq 1, \tag{8.30e}$$

$$\chi_{i,j}(t+1) - \chi_{i,j}(t) \leq 0, \tag{8.30f}$$

$$\chi_{i,j}(t) = \{0, 1\}, \tag{8.30g}$$

where the weight ϕ_v and ϕ_p are prioritization weights, where ϕ_v should be chosen to be much bigger than ϕ_p because the main objective is to reach the optimal flow set-point from the upper layer.

By solving the optimization problem (8.30) for each pumping station, the pump scheduling $\chi_{i,j}^*(t)$ for $\forall i \in \left[1, n_{c_j}\right] \subset \mathbb{Z}_+$, $\forall j \in [1, n_s] \subset \mathbb{Z}_+$, $\forall t \in [1, n_s] \subset \mathbb{Z}_+$ can be obtained for the lower layer.

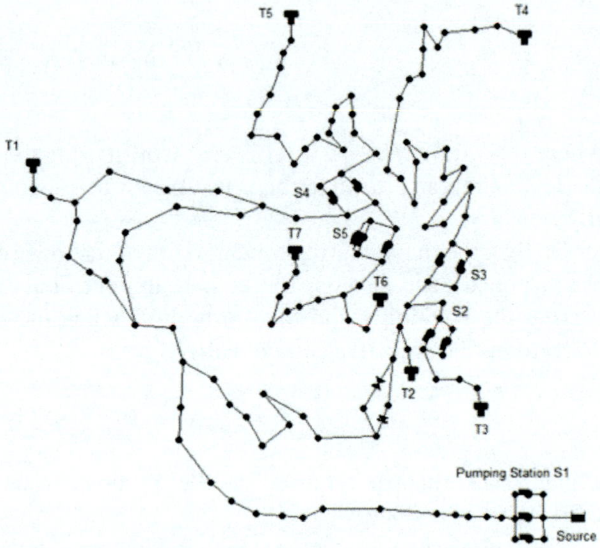

Fig. 8.1 Aggregate topology of the D-Town WDN

8.1.4 Application: The D-Town WDN

8.1.4.1 Description

The benchmark of D-Town network contains 388 water demand sectors, 405 links (pipes), 7 tanks, which contains multiple unidirectional and bidirectional links. The required pressure for all the water demand sectors is selected to be equal to 20 m. The unidirectional pipes and water demand sectors inside can be aggregated into its root node. Therefore, the aggregate topology of the D-Town water network is shown in Fig. 8.1. The required water demands in the root nodes are modified by aggregating the demands from a branch of unidirectional pipes and nodes while the required head of a branch is equivalent to the maximum head in this branch taking the head-loss through the pipes into account as well.

The required hydraulic head at each demand node is time-varying during one day since the head-loss through the pipes has been taken into account and the head depends on the water flow. For the control objectives associated to management of this case study, the prioritization is determined considering the economic objective is the most important and then the safety objective is more significant than the smoothness objective.

The online simulation has been carried out in a PC with the CPU of Intel (R) Core (TM) i7-5500U 2.4GHz, the memory of 12GB and MATLAB R2014a. The NEMPC strategy is implemented by means of the GAMS[1] and the CONOPT3 nonlinear solver,

[1]**General Algebraic Modelling System** (GAMS) is a programming language for mathematical optimization and able to solve the complex, large-scale and nonlinear optimization problems [12].

Fig. 8.2 Online simulation platform

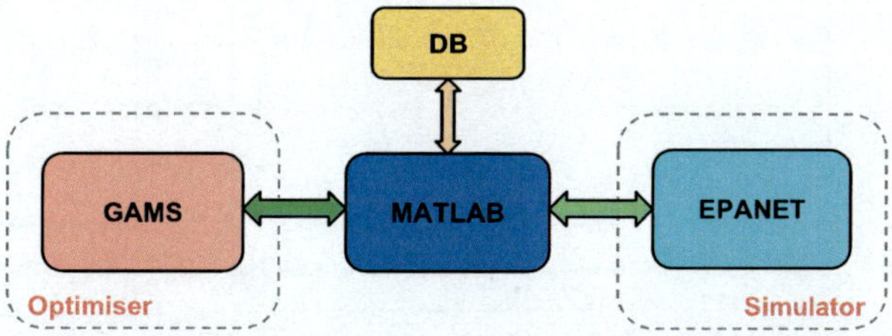

the EPANET[2] hydraulic simulator and MATLAB that is used for the communication between the GAMS model of the NEMPC controller and the EPANET hydraulic simulator. Besides, the proposed pump scheduling approach is also implemented in the MATLAB environment. The mixed-integer optimization problem of the pump scheduling approach is solved by using the MOSEK solver [14]. The topological graph of the communication is shown in Fig. 8.2. The database includes the water demands data and electrical tariff data.

8.1.4.2 Simulation Results

The optimization problem (8.21) is solved using GAMS programming while the optimization problem (8.30) is solved using linear programming. The MPC prediction horizon is chosen as 24 with the sampling time of 1 h in the upper layer. In the lower layer, the computational horizon of the optimization problem is chosen as 60 with the sampling time of 1 min. In the upper layer, the prioritization weights for economics, safety and smoothness objectives are chosen as 10, 1 and 0.1, respectively. In the lower layer, the prioritization weights ϕ_v and ϕ_p are chosen to be 10 : 1. For the pumping station having three parallel pumps, the weights of μ_1, μ_2 and μ_3 are chosen as 1.5, 1 and 0.5. For the pumping station having two parallel pumps, the weights of μ_1 and μ_2 are chosen as 1 and 0.5. The tolerance of the nonlinear solver is set as 10^{-4}.

The average single-step computation time of solving the upper layer nonlinear optimization problem is 53.21 seconds, being considerably smaller than the sampling time of 1 hour used in this layer. On the other hand, the average single-step computation time of solving the lower layer mixed-integer optimization problem is 4.19 seconds, being also smaller than the sampling time of 1 min used in this layer. Thus, the proposed strategy can be applied in real-time.

Figure 8.3 shows the head evolution of selected storage tanks. The dash blue line denotes the optimal hydraulic heads of the storage tanks. It is obvious that the head has daily quasi-periodic feature mainly because of the daily water demands and electricity tariffs. Moreover, results from the EPANET hydraulic simulator are

[2]**EPANET** software is a hydraulic simulator used for the hydraulic behavior analysis of a WDN. The WDN is built in EPANET consisting of water storage tanks/reservoirs, pumps, valves, pipes and nodes [13].

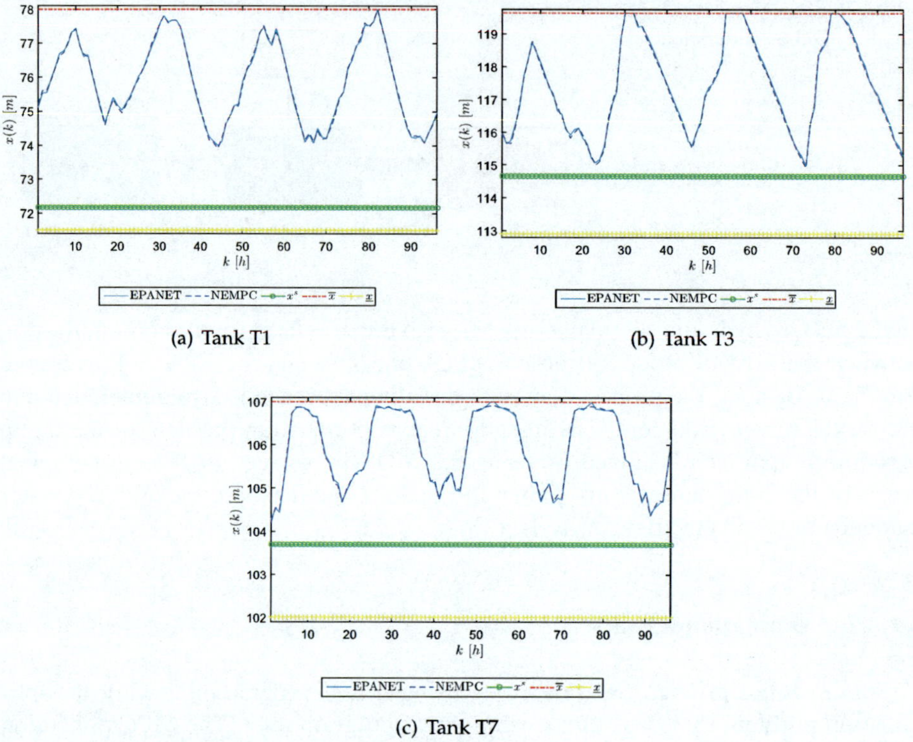

Fig. 8.3 Results of the head evolution of storage tanks

plotted in the cyan lines. By means of this state comparison between the EPANET simulator and optimizer, it is clear that the optimal system trajectories can be reached with the two-layer control strategy.

The selected average hourly water flows of the pumping stations are shown in the Fig. 8.4 in the magenta lines. The average hourly water flow of a pumping station can be calculated by

$$\bar{Q}_j = \frac{\sum_{i=1}^{N_q} \bar{q}_{i,j}}{60}. \tag{8.31}$$

The water flow in Fig. 8.4a is associated to the pumping station S1 and the average water flow is approximately similar and sometimes below the optimal flow set-point because the actual pumping flow is varying during each control interval depending on the boundary heads. Furthermore, the patterns of electrical tariff are added for reference in all the plots in Fig. 8.4. In general, the optimal flows are small when the electricity price is expensive. On the contrary, the flows are increasing when the electricity price becomes cheaper.

Figure 8.5 shows the optimal flow set-point and actual flow evolution of the valve in D-Town water network. The type of the valve is flow-controlled. The simulation flow is approximately tracking the optimal flow set-point. Hence, there is single-layer

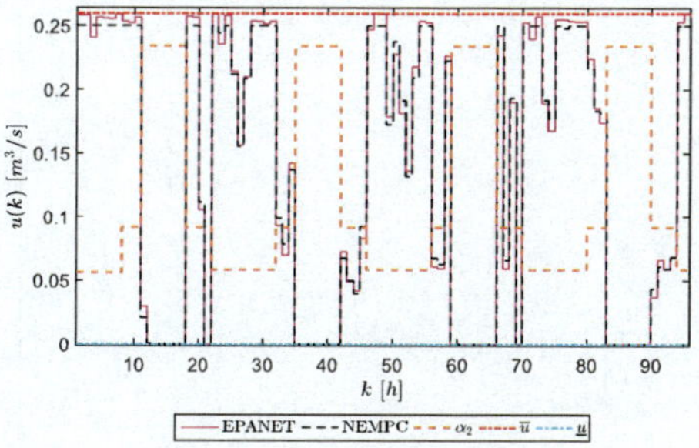

(a) Pumping station S1

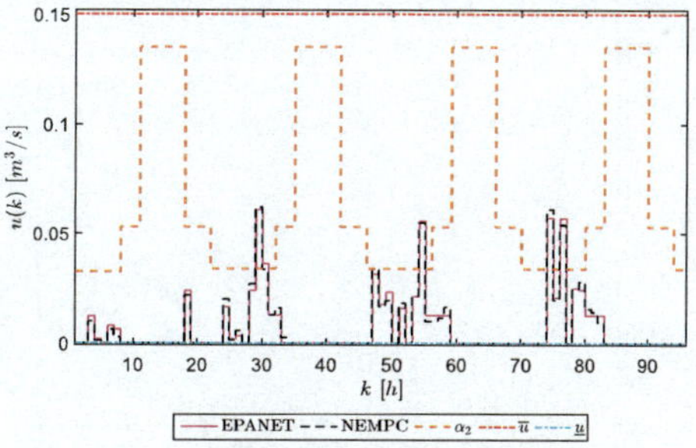

(b) Pumping station S2

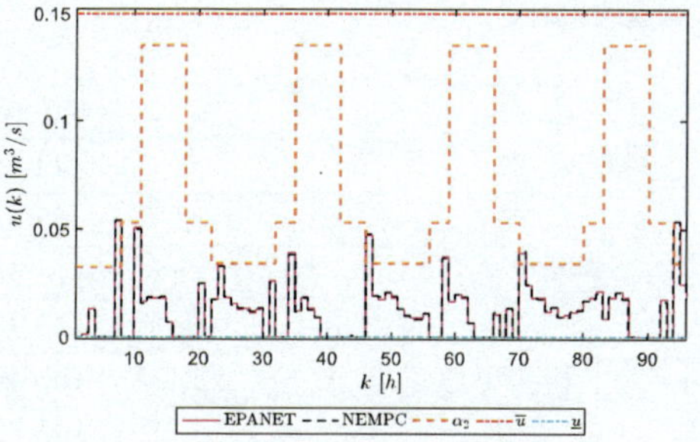

(c) Pumping station S4

Fig. 8.4 Results of the flows through pumping stations

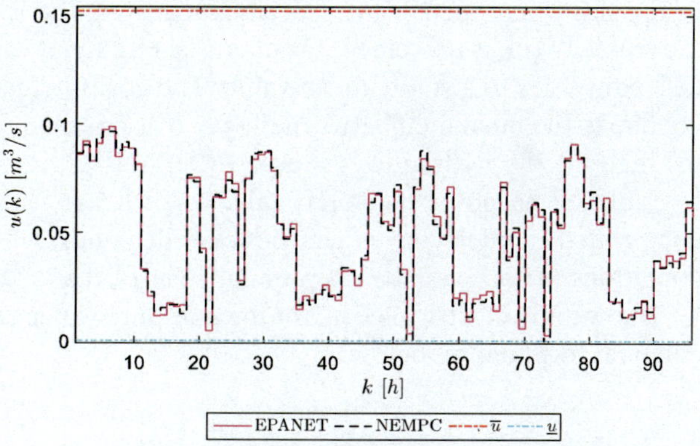

Fig. 8.5 Result of the flow through the valve V2

control for the valve and the optimal flow set-point is applied to the valve during one hour at an MPC step from the upper layer. But from this plot, there are small differences between the actual flow and optimal flow set-point.

Figure 8.6 presents the economic cost achieved by the EMPC controller at each sampling time (EMPC cost). It can be observed, that after a transient, the cost converges to a stable small cost. These results are confirmed with the results presented

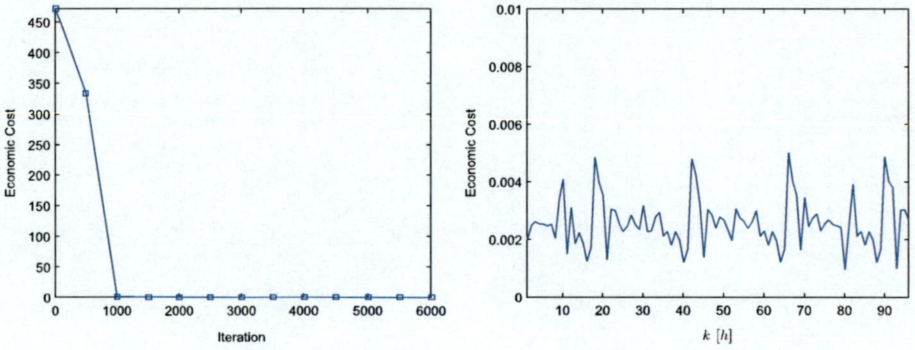

(a) EMPC cost evaluated at each iteration (b) EMPC cost with EPANET as virtual reality

Fig. 8.6 Economic costs with NEMPC

Table 8.2 Safety heads of storage tanks

Tank	Safety head (m)
T1	72.25
T2	65.42
T3	114.38
T4	133.90
T5	106.41
T6	102.83
T7	102.97

in Fig. 8.6 where the EMPC cost obtained using the EPANET simulator to emulate the real network is presented. From this last figure, it can be observed that the EMPC cost converges to a stable mean value. The cost fluctuations around this mean value are due to the mismatch between the control-oriented model used by the EMPC and the high fidelity hydraulic cost used by EPANET.

Table 8.2 proposes the safety tank water heads used in the online simulation to cope with the underlying stochastic water demands, which is considered as the initial conditions to compute the operational costs of the WDN.

The weekly electrical costs for the pumping water can be calculated by the mathematical formulation below:

$$\kappa^w := \sum_{i=1}^{168} \sum_{j=1}^{\Lambda} \frac{\rho_w g \Delta H_j^{ave}(i) \varphi_e(i) Q_j^{ave}(i)}{\eta(i)}, \tag{8.32}$$

where κ^w denotes the average weekly electricity costs for the utilization of the total pumping stations. ΔH^{ave} denotes mean head supplied by the pump and Q^{ave} represents the produced flows by the pump.

The annual electrical cost can be calculated by

$$\kappa^a := \frac{52 \times \kappa^w}{\varrho}, \tag{8.33}$$

where ϱ is the peak-day factor of 1.3 because the variability of the electric tariff, of the demand, and of any other design variable, during the year and the lifetime of this case study is not considered.

As a result, the annual operational cost of the D-Town water network is approximated to be 104, 482€. Compared to previous similar results for the D-Town network 117, 740€ by using the successive linear programming in [15] and approximately 168, 118€ by using the pseudo-genetic algorithm in [16], the two-layer NEMPC strategy is able to bring less operational costs for the management of the D-Town water network.

8.2 EMPC with Nonlinear Constraint Relaxation for WDNs

The periodicity in WDNs can be observed as in Fig. 8.3. In order to implement the EMPC controller proposed in Sect. 7.1, a linear descriptor model of WDNs is required. We now propose an iterative algorithm to relax the nonlinear algebraic equation (8.10c) by linear inequality constraints. For the relaxation of nonlinear constraints (8.10c), two cases are considered as follows.

Let us denote v_i, $i = 1, \ldots, n_v$ as the water flow for the pipe i and Δh_i, $i = 1, \ldots, n_v$ as the i-th row of $P_x x + P_z z$. According to [13], $\psi(v_i) = \alpha_i v_i |v_i|^{\beta-1}$. Therefore, the head-flow relation in the nonlinear algebraic equation (8.10c) can be explicitly written as

$$0 = \alpha_i v_i |v_i|^{\beta-1} + \Delta h_i, \tag{8.34}$$

where $\alpha_i \in \mathbb{R}_+$ is the pipe resistance coefficient for the i-th nonlinear equation due to friction with the pipe, and β is flow exponent that depends on the particular approximation, such as in Hazen-Williams, Darcy-Weisbach and Chezy-Manning formulas but, in all cases $\beta > 1$ according to ([13], Table 3.1).

The interconnected pipes in WDNs may be unidirectional or bidirectional. For the unidirectional pipe with a chosen positive direction, (8.34) becomes

$$0 = \alpha_i v_i^{\beta} + \Delta h_i, \tag{8.35}$$

with

$$0 \leq v_i \leq \overline{v}_i, \tag{8.36}$$

where $\overline{v}_i$ denotes the upper bound of the i-th flow.

The goal of dealing with these nonlinear algebraic equations in (8.34) is to relax them obtaining a set of linear inequality constraints using an iterative over-bounding algorithm. Note that finding these linear constraints with a proper constraint relaxation method is different than the traditional linearization method with a chosen operating point.

8.2.1 Nonlinear Constraint Relaxation for Unidirectional Pipes

We first discuss the relaxation for unidirectional pipes. By choosing a positive direction of the flow v_i, the nonlinear term $v_i\,|v_i|^{\beta-1}$ becomes $v_i\,|v_i|^{\beta-1} = v_i^{\beta}$ with $v_i \geq 0$. As shown in Fig. 8.7, the objective is to find a set of upper and lower linear bounds for over-bounding this term (shown in blue solid line).

The nonlinear algebraic equation (8.35) is equivalent to the satisfaction of the following inequalities

$$\alpha_i v_i^{\beta} + \Delta h_i \geq 0, \tag{8.37a}$$

$$\alpha_i v_i^{\beta} + \Delta h_i \leq 0, \tag{8.37b}$$

in which v_i^{β} is a convex function due to $\beta > 1$. From $0 \leq v_i \leq \overline{v}_i$, we have that $v_i^{\beta} \leq \overline{v}_i^{\beta-1} v_i$. Therefore, the inequality (8.37a) can be relaxed considering (8.36) as

$$\alpha_i \overline{v}_i^{\beta-1} v_i + \Delta h_i \geq 0. \tag{8.38}$$

On the other hand, from the convex nature of v_i^{β}, we have that every linearization constitutes a lower bound (dashed dotted lines in Fig. 8.7). The constraint (8.37b) can be approximated by considering N_a sampled operating points $v_{i,j}^{\star}$ for $j = 1, 2, \ldots, N_a$ as

$$0 \geq \alpha_i v_i^{\beta} + \Delta h_i \geq \alpha_i (a_j v_i + b_j) + \Delta h_i, \tag{8.39}$$

in which parameters a_j and b_j are given by

$$a_j = \beta {v_{i,j}^{\star}}^{\beta-1},$$

$$b_j = (1-\beta) {v_{i,j}^{\star}}^{\beta}.$$

In general, for a unidirectional pipe, the nonlinear algebraic equation (8.35) can be relaxed by using $N_a + 1$ inequality constraints as presented in (8.38) and (8.39). Figure 8.7 shows graphically the obtained relaxation. As a potential improvement, this relaxation can be refined iteratively. The iterative algorithm of nonlinear constraint relaxation can be improved by adding a penalty term in order to refine the region of v_i. As shown in Fig. 8.8, the upper bound can be moved by a scalar $\tau_i > 0$. Therefore, the region of v_i is refined to be $\left[v_i^a, v_i^b\right] \subseteq [0, \overline{v}_i]$.

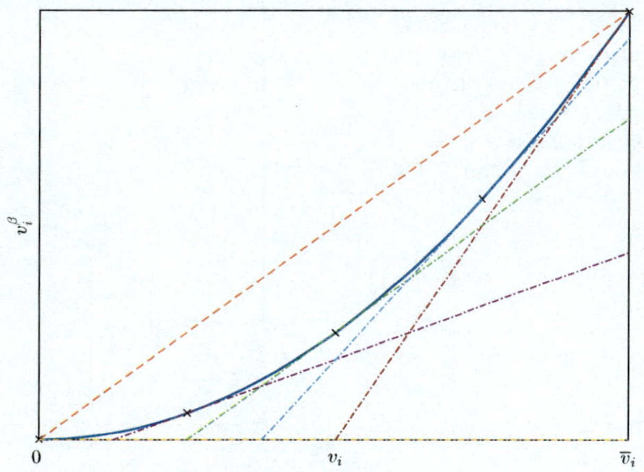

Fig. 8.7 Relaxation for v_i^β in unidirectional pipes: original function v_i^β is plotted in blue bold line, its upper bound is in dashed line and its lower bounds are in dashed dotted lines

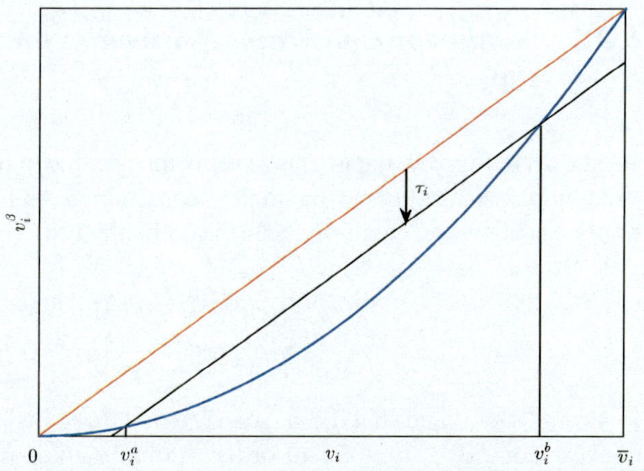

Fig. 8.8 Improving nonlinear constraint relaxation for unidirectional pipes

Considering a slack decision variable τ_i, (8.38) can be replaced by

$$\alpha_i \overline{v}_i^{\beta-1} v_i + \Delta h_i - \tau_i \geq 0, \tag{8.40a}$$

$$\tau_i \geq 0, \tag{8.40b}$$

where a small positive τ_i can be found in the MPC optimization loop. Hence, the cost function for the scalar $\tau_i(j)$ varying in the MPC prediction horizon $H_p = T$ can be penalized as

$$\ell^e(\tau_i(j)) := \lambda^e(j)\tau_i(j),$$

where $\lambda^e(j)$, $j = 1, 2, \ldots, T$ is a weight that can be set as a forgetting (monotonically decreasing) factor along the MPC prediction horizon T.

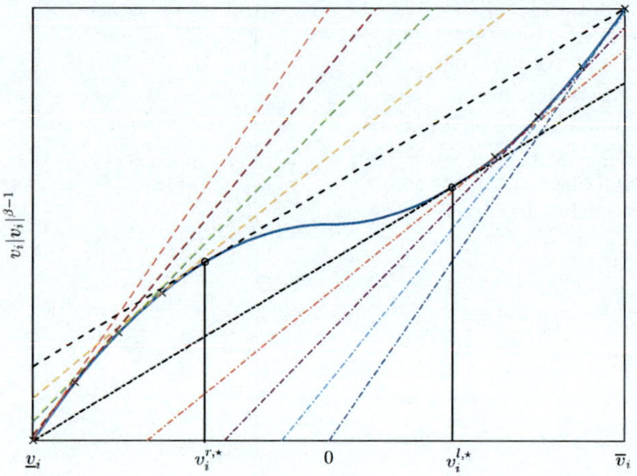

Fig. 8.9 Relaxation for $v_i\,|v_i|^{\beta-1}$ in bidirectional pipes: original constraint is plotted in blue bold line, upper bounds are shown in dashed line and lower bounds are shown in dashed dotted lines

8.2.2 *Nonlinear Constraint Relaxation for Bidirectional Pipes*

As shown in Fig. 8.9, the goal is to relax the nonlinear algebraic equation for bidirectional pipes also by linear inequality constraints. As in the unidirectional case, the nonlinear algebraic equation (8.34) is equivalent to

$$\alpha_i v_i\,|v_i|^{\beta-1} + \Delta h_i \le 0, \tag{8.41a}$$

$$\alpha_i v_i\,|v_i|^{\beta-1} + \Delta h_i \ge 0. \tag{8.41b}$$

From (8.41a) and (8.41b), we can see that these two inequality constraints are not convex along $\underline{v}_i \le v_i \le \overline{v}_i$. In order to obtain a convex relaxation for (8.41a), we consider lower bounds for $|v_i|^{\beta-1}$ with $\underline{v}_i \le v_i \le \overline{v}_i$ in the following form

$$a^l_j v_i + b^l_j \le v_i\,|v_i|^{\beta-1}\,,\ j = 1, \ldots, N_b + 1,$$

where a^l_j and b^l_j for $j = 1, \ldots, N_b + 1$ are two scalars. With a given a_j, the condition for the parameter b_j should be satisfied

$$b^l_j \le \min_{\underline{v}_i \le v_i \le \overline{v}_i} (v_i\,|v_i|^{\beta-1} - a^l_j v_i), \tag{8.42}$$

and let us consider the right side of (8.42)

$$M := \min_{\underline{v}_i \le v_i \le \overline{v}_i} (v_i\,|v_i|^{\beta-1} - a^l_j v_i) = \min\left\{M_+, M_-\right\},$$

where

$$M_+ = \min_{0 \leq v_i \leq \overline{v}_i} (v_i^\beta - a_j^l v_i),$$

$$M_- = \min_{\underline{v}_i \leq v_i < 0} (v_i(-v_i)^{\beta-1} - a_j^l v_i).$$

We now summarize the way to find a_j^l and b_j^l for $j = 1, \ldots, N_b + 1$. The minimum a_1^l along $\underline{v}_i \leq v_i \leq \overline{v}_i$ can be determined by

$$a_1^l = \beta \left(v_i^{l,\star}\right)^{\beta-1},$$

where $v_i^{l,\star}$ can be obtained by satisfying the following condition

$$\beta \left(v_i^{l,\star}\right)^{\beta-1} = \frac{\left(v_i^{l,\star}\right)^\beta + \underline{v}_i^\beta}{v_i^{l,\star} - \underline{v}_i}.$$

Denote $v_{i,1}^l = v_i^{l,\star}$. By choosing N_b values of $v_{i,j}^l$ in the interval $v_i^{l,\star} \leq v_{i,j}^l \leq \overline{v}_i$, we obtain $a_j^l = \beta \left(v_{i,j}^l\right)^{\beta-1}$ for $j = 2, \ldots, N_b + 1$. Therefore, the parameter b_j^l can be obtained by

$$b_j^l = \left(v_{i,j}^l\right)^\beta - a_j^l v_{i,j}^l, \ \ j = 1, \ldots, N_b + 1.$$

Furthermore, the upper and lower bounds are symmetric as shown in Fig. 8.9. Therefore, we can find the upper bounds in a similar way. Let us consider upper bounds of (8.41b) in the following form

$$a_j^r v_i + b_j^r \geq v_i \left|v_i\right|^{\beta-1}, \ \ j = 1, \ldots, N_b + 1.$$

Because of symmetry, a_1^r can be determined by

$$a_1^r = \beta \left(v_i^{r,\star}\right)^{\beta-1},$$

where $v_i^{r,\star}$ can be obtained by satisfying the following condition

$$\beta \left(v_i^{r,\star}\right)^{\beta-1} = \frac{\overline{v}_i^\beta + \left(v_i^{r,\star}\right)^\beta}{\overline{v}_i - v_i^{r,\star}}.$$

Denote $v_{i,1}^r = v_i^{r,\star}$. Similarly, by choosing N_b values of $v_{i,j}^r$ in the interval $\underline{v}_i \leq v_i \leq v_i^{r,\star}$, we obtain $a_j^r = \beta \left(v_{i,j}^r\right)^{\beta-1}$ for $j = 2, \ldots, N_b + 1$. Therefore, the parameter b_j^r can be computed as

$$b_j^r = v_{i,j}^r(-v_{i,j}^r)^{\beta-1} + a_j^r v_{i,j}^r, \ \ j = 1, \ldots, N_b + 1.$$

As a result, (8.34) for bidirectional pipes can be relaxed as $2N_b + 2$ linear inequalities. From (8.41a) and (8.41b), we obtain the relaxed linear inequality constraints

$$0 \geq \alpha_i v_i |v_i|^{\beta-1} + \Delta h_i \geq \alpha_i \left(a^l_j v_i + b^l_j\right) + \Delta h_i, \tag{8.43a}$$
$$0 \leq \alpha_i v_i |v_i|^{\beta-1} + \Delta h_i \leq \alpha_i \left(a^r_j v_i + b^r_j\right) + \Delta h_i, \tag{8.43b}$$

both for $j = 1, \ldots, N_b + 1$.

8.2.3 EMPC with Nonlinear Constraint Relaxation for WDNs

From the above results, we can obtain the relaxed linear inequality constraints in the MPC prediction horizon $H_p = T$ as follows:

$$\widetilde{P}^l_x(j)x(k+j) + \widetilde{P}^l_z(j)z(k+j) + \widetilde{P}^l_v(j)v(k+j) + \widetilde{P}^l_b(j) \leq 0, \tag{8.44a}$$
$$\widetilde{P}^r_x(j)x(k+j) + \widetilde{P}^r_z(j)z(k+j) + \widetilde{P}^r_v(j)v(k+j) + \widetilde{P}^r_b(j) \geq 0, \tag{8.44b}$$

for $j = 1, \ldots, T$. Taking into account the proposed iterative algorithm for the nonlinear constraint relaxation, the nonlinear algebraic constraint (8.10c) can be replaced by (8.44a) and (8.44b) along the MPC prediction horizon. We now formulate the optimization problem for implementing the economic MPC with nonlinear constraint relaxation as follows

$$\underset{\substack{x(0),\ldots,x(T)\\ u(0),\ldots,u(T-1)}}{\text{minimize}} \quad J_T(x, u, p) = \sum_{i=0}^{T-1} \sum_{j=1}^{\Gamma} \lambda_j \ell_j(i), \tag{8.45a}$$

subject to

$$x(i+1) = Ax(i) + B_u u(i) + B_v v(i) + B_d d(i), \tag{8.45b}$$
$$0 = E_u u(i) + E_v v(i) + E_d d(i), \tag{8.45c}$$
$$\widetilde{P}^l_x(i)x(i) + \widetilde{P}^l_z(i)z(i) + \widetilde{P}^l_v(i)v(i) + \widetilde{P}^l_b(i) \leq 0, \tag{8.45d}$$
$$\widetilde{P}^r_x(i)x(i) + \widetilde{P}^r_z(i)z(i) + \widetilde{P}^r_v(i)v(i) + \widetilde{P}^r_b(i) \geq 0, \tag{8.45e}$$
$$\underline{x} \leq x(i) \leq \overline{x}, \tag{8.45f}$$
$$\underline{u} \leq u(i) \leq \overline{u}, \tag{8.45g}$$
$$x(0) = x(T), \tag{8.45h}$$
$$x(j) = x(k), \quad j = \bmod(k, T). \tag{8.45i}$$

Let us denote the optimal solution of the optimization problem (8.45) as $u^*(j)$. Based on the receding horizon strategy, the optimal control action $u(k)$ at time step k is chosen as

$$u(k) = u^*(j), \; j = \text{mod}\,(k, T). \tag{8.46}$$

8.2.4 Application: The Richmond WDN

8.2.4.1 Description

The topology and layout of the Richmond WDN is shown in Fig. 8.10. The Richmond WDN has 6 water storage tanks, 7 booster pumps and 11 water demand sectors. Besides, there are 41 non-storage nodes and 41 pressurized pipes connected in this network. The demand pattern is also given for a 24-h period, that is $T = 24$. We use the *Chezy-Manning* head-flow formula to obtain (8.10c) as follows [13]:

$$z_i - z_j = R_{i,j} v_{i,j} \left| v_{i,j} \right|, \tag{8.47}$$

where z_i and z_j correspond to the hydraulic heads at any two adjacent nodes, and $v_{i,j}$ is the corresponding water flow. The parameter $R_{i,j}$ in the *Chezy-Manny* formula is given by

$$R_{i,j} = \frac{10.29 L_{i,j} C_{i,j}^2}{D_{i,j}^{5.33}}, \tag{8.48}$$

where $L_{i,j}$, $D_{i,j}$ and $C_{i,j}$ are the length, diameter and roughness coefficient of the corresponding pipe, respectively. $L_{i,j}$ and $D_{i,j}$ are given in the EPANET model of the Richmond WDN.

As shown in Fig. 8.10, this WDN has two bidirectional pipes in (8.34) and 39 unidirectional pipes in (8.35). In addition to the economic cost function defined in (8.45a), for the relaxed linear constraints (8.44a) and (8.44b) with the setting in (8.40), a penalty term $\lambda^e(j)$ is set to be a forgetting factor as

$$\lambda^e(j) = \max\left\{\lambda^e(j-1) - \epsilon, 0\right\}, \tag{8.49a}$$
$$\lambda^e(0) = \lambda^e, \tag{8.49b}$$

where ϵ denotes the relaxed step and λ^e is the initial value of this weight.

Therefore, the total MPC cost function is set to be

$$\bar{J}_T(x, u, p) = \sum_{i=0}^{T-1} \left(\lambda_1 \ell_1(i) + \lambda_2 \ell_2(i) + \lambda^e(i)\tau(i)\right), \tag{8.50}$$

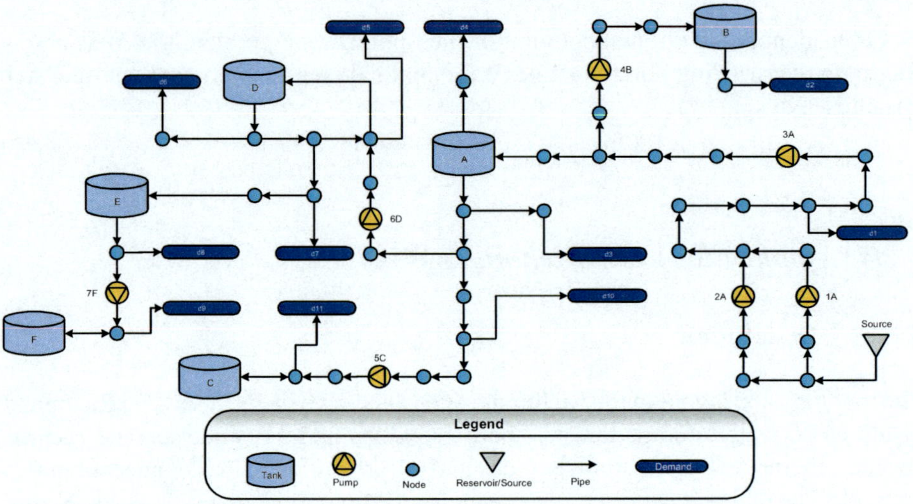

Fig. 8.10 Topology of the Richmond WDN

where τ denotes a slack variable for all the constraints in (8.40). In this simulation, we select the weights as $\lambda_1 = 10$, $\lambda_2 = 1$, $\lambda^e = 0.1$ and $\epsilon = 0.01$.

For the Richmond network, the period T is considered to be $T = 24$ (24 h) with the sampling time of an hour because of the periodicities of the water demand and electricity price considering the variations in the daily tariff. Hence, the prediction horizon of the proposed economic MPC strategy is also chosen to be $T = 24$. The minimal pressure at all the demand sectors is set to be 10 m. Furthermore, for the implementation of the proposed nonlinear constraint relaxation, we choose $N_a = N_b = 10$. Therefore, there are 11 relaxed constraints replacing (8.35) and 22 relaxed constraints replacing (8.34) for each pipe.

The simulations have been carried out with the MATLAB R2015a and the EPANET simulator [13] for seven days (168 h) in a PC of Intel i7-5500U CPU and 12GB RAM. The linear optimization problems are solved using the YALMIP toolbox [17] and the MOSEK solver [14]. The nonlinear optimization problems are solved using the nonlinear programming through the YALMIP toolbox and the IPOPT solver available in the OPTI toolbox [18]. The Richmond network is given in the EPANET simulator as the simulation model.

Recall the price signal $p = \alpha_1 + \alpha_2$. To compare the performance of the proposed economic MPC with the nonlinear economic MPC, we define the following key performance indicators (KPIs)

$$KPI_E := \frac{1}{n_s} \sum_{k=1}^{n_s} \left(p_j^\top u(k) \right), \; j = \mathrm{mod}(k, T), \tag{8.51a}$$

$$KPI_S := \frac{1}{n_s} \sum_{k=1}^{n_s} \sum_{i=1}^{n_x} \max \left\{ 0, \left(x_i^s - x_i(k) \right) \right\}, \tag{8.51b}$$

Table 8.3 Hydraulic heads at storage tanks to assess safety constraints

Tank	Elevation (m)	Volume (m)	Hydraulic Head x^s (m)
A	184.13	1.02	185.15
B	216	2.03	218.03
C	258.9	0.5	259.40
D	241.18	1.1	242.28
E	203.01	0.01	203.03
F	235.71	0.19	235.90

$$KPI_M := \frac{1}{n_s}\sum_{k=1}^{n_s}\sum_{i=1}^{n_x}\left(x_i(k) - x_i^s\right), \tag{8.51c}$$

where KPI_E is the economic KPI that measures the operational costs at each time step. KPI_S is the safety KPI that computes the average differences of the water storage that are lower than safety hydraulic head x_i^s given in Table 8.3. KPI_M is the measurement KPI that represents the additional water reserved in storage tanks. Based on the original benchmark available online, all the tanks are cylindrical and the relationship between water level and stored volume is considered to be linear and constant.

On the other hand, with the optimal solutions of the optimization problem (8.45), we would like to check whether all the nonlinear algebraic equations in (8.10c) are satisfied. To assess the relaxation algorithm for 40 nonlinear algebraic equations in the Richmond WDN, the error measurements for (8.10c) including MSE, mean absolute error (MAE) and symmetric mean absolute percent error (SMAPE) are introduced as follows:

$$MSE(k) := \frac{1}{n_e}\sum_{j=1}^{n_e}(P_x^j x(k) + P_z^j z(k) + \psi^j(v(k)))^2, \tag{8.52a}$$

$$MAE(k) := \frac{1}{n_e}\sum_{j=1}^{n_e}\left|P_x^j x(k) + P_z^j z(k) + \psi^j(v(k))\right|, \tag{8.52b}$$

$$SMAPE(k) := \frac{100}{n_e}\sum_{j=1}^{n_e}\frac{\left|P_x^j x(k) + P_z^j z(k) + \psi^j(v(k))\right|}{P_x^j x(k) + P_z^j z(k) + \psi^j(v(k))}, \tag{8.52c}$$

where P_x^j, P_z^j and $\psi^j(\cdot)$ denote the j-th row of P_x, P_z and $\psi(\cdot)$, respectively. n_e denotes the total number of nonlinear algebraic equations. In terms of MSE and MAE, they represent the violation of nonlinear algebraic equations. SMAPE is an indicator based on percentage errors.

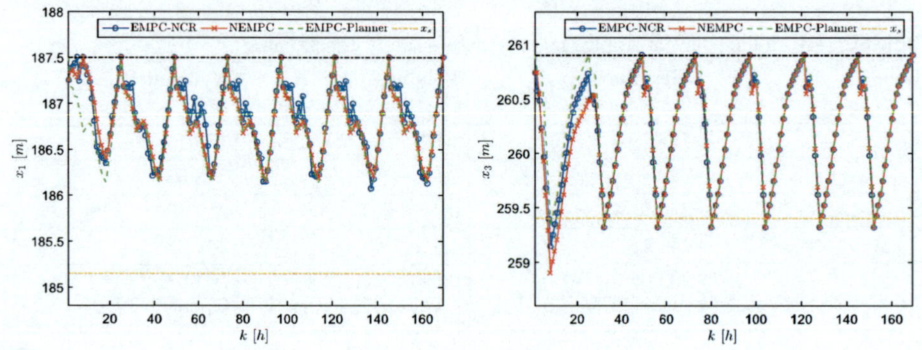

(a) Tank A (b) Tank C

Fig. 8.11 Results of system states using EMPC-NCR and NEMPC

8.2.4.2 Simulation Results

For the notation simplicity, we denote the simulation results of applying the proposed economic MPC with nonlinear constraint relaxation as EMPC-NCR, while the comparison results with the solutions of nonlinear planner in (7.4) and NEMPC controller in (7.5) both with updated nonlinear prediction model in (8.10) are denoted by EMPC-Planner and NEMPC, respectively. The closed-loop simulation results of system states and control inputs are shown in Figs. 8.11, 8.12, 8.13 and 8.14. In Figs. 8.11 and 8.12, the state trajectories obtained from applying the proposed EMPC-NCT are in solid lines with circles. Due to the convexity, the steady states can be obtained from the solution of the optimization problem (7.4) shown in dashed line. As a comparison, the state trajectories of NEMPC are also shown in solid lines with cross marks. From these results, we can see that the closed-loop trajectories obtained using the EMPC-NCR and NEMPC strategies are similar to those of the optimal planner trajectories (both states and control inputs). The NEMPC results are smoother and closer to the planner trajectories since a more accurate nonlinear model is used in the NEMPC optimization problem. Similarly, in terms of control inputs, as shown in Figs. 8.13 and 8.14, three trajectories of EMPC-NCR, NEMPC and EMPC-Planner are plotted. The input trajectories of EMPC-NCR are approaching the ones of EMPC-Planner.

To assess the performance of different control strategies, the comparison is also provided based on the defined KPIs. The computation results using the defined KPIs are shown in Table 8.4. In general, the performances of both MPC strategies are similar. Specifically, from the KPI_E results, the pure economic cost of EMPC-NCR is slightly cheaper than the one of NEMPC. According to KPI_S and KPI_M results, small differences between the reserved water in the storage tanks can be seen for both MPC strategies. This is because in the EMPC-NCR we use the pressure constraint on the variable z to guarantee the safety objective, which implies the water levels in the storage tanks should be greater than some certain values.

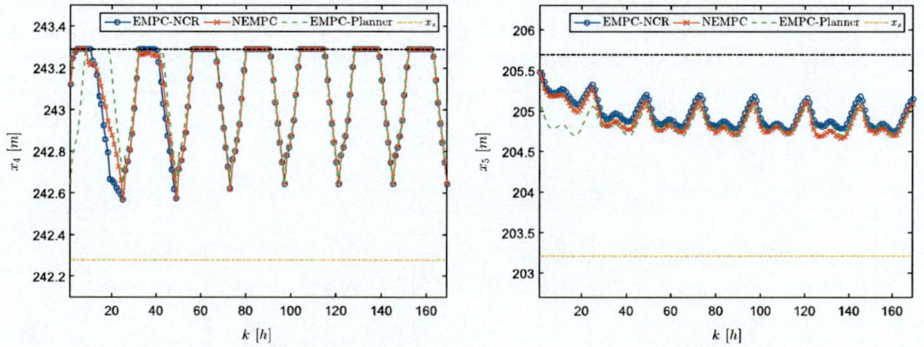

(a) Tank D (b) Tank E

Fig. 8.12 Results of system states using EMPC-NCR and NEMPC

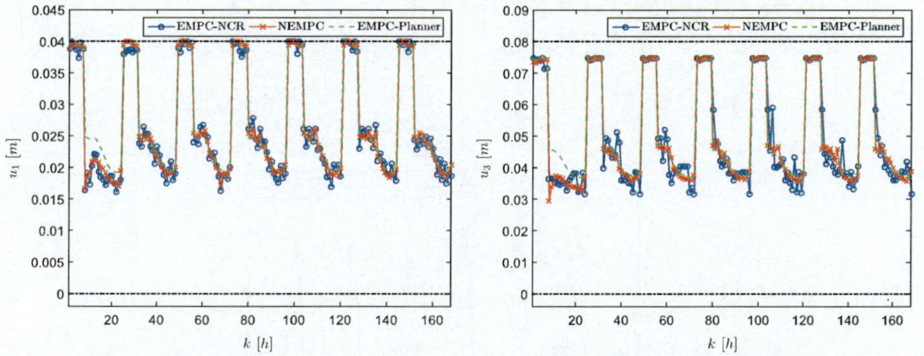

(a) Pump 1A (b) Pump 3A

Fig. 8.13 Results of control inputs (Pumps 1A and 3A) using EMPC-NCR and NEMPC

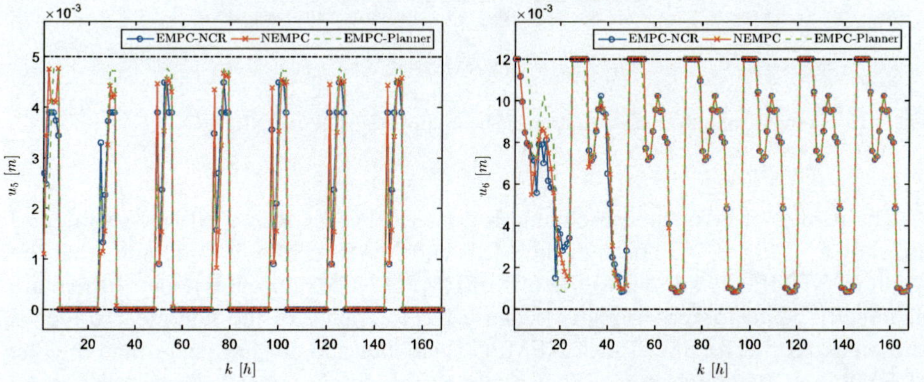

(a) Pump 5C (b) Pump 6D

Fig. 8.14 Results of control inputs (Pumps 5C and 6D) using EMPC-NCR and NEMPC

Table 8.4 KPI results using EMPC-NCR and NEMPC

MPC strategy	KPI_E	KPI_S	KPI_M
EMPC-NCR	0.6992	0.2604	6.7078
NEMPC	0.7028	0.1914	6.5249

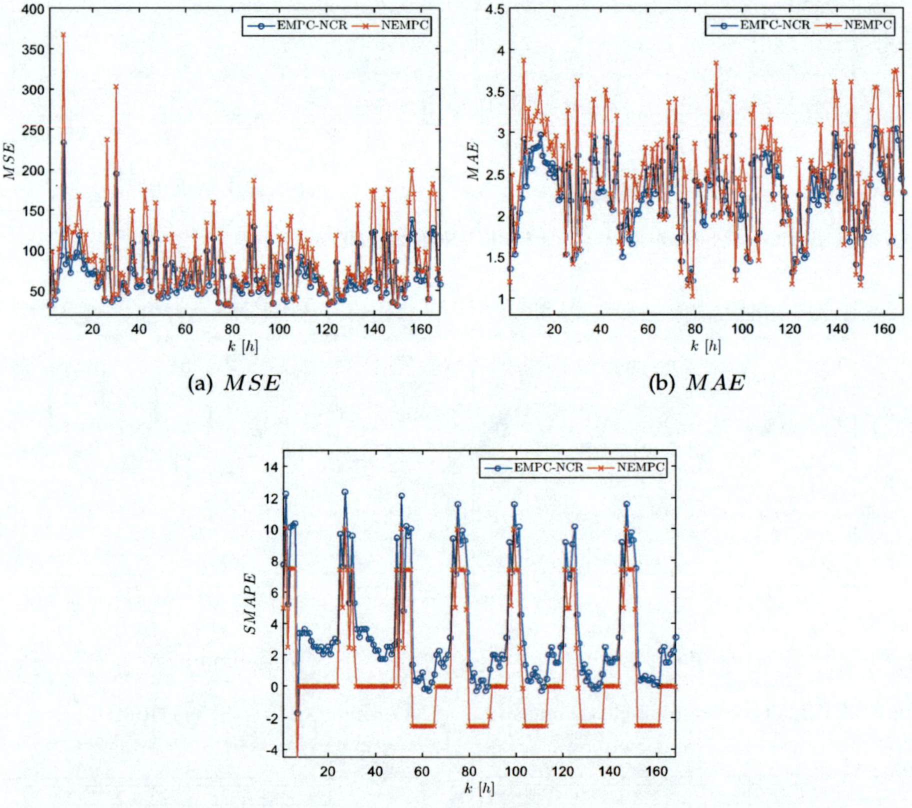

(a) MSE (b) MAE

(c) $SMAPE$

Fig. 8.15 Comparison of error measurements using EMPC-NCR and NEMPC

The results of error measurements for the EMPC-NCR and NEMPC strategies are shown in Fig. 8.15. Through the MSE and MAE results, it is obvious that the result of NEMPC is similar to the one of EMPC-NCR, although none of them are identically equal to zero. This is because the tolerance of the nonlinear solver is chosen as 10^{-5}. The SMAPE of NEMPC is smaller and closer to zero than the one of EMPC-NCR, which means that nonlinear algebraic equations are satisfied by NEMPC better than EMPC-NCR since the nonlinear programming technique is able to solve hard constraints. However, the EMPC-NCR strategy is able to produce a similar performance according to three error measurement results.

For the comparison of simulation time in a scenario of 168 h, it takes 62.86 min for NEMPC while 1.43 min for EMPC-NCR. Hence, the EMPC-NCR strategy has a significant improvement in the reduction of computational load and meanwhile based on the above comparison result, the performance of the EMPC-NCR strategy is similar to the NEMPC strategy. This reduction in the computation time would be more relevant in larger networks.

8.3 REMPC of SGs

In the following, let us consider discrete-time descriptor system affected by additive disturbances as

$$x(k+1) = Ax(k) + Bu(k) + B_d d(k) + B_w w(k), \tag{8.53a}$$

$$0 = E_x x(k) + E_u u(k) + E_d d(k) + E_w w(k), \tag{8.53b}$$

where $x \in \mathbb{R}^{n_x}$, $u \in \mathbb{R}^{n_u}$, $d \in \mathbb{R}^{n_d}$, $w \in \mathbb{R}^{n_w}$ denote the vectors of system states, control inputs, endogenous demands and unknown disturbances, respectively. Besides, $A \in \mathbb{R}^{n_x \times n_x}$, $B \in \mathbb{R}^{n_x \times n_u}$, $B_d \in \mathbb{R}^{n_x \times n_d}$, $B_w \in \mathbb{R}^{n_x \times n_w}$, $E_x \in \mathbb{R}^{n_r \times n_x}$, $E_u \in \mathbb{R}^{n_r \times n_u}$, $E_d \in \mathbb{R}^{n_r \times n_d}$ and $E_w \in \mathbb{R}^{n_r \times n_w}$ are system matrices.

Assumption 8.1 The disturbance vector $w \in \mathbb{R}^{n_w}$ is assumed to be unknown but bounded in the set $\mathcal{W}$.

We now extend REMPC proposed in Sect. 7.2 for the descriptor system (8.53) [6]. The goal is to use a robust tube-based technique based on nominal predictions that drives the closed-loop system trajectory to a neighborhood of an optimal periodic steady trajectory. A local controller is used to reduce the difference between nominal predictions and closed-loop trajectory. This local controller is designed to stabilize the nominal dynamical model of (8.53a) and simultaneously to satisfy the algebraic equation (8.53b).

8.3.1 Refined State and Input Constraints

Based on [19], an auxiliary input signal $v \in \mathbb{R}^{n_v}$ with $n_v = n_u - n_r$ is used. Therefore, the control input $u(k)$ is structured from the solution of (8.53) satisfying the algebraic equation (8.53b) for any $w(k)$ as

$$u(k) = M_x x(k) + M_d d(k) + M_w w(k) + M_v v(k), \tag{8.54}$$

where the matrices $M_x \in \mathbb{R}^{n_u \times n_x}$, $M_d \in \mathbb{R}^{n_u \times n_d}$, $M_w \in \mathbb{R}^{n_u \times n_w}$ and $M_v \in \mathbb{R}^{n_u \times n_v}$ should be designed.

By combining (8.53b) and (8.54), we have that the following condition holds:

$$E_x x(k) + E_d d(k) + E_w w(k) = -E_u M_x x(k) - E_u M_d d(k) \\ - E_u M_w w(k) - E_u M_v v(k).$$

which gives

$$E_x + E_u M_x = 0, \tag{8.55a}$$
$$E_d + E_u M_d = 0, \tag{8.55b}$$
$$E_w + E_u M_w = 0, \tag{8.55c}$$
$$E_u M_v = 0. \tag{8.55d}$$

From the condition (8.55), we can obtain matrices M_x, M_d and M_w. Note that there are infinite solutions of these matrices M_x, M_d and M_w. Specifically, M_v is the null space (kernel) of E_u, and M_x, M_d and M_w can be obtained in a generalized solution with pseudo-inverse matrices.

Besides, by combining (8.53a) and (8.54), we have

$$x(k+1) = \tilde{A}x(k) + \tilde{B}_v v(k) + \tilde{B}_d d(k) + \tilde{B}_w w(k), \tag{8.56}$$

where $\tilde{A} = A + BM_x$, $\tilde{B}_v = BM_v$, $\tilde{B}_d = B_d + BM_d$ and $\tilde{B}_w = B_w + BM_w$.

Let us define the nominal dynamical model of (8.56) as

$$\bar{x}(k+1) = \tilde{A}\bar{x}(k) + \tilde{B}_v \bar{v}(k) + \tilde{B}_d d(k), \tag{8.57}$$

where $\bar{x} \in \mathbb{R}^{n_x}$ and $\bar{v} \in \mathbb{R}^{n_v}$, and define the error between closed-loop states and state predictions as $e(k) = x(k) - \bar{x}(k)$. With (8.56) and (8.57), the error dynamics is given by

$$e(k+1) = \tilde{A}e(k) + \tilde{B}_v \left(v(k) - \bar{v}(k)\right) + \tilde{B}_w w(k). \tag{8.58}$$

To attenuate the effect of this error along the prediction horizon, a local control law $K \in \mathbb{R}^{n_v \times n_x}$ is proposed for $v(k) = Kx(k)$ and $\bar{v}(k) = K\bar{x}(k)$ such that the matrix $\tilde{A}_K = \tilde{A} + \tilde{B}_v K$ is Schur stable. Therefore, (8.58) is simplified as

$$e(k+1) = \tilde{A}_K e(k) + \tilde{B}_w w(k). \tag{8.59}$$

Since $w(k) \in \mathcal{W}$, by applying Lemma 6.1, we can obtain the set $\mathcal{Z}$ is RPI satisfying $\tilde{A}_K \mathcal{Z} \oplus \tilde{B}_w \mathcal{W} \subseteq \mathcal{Z}$. Therefore, for any $k \in \mathbb{N}$, $e(k) \in \mathcal{Z}$ is equivalent to $x(k) - \bar{x}(k) \in \mathcal{Z}$. Considering the state constraint set $x(k) \in \mathcal{X}$, we have

$$\bar{x}(k) \in \mathcal{X} \ominus \mathcal{Z}. \tag{8.60}$$

Based on (8.54), let us also denote the nominal input $\bar{u} \in \mathbb{R}^{n_u}$ as

$$\bar{u}(k) = M_x \bar{x}(k) + M_d d(k) + M_v \bar{v}(k). \tag{8.61}$$

By combining (8.54) and (8.61), we derive

$$\begin{aligned} u(k) - \bar{u}(k) &= M_x \left(x(k) - \bar{x}(k)\right) + M_w w(k) + M_v v(k) - M_v \bar{v}(k) \\ &= M_x e(k) + M_w w(k) + M_v K e(k). \end{aligned}$$

Since $u(k) \in \mathcal{U}$, $e(k) \in \mathcal{Z}$ and $w(k) \in \mathcal{W}$, we obtain

$$\bar{u}(k) \in \mathcal{U} \ominus M_x \mathcal{Z} \ominus M_w \mathcal{W} \ominus M_v K \mathcal{Z}. \tag{8.62}$$

As a result, the constraints on the nominal state and input vectors are refined as in (8.60) and (8.62), which will be used in the robust economic MPC design.

8.3.2 REMPC Planner for Descriptor Systems

Based on the optimization problem (7.4), the optimal periodic trajectory can be obtained solving the following open-loop optimization problem.

$$\underset{\substack{\bar{x}(0),\ldots,\bar{x}(T),\\ \bar{u}(0),\ldots,\bar{u}(T-1)}}{\text{minimize}} \; J_T\left(\bar{x}, \bar{u}, p\right) = \sum_{i=0}^{T-1} \ell\left(\bar{x}(i), \bar{u}(i), p_i\right), \tag{8.63a}$$

subject to

$$\bar{x}(i+1) = A\bar{x}(i) + B\bar{u}(i) + B_d d(i), \tag{8.63b}$$

$$0 = E_x \bar{x}(i) + E_u \bar{u}(i) + E_d d(i), \tag{8.63c}$$

$$\bar{x}(i) \in \mathcal{X} \ominus \mathcal{Z}, \tag{8.63d}$$

$$\bar{u}(i) \in \mathcal{U} \ominus M_x \mathcal{Z} \ominus M_w \mathcal{W} \ominus M_v K \mathcal{Z}, \tag{8.63e}$$

$$\bar{x}(0) = \bar{x}(T). \tag{8.63f}$$

The feasible solutions of the optimization problem (8.63) define the optimal periodic steady trajectory of system states and control inputs as

$$\bar{x}^p = \left\{\bar{x}(0), \ldots, \bar{x}(T)\right\}, \tag{8.64a}$$

$$\bar{u}^p = \left\{\bar{u}(0), \ldots, \bar{u}(T)\right\}. \tag{8.64b}$$

The closed-loop system with the proposed controller will be driven close to a neighborhood of this trajectory.

8.3.3 REMPC Controller for Descriptor Systems

In general, the REMPC controller is proposed by implementing the following optimization problem

$$\underset{\substack{\bar{x}(0),\ldots,\bar{x}(T),\\ \bar{u}(0),\ldots,\bar{u}(T-1)}}{\text{minimize}} \; J_T(\bar{x}, \bar{u}, p)\,, \tag{8.65a}$$

subject to

$$\bar{x}(i+1) = A\bar{x}(i) + B\bar{u}(i) + B_d d(i), \tag{8.65b}$$
$$0 = E_x\bar{x}(i) + E_u\bar{u}(i) + E_d d(i), \tag{8.65c}$$
$$\bar{x}(i) \in \mathcal{X} \ominus \mathcal{Z}, \tag{8.65d}$$
$$\bar{u}(i) \in \mathcal{U} \ominus M_x\mathcal{Z} \ominus M_w\mathcal{W} \ominus M_v K\mathcal{Z}, \tag{8.65e}$$
$$\bar{x}(0) = \bar{x}(T), \tag{8.65f}$$
$$x(k) - \bar{x}(j) \in \mathcal{Z}, \quad j = \text{mod}\,(k, T)\,. \tag{8.65g}$$

From the feasible solutions of the optimization problem (8.65), the control action at time k is chosen to be

$$u(k) = \bar{u}^*(j) + M_w w(k) + (M_x + M_v K)\,(x(k) - \bar{x}^*(j)), \tag{8.66}$$

with $j = \text{mod}\,(k, T)$. Note that in the closed-loop simulation, at time $k \in \mathbb{N}$, the current state $x(k)$ and the disturbance $w(k)$ are measurable to implement the control action $u(k)$ chosen in (8.66).

8.3.4 Application: The Smart Micro-Grid

We now apply the proposed robust control strategy into a smart micro-grid chosen from [19]. Periodic operation has been proved to be suitable for this system taking into account the potential periodicity of signals in the system.

8.3.4.1 Description

The micro-grid system includes three nano-grids placed in parallel. Each nano-grid consists of a cluster of batteries and a fuel cell. These batteries are used to compensate the voltage peaks from the fast system dynamics. Therefore, two system state variables are chosen as the state of charge of batteries and the storage level of hydrogen in the metal hydride tank. Besides, control inputs in each nano-grid are the power of exchange with the electric utility, the power of exchange with the hydrogen and the load power. The control-oriented model of this micro-grid is built by difference-algebraic equations in the form of (8.53) with the sampling time of 30 min, where system matrices are defined by

$$A = \begin{bmatrix} 1&0&0&0&0&0 \\ 0&1&0&0&0&0 \\ 0&0&1&0&0&0 \\ 0&0&0&1&0&0 \\ 0&0&0&0&1&0 \\ 0&0&0&0&0&1 \end{bmatrix}, \quad B_d = \begin{bmatrix} 5.5847 & 0 & 0 & 0 \\ 0 & 0 & 0 & 0 \\ 0 & 5.5847 & 0 & 0 \\ 0 & 0 & 0 & 0 \\ 0 & 0 & 5.5847 & 0 \\ 0 & 0 & 0 & 0 \end{bmatrix},$$

and $B = \text{diag}\,(B_n, B_n, B_n)$, $B_w = B_d$, with $B_n = \begin{bmatrix} 5.5847 & 5.5847 & -5.5847 \\ -3.4495 & 0 & 0 \end{bmatrix}$. And $E_x = 0$, $E_u = [0, 0, 1, 0, 0, 1, 0, 0, 1]$, $E_d = [0, 0, 0, -1]$, $E_w = E_d$. The constraint sets on the state vector $x(k) \in \mathcal{X}$ and the input vector $u(k) \in \mathcal{U}$, $\forall k \in \mathbb{N}$ are considered as follows:

$$\mathcal{X} = \left\{ x \in \mathbb{R}^6 : \begin{bmatrix} 40 \\ 20 \\ 40 \\ 20 \\ 40 \\ 20 \end{bmatrix} \le x \le \begin{bmatrix} 95 \\ 95 \\ 95 \\ 95 \\ 95 \\ 95 \end{bmatrix} \right\}, \mathcal{U} = \left\{ u \in \mathbb{R}^9 : \begin{bmatrix} -0.9 \\ -1.5 \\ 0 \\ -0.9 \\ -1.5 \\ 0 \\ -0.9 \\ -1.5 \\ 0 \end{bmatrix} \le u \le \begin{bmatrix} 0.9 \\ 1 \\ 2 \\ 0.9 \\ 1 \\ 2 \\ 0.9 \\ 1 \\ 2 \end{bmatrix} \right\}.$$

The patterns of the periodic signal $d(k)$ with the period $T = 48$ are shown in Fig. 8.16. This periodic signal is repeated along the simulation time. The variance matrix Σ_w for the Gaussian white disturbance $w(k)$ is given by $\Sigma_w = \text{diag}([0.0339, 0.0264, 0.0189, 0.0532])$ and with the 95% confidence level, the set $w(k) \in \mathcal{W}$, $\forall k \in \mathbb{N}$ can be obtained. The initial state $x(0)$ is chosen as $x(0) = [67.2513, 47.4267, 67.0940, 47.4985, 67.3972, 47.0535]^\top$.

According to [19], the main control objectives for the management of this micro-grid are considered:

- To optimize the economic costs by maximizing the benefit of the energy exchange taken into account a time-varying electricity prices presented in c;
- To minimize the usage damages of equipment.

Based on these two objectives, the cost functions are defined as follows:

$$\ell_1(\bar{u}(i), p_i) = \lambda_1 (p_i - P_1 \bar{u}(i))^2,$$
$$\ell_2(\bar{u}(i)) = \lambda_2 \bar{u}_1(i)^2 + \lambda_2 \bar{u}_2(i)^2 + \lambda_2 \bar{u}_4(i)^2 + \lambda_2 \bar{u}_5(i)^2 + \lambda_2 \bar{u}_7(i)^2 + \lambda_2 \bar{u}_8(i)^2,$$

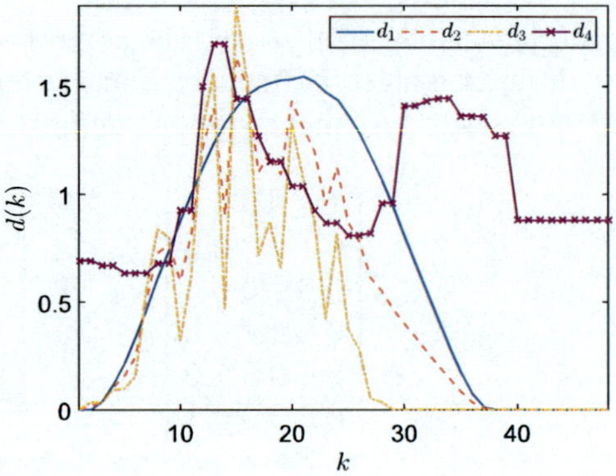

Fig. 8.16 Pattern of the periodic signal $d(k)$

where $P_1 = [0, 1, 0, 0, 1, 0, 0, 1, 0]$. Moreover, λ_1 and λ_2 are prioritization weights. In this simulation, they are chosen as $\lambda_1 = 10$ and $\lambda_2 = 40$.

8.3.4.2 Simulation Results

By satisfying the condition in (8.55), we can obtain M_v as the null space of E_u and M_d is equal to M_w. For the design of the local control law K, the LQR technique is used with the weighing matrices $Q = \text{diag}([0.0182, 0.0133, 0.0182, 0.0133, 0.0182, 0.0133])$, $R = \text{diag}([0.5556, 0.4, 0.5, 0.5556, 0.4, 0.5, 0.5556, 0.4, 0.5])$ and an mRPI set can be obtained using Lemma 6.1.

The simulation has been carried out for 2 h (192 sampling time steps). The optimization problems (8.63) and (8.65) are solved using the linear programming technique. Note that the planner implemented (8.63) is only solved once to find the optimal periodic steady trajectory (8.64). And the closed-loop simulation considers the system (8.53) with the REMPC controller in (8.65). The Gaussian white disturbances $w(k) \in \mathcal{W}$ are sampled in a customized way as shown in Fig. 8.17c. With these disturbances, the closed-loop system is recursively feasible. Besides, some simulation results of the closed-loop state and input trajectories are shown in Figs. 8.17 and 8.18.

Since this micro-grid consists of three nano-grids, we show the results of the first nano-grid. As shown in Fig. 8.17, the blue lines represent the closed-loop states $x_1(k)$ and $x_2(k)$ while the red dashed lines represent the optimal periodic steady states x_1^p and x_2^p. When the disturbances are present in Fig. 8.17c, both closed-loop states reach a neighborhood of the optimal periodic steady states and when the disturbances vanish, they converge to their optimal periodic steady states. For this nano-grid, as also shown in Fig. 8.18 control inputs are close to a neighborhood of the optimal periodic steady inputs.

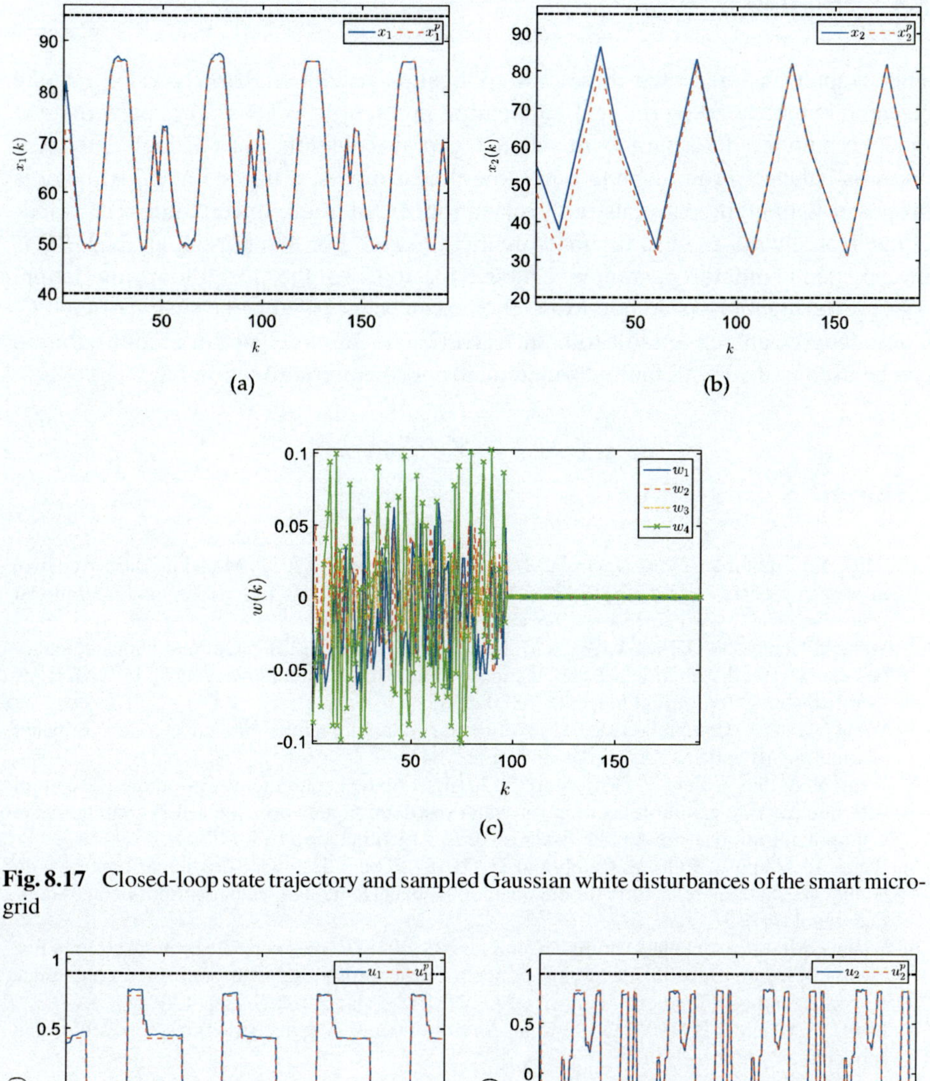

Fig. 8.17 Closed-loop state trajectory and sampled Gaussian white disturbances of the smart micro-grid

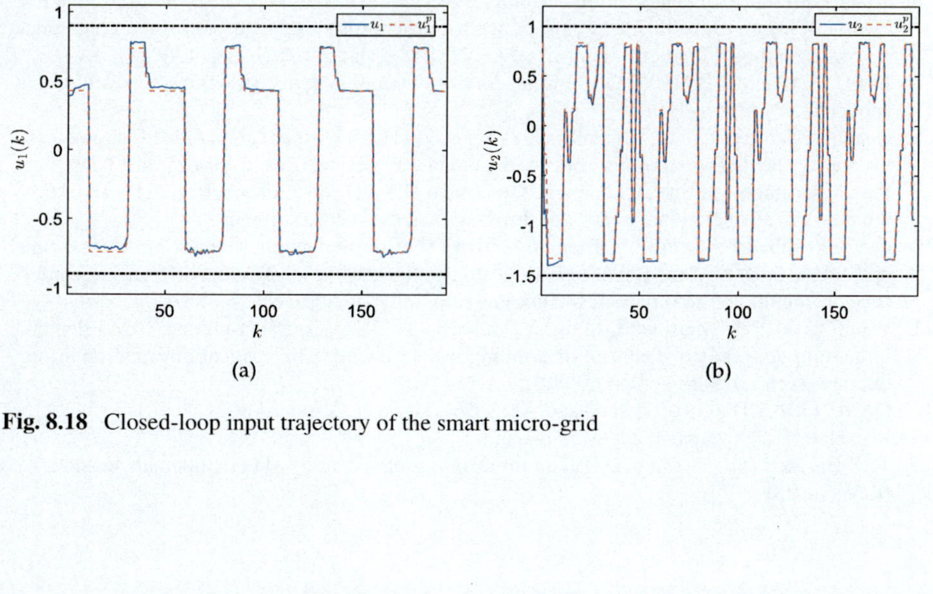

Fig. 8.18 Closed-loop input trajectory of the smart micro-grid

8.4 Summary

This chapter has presented three real application results of EMPC strategies. The research is motivated by the real application problem in WDNs. The mathematical model is built in a descriptor form, where algebraic equations must be included based on mass balance. From dealing with a nonlinear model, a linear EMPC strategy is proposed with nonlinear constraint relaxation. In addition, uncertainties, i.e. modeling error, always exist in the mathematical model. For this reason, an REMPC is investigated in order to guarantee recursive feasibility of the closed-loop simulation. As a future direction, fault diagnosis scheme can be included in the design of EMPC. Consequently, an active fault-tolerant EMPC with suitable system reconfiguration can be used to deal with faults from actuator or sensor malfunctioning.

References

1. Wang Y, Cembrano G, Puig V, Urrea M, Romera J, Saporta D (2017) Model predictive control of water networks considering flow and pressure. In: Real-time monitoring and operational control of drinking-water systems. Springer, pp 251–267
2. Wang Y, Cembrano G, Puig V, Urrea M, Romera J, Saporta D (2017) Optimal management of barcelona water distribution network using non-linear model predictive control. In: 20th IFAC world congress, Toulouse, France, pp 5380–5385
3. Wang Y, Puig V, Cembrano G (2017) Non-linear economic model predictive control of water distribution networks. J Process Control 56:23–34
4. Wang Y, Alamo T, Puig V, Cembrano G (2016) Periodic economic model predictive control with nonlinear-constraint relaxation for water distribution networks. In: IEEE conference on control application (IEEE-CCA), Buenos Aires, Argentina, pp 1137–1172
5. Wang Y, Alamo T, Puig V, Cembrano G (2018) Economic model predictive control with nonlinear constraint relaxation for the operational management of water distribution networks. Energies 11(4):991
6. Wang Y, Muñoz de la Peña D, Puig V, Cembrano G (2018) Robust periodic economic predictive control based on probabilistic set invariance for descriptor systems. In: 6th IFAC conference on nonlinear model predictive control (IFAC-NMPC), Madison, USA, pp 436–441
7. Brdys M, Ulanicki B (1994) Operational control of water systems: structures, algorithms and applications. Prentice-Hall
8. Ocampo-Martinez C, Puig V, Cembrano G, Quevedo J (2013) Application of MPC strategies to the management of complex networks of the urban water cycle. IEEE Control Syst 33(1):15–41
9. Puig V, Ocampo-Martinez C, Pérez R, Cembrano G, Quevedo J, Escobet T (2017) Real-time monitoring and operational control of drinking-water systems. Springer
10. Quevedo J, Saludes J, Puig V, Blanch J (2014) Short-term demand forecasting for real-time operational control of the Barcelona water transport network. In: 22nd mediterranean conference on control and automation (MED), Palermo, Italy, pp 990–995
11. Wang Y, Ocampo-Martinez C, Puig V, Quevedo J (2016) Gaussian-process-based demand forecasting for predictive control of drinking water networks. In: Critical information infrastructures security. Springer, pp 69–80
12. GAMS (2016) The solver manuals. GAMS Development Corporation
13. Rossman L (2000) Epanet 2.0 users manual
14. MOSEK ApS (2015) The MOSEK optimization toolbox for MATLAB manual. Version 7.1 (Revision 28)

15. Price E, Ostfeld A (2014) Battle of background leakage assessment for water networks using successive linear programing. In: 16th water distribution system analysis conference (WDSA), vol 89. Bari, Italy, pp 45–52
16. Iglesias-Rey P, Martínez-Solano F, Meliá DM, Martínez-Solano P (2014) BBLAWN: a combined use of best management practices and an optimization model based on a pseudo-genetic algorithm. In: 16th water distribution system analysis conference (WDSA), vol 89. Bari, Italy, pp 29–36
17. Löfberg J (2004) YALMIP: a toolbox for modeling and optimization in MATLAB
18. Currie J, Wilson D (2012) OPTI: lowering the barrier between open source optimizers and the industrial MATLAB user. Savannah, USA
19. Pereira M, Limon D, Muñoz de la Peña D, Limon D (2017) Robust economic model predictive control of a community micro-grid. Renew Energy 100:3–17

Chapter 9
Fault-Tolerant Control of Discrete-Time Descriptor Systems Using Virtual Actuator and Virtual Sensor

Due to the increasing complexity of modern control systems, the possibility of actuator and sensor faults has increased dramatically. These faults may degrade the performance, leading to unsatisfactory behavior, or in the worst cases to instability, thus bearing catastrophic consequences for the system itself and for the safety of living beings around them. Motivated by the increasing need for safety and reliability, FTC techniques have attracted a lot of interest in the control community, since they allow to maintain the system performance close to the desired one while preserving stability in spite of the faults [1, 2]. Following a well-established nomenclature, the existing FTC techniques can be classified into *passive* and *active* [3]. In passive FTC, faults are considered as uncertainties and the controller is designed using robust control, leading to simpler implementation but more conservativeness in terms of performance and fault-tolerance [4–6]. On the other hand, active FTC reacts actively to faults by performing some reconfiguration action based on online information coming from a fault diagnosis module [7–9].

Among the successful active FTC paradigms proposed to achieve fault tolerance, this chapter presents an FTC framework for discrete-time descriptor systems using VA and VS with delayed state feedback. As discussed in Chap. 2, the descriptor observer is based on the Luenberger structure used in (2.27). In the case of discrete-time descriptor systems, an algebraic loop could exist in the observer-based state feedback, which prevents the implementation. To overcome this issue, an observer-based delayed controller is designed for discrete-time descriptor systems. A separation principle is formulated and proved in a general manner. With this observer-based delayed control scheme, VA and VS are defined for discrete-time descriptor systems. The closed-loop system with reconfiguration is presented. Based on the presented separation principle, the gains of VA, VS, descriptor observer and delayed controller can be designed independently.

Y. Wang, *Advances in State Estimation, Diagnosis and Control of Complex Systems*, Springer Theses, https://doi.org/10.1007/978-3-030-52440-1_9

9.1 Observer-Based Delayed Control of Discrete-Time Descriptor Systems

In [10, 11], the solution of state-feedback control design for descriptor system has been presented. However, so far, the problem of observer-based state-feedback control of descriptor systems has only been addressed in the case of continuous-time systems in [12], but not in the case of discrete-time systems. In particular, when formulating the problem in discrete-time, an algebraic loop appears in the implementation, preventing the use of a classical state feedback law. In fact, let us consider the discrete-time descriptor system

$$Ex(k+1) = Ax(k) + Bu(k), \tag{9.1a}$$
$$y(k) = Cx(k), \tag{9.1b}$$

where $x \in \mathbb{R}^{n_x}$, $u \in \mathbb{R}^{n_u}$ and $y \in \mathbb{R}^{n_y}$ denote the system state, control input and measurement output vectors, respectively. $A \in \mathbb{R}^{n_x \times n_x}$, $B \in \mathbb{R}^{n_x \times n_u}$, $C \in \mathbb{R}^{n_y \times n_x}$ and $E \in \mathbb{R}^{n_x \times n_x}$ are the state space matrices, with E possibly singular, such that $\text{rank}(E) = r \leq n_x$.

Under the assumption that matrices E and C satisfy rank $\left[E^\top, C^\top\right] = n_x$, based on (2.27), we use the descriptor observer as follows:

$$z(k+1) = (TA - LC)\,\hat{x}(k) + TBu(k) + Ly(k), \tag{9.2a}$$
$$\hat{x}(k) = z(k) + Ny(k), \tag{9.2b}$$

where $z \in \mathbb{R}^{n_x}$ and $\hat{x} \in \mathbb{R}^{n_x}$ denote the observer state and the estimated state, respectively, $L \in \mathbb{R}^{n_x \times n_y}$ is an observer gain. Besides, $T \in \mathbb{R}^{n_x \times n_x}$ and $N \in \mathbb{R}^{n_x \times n_y}$ are matrices such that

$$TE + NC = I, \tag{9.3}$$

whose existence is guaranteed by the above-mentioned observability and the rank condition.

To make the descriptor system (9.1) admissible, i.e. regular, causal and stable, using a feedback law fed by the estimated state $\hat{x}$, one might propose to use a standard state-feedback law, that is $u(k) = K\hat{x}(k)$, where $K \in \mathbb{R}^{n_x \times n_u}$ is a controller gain. However, in many cases, the choice $u(k) = K\hat{x}(k)$ creates an algebraic loop, which makes the practical implementation impossible. In fact, $u(k)$ depends on $\hat{x}(k)$, which is calculated using $y(k)$ through (9.2b). The output vector $y(k)$ depends on $x(k)$ through (9.1b), which might depend on $u(k)$ itself if rank $[E, B] \neq \text{rank}(E)$.

For this reason, we propose a delayed state feedback to perform the observer-based control of the descriptor system (9.1), as follows:

$$u(k) = K\hat{x}(k-1). \tag{9.4}$$

Thus, the closed-loop system can be modeled as a descriptor system with state delay. In the following, we will discuss about this choice by showing that the separation principle still holds, so that the controller design and the observer design can be performed independently. Then, by revisiting some preliminary results for discrete-time descriptor systems with state delay given in [11], we propose an improved admissibility condition for these systems taking into account a Lyapunov functional as in [13], which is later used to present a design procedure for the observer-based state-feedback control of discrete-time descriptor systems.

9.1.1 Separation Principle

Let us define the state estimation error $e(k) = x(k) - \hat{x}(k)$. From (9.1b) and (9.2b), we obtain $e(k) = x(k) - z(k) - NCx(k)$ which, taking into account (9.3), becomes $e(k) = TEx(k) - z(k)$. Then, the dynamics of $e(k)$ can be formulated as follows:

$$\begin{aligned} e(k+1) &= TEx(k+1) - z(k+1) \\ &= TAx(k) + TBu(k) - (TA - LC)\hat{x}(k) - TBu(k) - Ly(k) \\ &= (TA - LC)\big(x(k) - \hat{x}(k)\big) \\ &= (TA - LC)\,e(k). \end{aligned}$$

Besides, using (9.4) with $\hat{x}(k-1) = x(k-1) - e(k-1)$, the descriptor system (9.1) can be rewritten as

$$Ex(k+1) = Ax(k) + BKx(k-1) - BKe(k-1).$$

As a result, the augmented system can be expressed as the descriptor system with state delay

$$\bar{E}\begin{bmatrix} x(k+1) \\ e(k+1) \end{bmatrix} = \bar{A}\begin{bmatrix} x(k) \\ e(k) \end{bmatrix} + \bar{A}_d\begin{bmatrix} x(k-1) \\ e(k-1) \end{bmatrix}, \tag{9.5}$$

with

$$\bar{E} = \begin{bmatrix} E & 0 \\ 0 & I \end{bmatrix}, \bar{A} = \begin{bmatrix} A & 0 \\ 0 & (TA - LC) \end{bmatrix}, \bar{A}_d = \begin{bmatrix} BK & -BK \\ 0 & 0 \end{bmatrix}.$$

Consider the class of discrete-time descriptor systems with state delay

$$Ex(k+1) = Ax(k) + A_d x(k-1). \tag{9.6}$$

According to [11, pp. 178], we first recall the following definition for discrete-time descriptor systems with state delay.

Definition 9.1 The discrete-time descriptor system with state delay (9.6) is said to be

- *regular* if $\det\left(z^2 E - zA - A_d\right)$ is not identically zero;
- *causal* if it is regular and

$$\deg\left(z^{n_x}\det\left(zE - A - z^{-1}A_d\right)\right) = n_x + \text{rank}(E);$$

- *stable* if it is regular and $\max\left(|\nu|\right) < 1$, with $\nu \in \lambda\left(E, A, A_d\right)$, where

$$\lambda\left(E, A, A_d\right) = \left\{z : \det\left(z^2 E - zA - A_d\right) = 0\right\};$$

- *admissible* if it is regular, causal and stable.

The following theorem establishes the separation principle for a discrete-time descriptor system with state delay in a block-triangular form, i.e. (9.5).

Theorem 9.1 *The following statements are equivalent:*

- *The descriptor systems with state delay*

$$E_1 x_1(k+1) = A_{11} x_1(k) + A_{d_{11}} x_1(k-1), \tag{9.7a}$$

$$E_2 x_2(k+1) = A_{22} x_2(k) + A_{d_{22}} x_2(k-1), \tag{9.7b}$$

where the states $x_1 \in \mathbb{R}^{n_1}$ and $x_2 \in \mathbb{R}^{n_2}$ are admissible.

- *The descriptor system with state delay*

$$\begin{aligned}\begin{bmatrix} E_1 & 0 \\ 0 & E_2 \end{bmatrix}\begin{bmatrix} x_1(k+1) \\ x_2(k+1) \end{bmatrix} &= \begin{bmatrix} A_{11} & A_{12} \\ 0 & A_{22} \end{bmatrix}\begin{bmatrix} x_1(k) \\ x_2(k) \end{bmatrix} \\ &+ \begin{bmatrix} A_{d_{11}} & A_{d_{12}} \\ 0 & A_{d_{22}} \end{bmatrix}\begin{bmatrix} x_1(k-1) \\ x_2(k-1) \end{bmatrix},\end{aligned} \tag{9.8}$$

where $A_{12} \in \mathbb{R}^{n_1 \times n_2}$ and $A_{d_{12}} \in \mathbb{R}^{n_1 \times n_2}$ is admissible.

Proof *(Regularity)* The following equality holds

$$\begin{aligned}&\det\left(z^2\begin{bmatrix} E_1 & 0 \\ 0 & E_2 \end{bmatrix} - z\begin{bmatrix} A_{11} & A_{12} \\ 0 & A_{22} \end{bmatrix} - \begin{bmatrix} A_{d11} & A_{d12} \\ 0 & A_{d22} \end{bmatrix}\right) \\ &= \det\left(\begin{bmatrix} z^2 E_1 - zA_{11} - A_{d11} & -zA_{12} - A_{d12} \\ 0 & z^2 E_2 - zA_{22} - A_{d22} \end{bmatrix}\right) \\ &= \det\left(z^2 E_1 - zA_{11} - A_{d11}\right)\det\left(z^2 E_2 - zA_{22} - A_{d22}\right).\end{aligned}$$

Hence, according to Definition 9.1, the regularity of the systems (9.7a) and (9.7b) is equivalent to that of the system (9.8).

(Causality) Let us denote

$$\Psi = \begin{bmatrix} zE_1 - A_{11} - z^{-1}A_{d11} & -A_{12} - z^{-1}A_{d12} \\ 0 & zE_2 - A_{22} - z^{-1}A_{d22} \end{bmatrix}.$$

According to Definition 9.1, we also have that

$$\begin{aligned} &\deg\left(z^{n_1+n_2}\det(\Psi)\right) \\ =&\deg\left(z^{n_1}\det\left(zE_1 - A_{11} - z^{-1}A_{d11}\right)z^{n_2}\det\left(zE_2 - A_{22} - z^{-1}A_{d22}\right)\right) \\ =&\deg\left(z^{n_1}\det\left(zE_1 - A_{11} - z^{-1}A_{d11}\right)\right) + \deg\left(z^{n_2}\det\left(zE_2 - A_{22} - z^{-1}A_{d22}\right)\right). \end{aligned}$$

From causality of the systems (9.7a) and (9.7b), it follows that

$$\begin{aligned} \deg\left(z^{n_1}\det\left(zE_1 - A_{11} - z^{-1}A_{d11}\right)\right) &= n_1 + \text{rank}(E_1), \\ \deg\left(z^{n_2}\det\left(zE_2 - A_{22} - z^{-1}A_{d22}\right)\right) &= n_2 + \text{rank}(E_2). \end{aligned}$$

Then, we know that

$$\begin{aligned} \deg\left(z^{n_1+n_2}\det(\Psi)\right) =& n_1 + \text{rank}(E_1) + n_2 + \text{rank}(E_2) \\ =& (n_1 + n_2) + \text{rank}\begin{bmatrix} E_1 & 0 \\ 0 & E_2 \end{bmatrix}, \end{aligned}$$

which implies causality of the system (9.8).

On the other hand, for the pairs $(E_1, A_{11}, A_{d_{11}})$ and $(E_2, A_{22}, A_{d_{22}})$, we know

$$\begin{aligned} \deg\left(z^{n_1}\det\left(zE_1 - A_{11} - z^{-1}A_{d11}\right)\right) &\le n_1 + \text{rank}(E_1), \\ \deg\left(z^{n_2}\det\left(zE_2 - A_{22} - z^{-1}A_{d22}\right)\right) &\le n_2 + \text{rank}(E_2). \end{aligned}$$

From causality of the system (9.8), it follows

$$\deg(z^{n_1+n_2}\det(\Psi)) = (n_1 + n_2) + \text{rank}(E_1) + \text{rank}(E_2),$$

which implies $\deg\left(z^{n_1}\det\left(zE_1 - A_{11} - z^{-1}A_{d11}\right)\right) = n_1 + \text{rank}(E_1)$ and $\deg\left(z^{n_2}\det\left(zE_2 - A_{22} - z^{-1}A_{d22}\right)\right) = n_2 + \text{rank}(E_2)$, and therefore causality of the systems (9.7a) and (9.7b).

(Stability) Following the proof of regularity, we know

$$\lambda\left(\begin{bmatrix} E_1 & 0 \\ 0 & E_2 \end{bmatrix}, \begin{bmatrix} A_{11} & A_{12} \\ 0 & A_{22} \end{bmatrix}, \begin{bmatrix} A_{d_{11}} & A_{d_{12}} \\ 0 & A_{d_{22}} \end{bmatrix}\right) = \lambda\left(E_1, A_1, A_{d11}\right) \cup \lambda\left(E_2, A_2, A_{d22}\right),$$

which implies the equivalence of the stability according to Definition 9.1.

(Admissibility) Since we have proved the equivalence of regularity, causality and stability in systems (9.7a), (9.7b) and (9.8), we can conclude the equivalence of admissibility of systems (9.7a) and (9.7b) and the system (9.8). □

Theorem 9.1 is crucial, since it states that the admissibility of (9.5) can be enforced by considering independently the systems

$$Ex(k+1) = Ax(k) + BKx(k-1), \tag{9.9a}$$

$$e(k+1) = (TA - LC)e(k), \tag{9.9b}$$

where (9.9b) is in a dynamical form without state delay and the design of a stabilizing observer gain L is available in literature. We focus on the design of a delayed controller gain K to guarantee the descriptor system (9.9a) admissible.

9.1.2 Improved Admissibility Analysis and Controller Design

In the following, we present an improved admissibility condition for the descriptor system with state delay (9.6). Then, we propose the design condition of the delayed controller using matrix inequalities.

9.1.2.1 Improved Admissibility Condition

Let us recall an admissible result in [11, Theorem 9.3], which provides a sufficient condition for the admissibility of the system (9.6).

Proposition 9.1 *The discrete-time descriptor system with state delay (9.6) is admissible if there exist matrices $P \in \mathbb{S}^{n_x}$ and $Q \in \mathbb{S}^{n_x}_{\succeq 0}$ such that*

$$E^\top PE \succeq 0, \tag{9.10a}$$

$$\begin{bmatrix} A^\top PA - E^\top PE + Q & A^\top PA_d \\ A_d^\top PA & A_d^\top PA_d - Q \end{bmatrix} \prec 0. \tag{9.10b}$$

Proof Based on [11, Theorem 9.3] with $\tau = 1$, we can obtain (9.10) with $Q \succeq 0$. Since $\tau = 1$, following the proof of [11, Theorem 9.3], we know the matrix $\hat{P} = \text{diag}(P, Q)$ and Q does not appear in [11, Eq. (9.33)] and hence can be positive semi-definite. □

However, as stated by [13], conditions as the ones provided by Proposition 9.1 may lead to conservativeness, since the considered Lyapunov functional is of the type

$$V(k) = x(k)^\top E^\top PEx(k) + x(k-1)^\top Qx(k-1),$$

and the possibility of introducing an additional term related to $(x(k) - x(k-1))$ is ignored. Inspired by the choice of the Lyapunov functional in [13, Eq. (6)], we now present an improved admissibility condition in the following theorem.

Theorem 9.2 *The discrete-time descriptor system with state delay (9.6) is admissible if there exist matrices $P \in \mathbb{S}^{n_x}$, $Q \in \mathbb{S}^{n_x}$ and $S \in \mathbb{S}^{n_x}$ such that*

$$\begin{bmatrix} E^\top (P+S) E & -E^\top SE \\ -E^\top SE & Q + E^\top SE \end{bmatrix} \succeq 0, \tag{9.11a}$$

$$\begin{bmatrix} \phi_1 & \phi_2 \\ \phi_2^\top & \phi_3 \end{bmatrix} \prec 0, \tag{9.11b}$$

with

$$\phi_1 = A^\top (P+S)A + Q - E^\top PE - \mathbf{He}\left(E^\top SA\right) \tag{9.12a}$$

$$\phi_2 = A^\top (P+S)A_d - E^\top SA_d + E^\top SE, \tag{9.12b}$$

$$\phi_3 = A_d^\top (P+S)A_d - Q - E^\top SE. \tag{9.12c}$$

Proof Define the variable

$$\xi(k) = \left[x(k)^\top, \ x(k-1)^\top\right].$$

Then, it can be verified that the system (9.6) can be rewritten as

$$\hat{E}\xi(k+1) = \hat{A}\xi(k), \tag{9.13}$$

where

$$\hat{E} = \begin{bmatrix} E & 0 \\ 0 & I \end{bmatrix}, \hat{A} = \begin{bmatrix} A & A_d \\ I & 0 \end{bmatrix}. \tag{9.14}$$

Set

$$\hat{P} = \begin{bmatrix} P+S & -SE \\ -E^\top S & Q + E^\top SE \end{bmatrix}. \tag{9.15}$$

For the system (9.13), it can be shown that

$$\hat{A}^\top \hat{P} \hat{A} - \hat{E}^\top \hat{P} \hat{E} = \begin{bmatrix} \phi_1 & \phi_2 \\ \phi_2^\top & \phi_3 \end{bmatrix} \prec 0. \tag{9.16}$$

By noting that $\hat{E}^\top \hat{P} \hat{E} \succeq 0$ due to (9.11a), and employing [14, Theorem 2], we have that the system (9.13) is admissible. From regularity of (9.13), it follows that $\det(z\hat{E} - \hat{A})$ is not identically zero, and since $\det(z\hat{E} - \hat{A}) = \det(z^2 E - zA - A_d)$, regularity of (9.6) follows from Definition 9.1. Moreover, from the causality of (9.13), we have that

$$\deg\left(\det\left(z\hat{E} - \hat{A}\right)\right) = \operatorname{rank}(\hat{E}) = n_x + \operatorname{rank}(E),$$

which proves causality of (9.6) since $\det(z\hat{E} - \hat{A}) = z^{n_x} \det\left(zE - A - z^{-1}A_d\right)$. Finally, the stability of (9.13) implies the stability of (9.6) and, therefore, its admissibility. □

Remark 9.1 Note that Theorem 9.2 can be reduced to Proposition 9.1 when $S = 0$.

The condition of Theorem 9.2 includes non-strict inequalities due to (9.11a). Following the spirit of [11, Theorem 9.4], we next present the admissibility condition with strict inequalities.

Theorem 9.3 *The discrete-time descriptor system with state delay (9.6) is admissible if there exist matrices $P \in \mathbb{S}^{n_x}$, $Q \in \mathbb{S}^{n_x}$, $S \in \mathbb{S}^{n_x}$ and $W \in \mathbb{R}^{2n_x \times (n_x - r)}$ such that*

$$\hat{P} \succ 0, \tag{9.17a}$$

$$\begin{bmatrix} \phi_1 & \phi_2 \\ \phi_2^\top & \phi_3 \end{bmatrix} + \mathbf{He}\left(WE^\perp [A,\ A_d]\right) \prec 0, \tag{9.17b}$$

with $\hat{P}$ as in (9.15), and ϕ_1, ϕ_2 and ϕ_3 as in (9.12).

Proof Consider the matrix

$$\hat{E}^\perp = \left[E^\perp,\ 0\right],$$

which is of full row rank and satisfies $\hat{E}^\perp \hat{E} = 0$ with $\hat{E}$ defined as in (9.14). It is straightforward from (9.16) that

$$\begin{aligned} &\hat{A}^\top \hat{P} \hat{A} - \hat{E}^\top \hat{P} \hat{E} + \mathbf{He}\left(W\hat{E}^\perp \hat{A}\right) \\ &= \begin{bmatrix} \phi_1 & \phi_2 \\ \phi_2^\top & \phi_3 \end{bmatrix} + \mathbf{He}\left(WE^\perp [A,\ A_d]\right) \prec 0. \end{aligned}$$

Since $\hat{P} \succ 0$, according to [15, Theorem 1], the system (9.13) is admissible. Following a discussion similar to the proof of Theorem 9.2, the system (9.6) is shown to be admissible. □

Remark 9.2 Note that Theorem 9.3 can also be reduced to [11, Theorem 9.4] when $S = 0$.

9.1.2.2 Delayed Controller Design

Based on above results, we now present the condition for the design of a controller gain K, which is obtained by applying Theorem 9.3 taking into account that (9.9a) is in the form (9.6) with $A_d = BK$.

Theorem 9.4 *The discrete-time descriptor system with state delay (9.9a) is admissible if there exist matrices $P \in \mathbb{S}^n$, $Q \in \mathbb{S}^n$, $S \in \mathbb{S}^n$, $W_1 \in \mathbb{R}^{n\times(n-r)}$, $W_2 \in \mathbb{R}^{n\times(n-r)}$ and $K \in \mathbb{R}^{m\times n}$ such that (9.17a) and*

$$\begin{bmatrix} \psi_1 & \psi_2 & A^\top(P+S) \\ \psi_2^\top & \psi_3 & K^\top B^\top(P+S) \\ (P+S)A & (P+S)BK & -(P+S) \end{bmatrix} \prec 0, \tag{9.18}$$

with $\hat{P}$ as in (9.15) and

$$\begin{aligned} \psi_1 &= Q - E^\top P E + \mathbf{He}\left(W_1 E^\perp A - E^\top S A\right), \\ \psi_2 &= W_1 E^\perp BK + A^\top \left(E^\perp\right)^\top W_2^\top - E^\top SBK + E^\top SE, \\ \psi_3 &= \mathbf{He}\left(W_2 E^\perp BK\right) - Q - E^\top SE. \end{aligned}$$

Proof According to (9.17b), let us set $W = \left[W_1^\top,\ W_2^\top\right]^\top$ and $A_d = BK$. Taking into account that the positive definiteness of the matrix $(P+S)$ is ensured by (9.17a), applying the Schur complement to (9.17b), we obtain

$$\begin{bmatrix} \psi_1 & \psi_2 & A^\top \\ \psi_2^\top & \psi_3 & K^\top B^\top \\ A & BK & -(P+S)^{-1} \end{bmatrix} \prec 0.$$

Pre- and post-multiplying the above inequality by $\mathrm{diag}(I, P+S)$, we thus obtain (9.18). □

9.2 Problem Statement in FTC

Consider the following discrete-time descriptor system subject to actuator and sensor faults

$$Ex_f(k+1) = Ax_f(k) + B_f(\phi(k))\left(u_f(k) + f_a(k)\right), \tag{9.19a}$$

$$y_f(k) = C_f(\gamma(k))x_f(k) + f_s(k), \tag{9.19b}$$

where $x_f \in \mathbb{R}^{n_x}$, $u_f \in \mathbb{R}^{n_u}$ and $y_f \in \mathbb{R}^{n_y}$ denote the vectors of faulty system states, faulty control inputs and faulty measurement output vectors, respectively. $f_a \in \mathbb{R}^{n_u}$ and $f_s \in \mathbb{R}^{n_y}$ denote the vectors of *additive* actuator and sensor faults. $\phi \in \mathbb{R}^{n_u}$ and $\gamma \in \mathbb{R}^{n_y}$ denote the vectors of *multiplicative* actuator and sensor faults with

$$\phi(k) = \left[\phi_1(k), \ldots, \phi_{n_u}(k)\right]^\top, \quad 0 \le \phi_i(k) \le 1, \quad i = 1, \ldots, n_u, \tag{9.20a}$$

$$\gamma(k) = \left[\gamma_1(k), \ldots, \gamma_{n_y}(k)\right]^\top, \quad 0 \le \phi_i(k) \le 1, \quad i = 1, \ldots, n_y, \tag{9.20b}$$

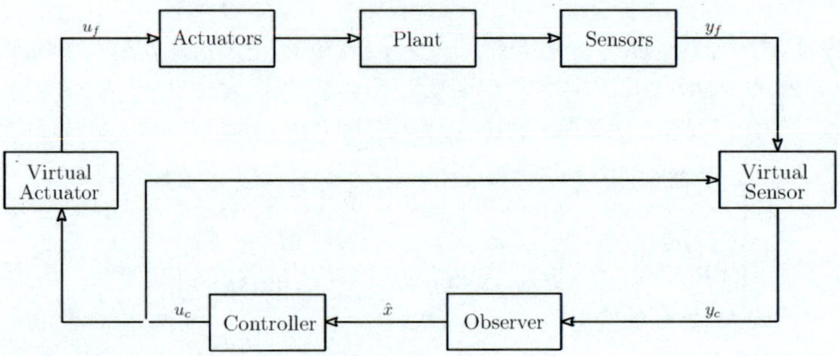

Fig. 9.1 FTC scheme using VA and VS

Besides, $A \in \mathbb{R}^{n_x \times n_x}$ and $B_f(\phi(k)) \in \mathbb{R}^{n_x \times n_u}$ and $C_f(\gamma(k)) \in \mathbb{R}^{n_y \times n_x}$ are defined in the following structure

$$B_f(\phi(k)) = B\text{diag}\left(\phi_1(k), \ldots, \phi_{n_u}(k)\right), \tag{9.21a}$$

$$C_f(\gamma(k)) = \text{diag}\left(\gamma_1(k), \ldots, \gamma_{n_y}(k)\right) C, \tag{9.21b}$$

where B and C are given in the nominal descriptor system (9.1).

Assumption 9.1 The additive and multiplicative actuator and sensor faults are assumed to be estimated as known variables.

As shown in Fig. 9.1, in this chapter, we focus on designing a VA and a VS for the reconfiguration of the faulty system (9.19) and its application to FTC using a nominal observer-based delayed state-feedback controller as in (9.4) with (9.2).

9.3 FTC of Descriptor Systems with Reconfiguration

We now propose a general FTC scheme for discrete-time descriptor systems using VA and VS [16]. In this FTC scheme, an observer-based delayed state-feedback controller defined in (9.4) is used. Then, we define the structure of a virtual delayed actuator (VDA) in a descriptor form as well as in a form that accommodates the delayed state feedback. On the other hand, as introduced in Chap. 2, the observer of discrete-time descriptor system is in a dynamical form. Hence, the duality between VDA and VS in traditional dynamical systems no longer hold. We now present the closed-loop system dynamics that proves separation principle can also be used.

9.3.1 Nominal Observer-Based Delayed Controller

Recall the observer-based delayed state-feedback controller in (9.2) and (9.4) as the nominal controller as follows:

$$u_c(k) = K\hat{x}(k-1), \tag{9.22a}$$
$$z(k+1) = (TA - LC)\hat{x}(k) + TBu_c(k) + Ly_c(k), \tag{9.22b}$$
$$\hat{x}(k) = z(k) + Ny_c(k), \tag{9.22c}$$

where $u_c \in \mathbb{R}^{n_u}$ and $y_c \in \mathbb{R}^{n_y}$ are nominal input and output provided by the VDA and the VS, respectively.

9.3.2 VDA and VS for Descriptor Systems

We now define the VDA and the VS for the descriptor system (9.19).

For designing the VDA, let us define the following matrices

$$N_{va}(\phi(k)) = B_f(\phi(k))^{\dagger} B, \tag{9.23a}$$
$$B^* = B_f(\phi(k))N_{va}(\phi(k)). \tag{9.23b}$$

Definition 9.2 (*Virtual delayed actuator for descriptor systems*) Given the descriptor system subject to actuator and sensor faults in (9.19), the VDA is defined as follows:

$$Ex_{va}(k+1) = Ax_{va}(k) + B^* M_{va}x_{va}(k-1) + \left(B - B^*\right) u_c(k), \tag{9.24a}$$
$$u_f(k) = N_{va}(\phi(k)) \left(u_c(k) - M_{va}x_{va}(k-1)\right) - f_a(k), \tag{9.24b}$$

where $x_{va} \in \mathbb{R}^{n_x}$ denotes the vector of VDA states. Moreover, $M_{va} \in \mathbb{R}^{n_u \times n_x}$ is a VDA gain.

For designing the VS, let us also define the following matrices:

$$N_{vs}(\gamma(k)) = CC_f(\gamma(k))^{\dagger}, \tag{9.25a}$$
$$C^* = N_{vs}(\gamma(k))C_f(\gamma(k)). \tag{9.25b}$$

Assumption 9.2 The pair (E, C^*) is assumed to be observable and hence matrices E and C^* satisfy the rank condition

$$\operatorname{rank}\begin{bmatrix} E \\ C^* \end{bmatrix} = n_x. \tag{9.26}$$

If Assumption 9.2 holds, there exist matrices $T_s \in \mathbb{R}^{n_x \times n_x}$ and $N_s \in \mathbb{R}^{n_x \times n_y}$ satisfying the following condition

$$T_s E + N_s C^* = I_{n_x}. \tag{9.27}$$

Definition 9.3 (*Virtual sensor of descriptor systems*) Given the descriptor system subject to actuator and sensor faults in (9.19), the VS is defined as follows:

$$\begin{aligned} z_{vs}(k+1) &= \left(T_s A - M_{vs} C^*\right) x_{vs}(k) + T_s B u_c(k) \\ &\quad + M_{vs} N_{vs}(\gamma(k)) \left(y_f(k) + C_f(\gamma(k)) x_{va}(k) - f_s(k)\right), \qquad (9.28a) \\ x_{vs}(k) &= z_{vs}(k) + N_s N_{vs}(\gamma(k)) \left(y_f(k) + C_f(\gamma(k)) x_{va}(k) - f_s(k)\right), \qquad (9.28b) \\ y_c(k) &= N_{vs}(\gamma(k)) \left(y_f(k) + C_f(\gamma(k)) x_{va}(k) - f_s(k)\right) + (C - C^*) x_{vs}(k), \qquad (9.28c) \end{aligned}$$

where $x_{vs} \in \mathbb{R}^{n_x}$ and $z_{vs} \in \mathbb{R}^{n_x}$ denote the vector of states and intermediate states of VS. Moreover, $M_{vs} \in \mathbb{R}^{n_x \times n_y}$ is a VS gain.

Remark 9.3 According to [17], B^* and C^* are independent to multiplicative actuator and sensor faults $\phi(k)$ and $\gamma(k)$.

9.3.3 The Closed-Loop Analysis and Designs

We now analyze the closed-loop dynamics of the faulty system (9.19) with the VDA and the VS in Definitions 9.2 and 9.3 as well as the nominal observer-based delayed controller (9.22).

Theorem 9.5 *Consider the faulty descriptor (9.19), the nominal controller in (9.22), the VDA in (9.24) and the VS (9.28). Let us define the variable* $\zeta(k) = \left[\zeta_1(k)^\top, \zeta_2(k)^\top, \zeta_3(k)^\top, \zeta_4(k)^\top\right]^\top$ *with*

$$\begin{aligned} \zeta_1(k) &= x_{va}(k), && (9.29a) \\ \zeta_2(k) &= x_f(k) + x_{va}(k), && (9.29b) \\ \zeta_3(k) &= \hat{x}(k) - x_{vs}(k), && (9.29c) \\ \zeta_4(k) &= x_{vs}(k) - x_f(k) - x_{va}(k). && (9.29d) \end{aligned}$$

Then, the closed-loop behavior is given by

$$\tilde{E}\zeta(k+1) = \tilde{A}\zeta(k) + \tilde{A}_d\zeta(k-1), \tag{9.30}$$

where

$$\tilde{E} = \begin{bmatrix} E & 0 & 0 & 0 \\ 0 & E & 0 & 0 \\ 0 & 0 & I_{n_x} & 0 \\ 0 & 0 & 0 & I_{n_x} \end{bmatrix},$$

$$\tilde{A} = \begin{bmatrix} A & 0 & 0 & 0 \\ 0 & A & 0 & 0 \\ 0 & 0 & (TA - LC) & \Xi \\ 0 & 0 & 0 & (T_s A - M_{vs} C^*) \end{bmatrix},$$

$$\tilde{A}_d = \begin{bmatrix} B^* M_{va} & (B - B^*)\, K & (B - B^*)\, K & (B - B^*)\, K \\ 0 & BK & BK & BK \\ 0 & 0 & 0 & -TBK \\ 0 & 0 & 0 & 0 \end{bmatrix},$$

with $\Xi = (TA - LC^*) - (TE + NC^*)(T_s A - M_{vs} C^*)$.

Proof From the definition of $\zeta(k)$ in (9.29), we know that

$$\begin{aligned} x_{va}(k) &= \zeta_1(k), \\ x_f(k) &= -\zeta_1(k) + \zeta_2(k), \\ \hat{x}(k) &= \zeta_2(k) + \zeta_3(k) + \zeta_4(k), \\ x_{vs}(k) &= \zeta_2(k) + \zeta_4(k). \end{aligned}$$

From (9.24) and (9.22), it follows

$$\begin{aligned} E\zeta_1(k+1) &= Ex_{va}(k+1) \\ &= Ax_{va}(k) + (B - B^*)K\hat{x}(k-1) + B^* M_{va} x_{va}(k-1) \\ &= A\zeta_1(k) + B^* M_{va} \zeta_1(k-1) + (B - B^*)K\zeta_2(k-1) \\ &\quad + (B - B^*)K\zeta_3(k-1) + (B - B^*)K\zeta_4(k-1). \end{aligned}$$

From (9.19), (9.24) and (9.22), we have that

$$\begin{aligned} E\zeta_2(k+1) &= Ex_f(k+1) + Ex_{va}(k+1) \\ &= A(x_f(k) + x_{va}(k)) + BK\hat{x}(k-1) \\ &= A\zeta_2(k) + BK\zeta_2(k-1) + BK\zeta_3(k-1) + BK\zeta_4(k-1). \end{aligned}$$

From (9.19), (9.24), (9.28) and (9.22), we have

$$\begin{aligned} \zeta_4(k+1) &= x_{vs}(k+1) - x_f(k+1) - x_{va}(k+1) \\ &= z_{vs}(k+1) + N_s C^* \big(x_f(k+1) + x_{va}(k+1)\big) - x_f(k+1) - x_{va}(k+1). \end{aligned}$$

Considering the condition (9.27), the above equation can be simplified as

$$\begin{aligned}\zeta_4(k+1) &= z_{vs}(k+1) - T_s E\left(x_f(k+1) + x_{va}(k+1)\right)\\ &= (T_s A - M_{vs}C^*)(x_{vs}(k) - x_f(k) - x_{va}(k))\\ &= (T_s A - M_{vs}C^*)\zeta_4(k).\end{aligned}$$

From (9.19), (9.22) and (9.28), we have

$$\begin{aligned}\zeta_3(k+1) &= \hat{x}(k+1) - x_{vs}(k+1)\\ &= z(k+1) + Ny_c(k+1) - z_{vs}(k+1) - N_s C^*(x_f(k+1) + x_{va}(k+1))\\ &= (TA - LC)\hat{x}(k) + TBu_c(k) + LC^*(x_f(k) + x_{va}(k)) + L(C - C^*)x_{vs}(k)\\ &\quad + NC^*(x_f(k+1) + x_{va}(k+1)) + N(C - C^*)x_{vs}(k+1)\\ &\quad - (T_s A - M_{vs}C^*)x_{vs}(k) - T_s Bu_c(k) - M_{vs}C^*(x_f(k) + x_{va}(k))\\ &\quad - N_s C^*(x_f(k+1) + x_{va}(k+1)).\end{aligned}$$

Considering the conditions (9.3) and (9.27), the above equation becomes

$$\begin{aligned}\zeta_3(k+1) = (TA - LC)\zeta_3(k) &+ \left((TA - LC^*) - (TE + NC^*)(T_s A - M_{vs}C^*)\right)\zeta_4(k)\\ &- TBK\zeta_4(k-1).\end{aligned}$$

As a result, according to formulations above, we can conclude that the closed-loop dynamics can be formulated as in (9.30). □

From the result of Theorem 9.5, based on the separation principle presented in Theorem 9.1, the admissibility of the closed-loop system in (9.30) is equivalent to admissibility of the following subsystems:

$$E\delta(k+1) = A\delta(k) + B^* M_{va}\delta(k-1), \tag{9.31a}$$

$$E\delta(k+1) = A\delta(k) + BK\delta(k-1), \tag{9.31b}$$

$$\delta(k+1) = (TA - LC)\delta(k), \tag{9.31c}$$

$$\delta(k+1) = (T_s A - M_{vs}C^*)\delta(k), \tag{9.31d}$$

where $\delta \in \mathbb{R}^{n_x}$ denotes an auxiliary state.

From (9.31a) and (9.31b), the designs of the gains of the VDA and the nominal controller correspond to the descriptor delay system form in (9.6). Thus, they can be designed using Theorem 9.4.

On the other hand, since that (9.31c) and (9.31d) are in a dynamical form, the designs of the gains of the descriptor observer and the VS can use the standard Lyapunov stability result, i.e. in [18].

9.4 Numerical Example

We use a numerical example to test the validity of the proposed FTC strategy. Consider the nominal descriptor system (9.1) with the following matrices

$$E=\begin{bmatrix}1 & 0.8 & 0\\ 0 & 1 & 0\\ 0 & 0.4 & 0\end{bmatrix},\quad A=\begin{bmatrix}1.05 & 0.68 & 0\\ 0 & 0.85 & 0.3\\ 0 & 0.34 & 1\end{bmatrix},$$
$$B=\begin{bmatrix}1 & 1.3 & 2\\ 0 & 1 & 0.3\\ 0 & 0.4 & 1\end{bmatrix},\quad C=\begin{bmatrix}1 & 0 & 0\\ 0 & 1 & 1\end{bmatrix},$$

and considering the faulty descriptor system (9.19) with known actuator and sensor faults

$$\begin{aligned}B_f &= B\mathrm{diag}\left([0, 0.3, 0.5]\right),\\ C_f &= \mathrm{diag}\left([0.3, 0.8]\right)C,\end{aligned}$$

and $f_a = 0$, $f_s = 0$. Then, for designing the VDA and the VS, according to (9.23), it comes

$$N_{va}=\begin{bmatrix}0 & 0 & 0\\ 0.0157 & 3.3333 & 0\\ 0.7798 & 0 & 2\end{bmatrix},\quad B^{*}=\begin{bmatrix}0.7859 & 1.3 & 2\\ 0.1217 & 1 & 0.3\\ 0.3918 & 0.4 & 1\end{bmatrix},$$

and according to (9.25), we can obtain

$$N_{vs}=\begin{bmatrix}3.3333 & 0\\ 0 & 1.25\end{bmatrix},\quad C^{*}=\begin{bmatrix}1 & 0 & 0\\ 0 & 1 & 1\end{bmatrix}.$$

Since $\operatorname{rank}\left(\left[E^{\top},\ C^{\top}\right]\right) = n_x = 3$, from the conditions (9.3), we choose the following matrices

$$T=\begin{bmatrix}0.0135 & -0.2162 & 0.5135\\ -0.1081 & 0.7297 & 0.8919\\ -0.6486 & -0.6216 & 0.3514\end{bmatrix},\quad N=\begin{bmatrix}0.9865 & 0\\ 0.1081 & 0\\ 0.6486 & 1\end{bmatrix}.$$

Moreover, in this example, it can be also verified that $\operatorname{rank}\left(\left[E^{\top},\ C^{*^{\top}}\right]\right) = n_x = 3$. Then, we can also obtain $T_s = T$ and $N_s = N$.

From this example, by computing the generalized eigenvalues $\lambda(E, A) = \{1.05, 0.85, \infty\}$, the open-loop system (9.1) is unstable and therefore not admissible. By means of the proposed FTC scheme as shown in Fig. 9.1, according to discussions

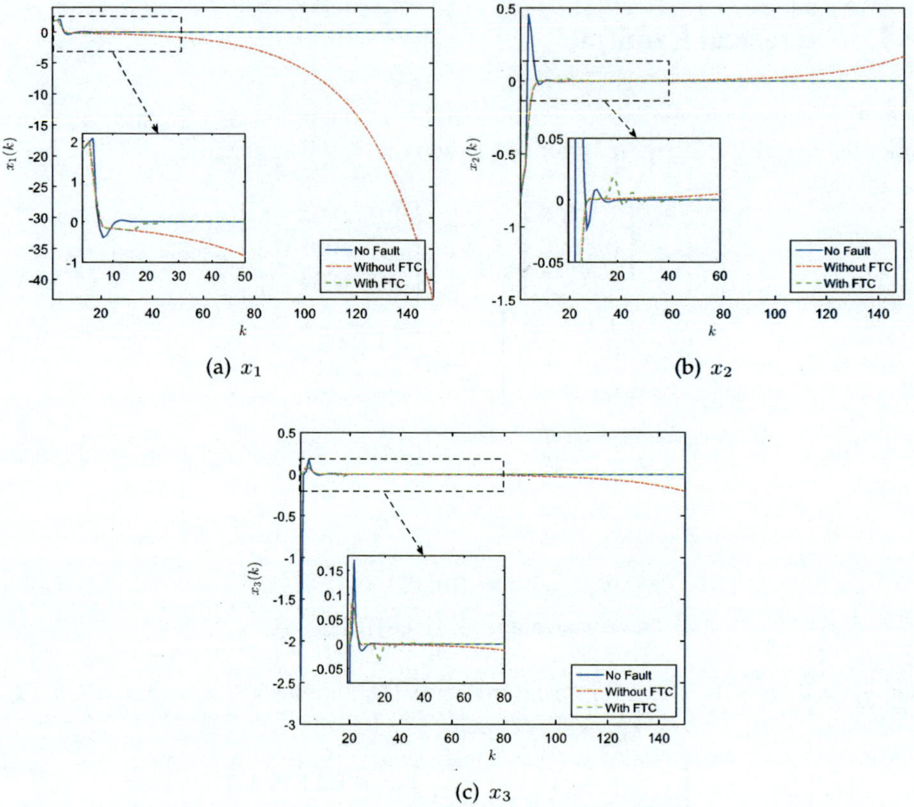

Fig. 9.2 Closed-loop FTC results

above, we can independently design the gains of the VDA, delayed controller, the descriptor observer and the VS. Then, we obtain

$$M_{va} = \begin{bmatrix} 0.4230 & 0.3259 & -0.0138 \\ 0.0054 & -0.2638 & -0.0048 \\ -0.3899 & -0.0835 & 0.0114 \end{bmatrix}, \quad K = \begin{bmatrix} -0.4608 & 0.1956 & 0.0014 \\ 0.0052 & -0.2842 & -0.0013 \\ -0.0034 & -0.0431 & 0.0008 \end{bmatrix},$$

$$L = \begin{bmatrix} 0.2871 & 0.3280 \\ -0.1047 & 1.0930 \\ -0.4827 & 0.0395 \end{bmatrix}, \quad M_{vs} = \begin{bmatrix} 0.2871 & 0.3280 \\ -0.1047 & 1.0930 \\ -0.4827 & 0.0395 \end{bmatrix}.$$

Given the following initial condition: $x(0) = [1.8, -0.8, -2.5]^{\top}$, $u(0) = 0$, $x_{va}(0) = 0$, $z_{vs}(0) = 0$ and $x_{vs}(0)$. With the designed gains above, the closed-loop simulation has been carried out for 150 sampling steps. The results of state trajectories are shown in Fig. 9.2. The faults were introduced from time step $k = 3$ and the FTC strategy was implemented from time step $k = 15$. For three states, three different scenarios have been provided, which are (i) no faults; (ii) faults without applying the proposed FTC; (iii) faults with applying the proposed FTC.

From this figure, it can be seen that from $k = 0$ to $k = 3$, state trajectories of three scenarios are the same since no faults occurred. From $k = 3$ to $k = 15$, since actuator and sensor faults have been introduced in the closed-loop system but no FTC has been implemented, the green and red dashed lines are starting to diverge simultaneously. From $k = 15$, state trajectories with using the proposed FTC (in green dashed lines) are converging to zero during a transient time. Thus, from the closed-loop simulation result, we can conclude that the closed-loop system is stable and hence admissible. And the proposed FTC strategy is effective.

9.5 Summary

This chapter has presented an FTC scheme for discrete-time descriptor systems using VDA and VS. Based on the discussions in this chapter, in order to solve the implementation problem of an observer-based state-feedback control of discrete-time descriptor systems, an observer-based delayed control for discrete-time descriptor systems has been proposed. For the admissibility analysis and the delayed controller design, an improved admissibility condition has been studied. Then, based on the delayed controller, a VDA and a VS for FTC reconfiguration have been defined. According to the separation principle for descriptor delay systems, the gains of nominal controller, descriptor observer, VDA and VS can be design independently. Through a simulation result, the proposed FTC strategy has been verified.

Future direction of the topic in this chapter can be summarized as follows:

- Discrete-time descriptor systems could be affected by uncertainties. Robustness should be addressed in the designs of VA and VS;
- Nonlinearities sometimes appear in descriptor systems. VA and VS could be extended into nonlinear systems using LPV embedding;
- Set-membership approach for discrete-time descriptor systems proposed in Chap. 2 could be linked with VS.

References

1. Blanke M, Kinnaert M, Lunze J, Staroswiecki M (2016) Diagnosis and fault-tolerant control. Springer, Berlin Heidelberg
2. Rotondo D (2018) Advances in gain-scheduling and fault tolerant control techniques. Springer
3. Zhang Y, Jiang J (2008) Bibliographical review on reconfigurable fault-tolerant control systems. Annu Rev Control 32(2):229–252

4. Guezmil A, Berriri H, Pusca R, Sakly A, Romary R, Mimouni M (2019) Experimental investigation of passive fault tolerant control for induction machine using sliding mode approach. Asian J Control 21(1):520–532
5. Nasiri A, Nguang S, Swain A, Almakhles D (2019) Passive actuator fault tolerant control for a class of MIMO nonlinear systems with uncertainties. Int J Control 92(3):693–704
6. Stefanovski J (2018) Passive fault tolerant perfect tracking with additive faults. Automatica 87:432–436
7. Rabaoui B, Rodrigues M, Hamdi H, BenHadj Braiek N (2018) A model reference tracking based on an active fault tolerant control for LPV systems. Int J Adapt Control Signal Process 32(6):839–857
8. Shen Q, Yue C, Goh C, Wang D (2018) Active fault-tolerant control system design for spacecraft attitude maneuvers with actuator saturation and faults. IEEE Trans Ind Electron 66(5):3763–3772
9. Xu F, Olaru S, Puig V, Ocampo-Martinez C, Niculescu S (2017) Sensor-fault tolerance using robust MPC with set-based state estimation and active fault isolation. Int J Robust Nonlinear Control 27(8):1260–1283
10. Duan G (2010) Analysis and design of descriptor linear systems. Springer, New York, USA
11. Xu S, Lam J (2006) Robust control and filtering of singular systems. Springer
12. Darouach M (2014) Observers and observer-based control for descriptor systems revisited. IEEE Trans Autom Control 59(5):1367–1373
13. Zhang B, Xu S, Zou Y (2008) Improved stability criterion and its applications in delayed controller design for discrete-time systems. Automatica 44(11):2963–2967
14. Hsiung K-L, Lee L (1999) Lyapunov inequality and bounded real lemma for discrete-time descriptor systems. IET Control Theory Appl 146(4):327–331
15. Xu S, Lam J (2004) Robust stability and stabilization of discrete singular systems: an equivalent characterization. IEEE Trans Autom Control 49(4):568–574
16. Wang Y, Rotondo D, Puig V, Cembrano G, Zhao Y (2019) Fault tolerant control of discrete-time descriptor systems using virtual actuators. In: International conference on control and fault-tolerant systems (Systol), Casablanca, Morocco, pp 183–188
17. Rotondo D, Nejjari F, Puig V (2014) A virtual actuator and sensor approach for fault tolerant control of LPV systems. J Process Control 24(3):203–222
18. Duan G, Yu H (2013) LMIs in control systems: analysis, design and applications. CRC Press

Chapter 10
Concluding Remarks

10.1 Conclusions

In this thesis, several theoretical contributions and application results on robust state estimation, set-based fault diagnosis, EMPC and FTC strategies for complex systems have been presented. Specifically, the conclusions are summarized with respect to the envisaged thesis objectives as follows:

(i) *Develop robust state estimation approaches based on set theory for descriptor systems;*
It has been shown in Chap. 2 that a set-based framework for discrete-time descriptor systems has been proposed. The set-membership approach based on zonotopes and zonotopic Kalman observer have been extended to descriptor systems subject to unknown-but-bounded uncertainties, which are able to achieve robust state estimation results. As a significant difference from classical state estimation approaches, the obtained results include not only a punctual value but also guaranteed estimation bounds that reflect the worst-case state estimation in presence of uncertainties. This framework has also been proved to be applicable in fault diagnosis.

(ii) *Improve the limitation of set-membership approach for complex systems;*
The disadvantage of classical set-membership approach derived from the geometrical complexity has been solved by investigating a distributed approach. It has been shown in Chap. 3 that considering a distributed model with coupled states, distributed state bounding zonotopes are built for agent by agent. Then, the correction from the consistency test is implemented between the system model and the individual measured output. In terms of communication among agents in the distributed model, each agent send its distributed zonotope to all its neighbors and the effects of neighbors are translated by the distributed model. This distributed approach has been demonstrated with a simple example and compared with a corresponding centralized approach. According to the simu-

Y. Wang, *Advances in State Estimation, Diagnosis and Control of Complex Systems*, Springer Theses, https://doi.org/10.1007/978-3-030-52440-1_10

lation results, the performance of the proposed distributed approach is slightly worse than the centralized one due to the fact that less information is used for the correction. However, for high dimensional complex systems, centralized set-based approaches cannot be directly applied. Hence, this distributed approach provides a potential way to deal with those systems.

(iii) *Investigate fault diagnosis strategies based on set theory for descriptor systems;* To achieve fault diagnosis of descriptor systems, the proposed set-based framework has been extended for FD, FI and FE in Chaps. 4 and 5. For the design, in addition to make the observer gain be robust against uncertainties, additional conditions are required for different objectives. For FD, two fault sensitivity criteria are defined. For FI, unknown input observers are employed to locate where the occurred fault is from. For FE, the identification condition is introduced to estimate faults from measured outputs. On the other hand, in Chap. 3, a general framework of set-invariance characterizations has also been proposed for discrete-time descriptor systems. Taking into account that occurred faults may lead to a different operating mode of the considered model. The active mode detection mechanism has been provided using set invariance. These contributions complete a set-based fault diagnosis for discrete-time descriptor systems.

(iv) *Contribute to EMPC strategies for periodic operation with applications to realistic complex systems;*
Motivated by WDNs, the exogenous and endogenous signals in these networks imply a periodic behavior and naturally an optimal periodic operation could be useful for management of these real systems. For this reason, as presented in Chap. 7, a novel EMPC formulation for periodic operation has been addressed as well as its robust case. This formulation has also been proved to be recursively feasible, converging to (a neighborhood of) a periodic steady trajectory. In addition, when a defined optimality certificate is satisfied, this steady trajectory is equivalent to the optimal trajectory obtained by the corresponding planner. From application point of view, to apply a control strategy into a real system, some issues could happen. As shown in Chap. 8, a two-layer control strategy is used for translate the optimal set-points produced from the upper-layer EMPC controller into a sequence of ON/OFF operations in WDNs. A nonlinear constraint relaxation approach is used for dealing with nonlinear algebraic equation in the control-oriented model of WDNs. Besides, a robust technique is used for satisfying disturbances in descriptor model of SG. These contributions provide an insight on finding an optimal management of complex systems.

(v) *Include fault-tolerant capability in the controller design for descriptor systems.* In terms of the critical nature, the fault-tolerant capability has been introduced by means of defined VA and VS for descriptor systems as shown in Chap. 9. Taking into account that occurred faults can be estimated, these faults are hidden using VA and VS for descriptor systems. The advantage of this approach is to make use of the nominal controller in the reconfiguration loop in presence of fault occurrence.

10.2 Future Work

There are still some open issues regarding the presented problems in Chaps. 2–9. From the summary of each chapter, several improvements have been introduced. Generally speaking, some interesting ideas for future directions derived from this thesis are suggested as follows:

- Applications of all the presented theoretical results to real case studies including large-scale complex systems are interesting. Among them, new challenges in the implementations of these approaches could be met and useful solutions could be demonstrated.
- The set-based framework deserves to be extended to nonlinear descriptor systems. The DC programming could be an option. Considering a nonlinear function can be a difference between two convex functions, uncertain system states can be bounded by convex analysis.
- It is interesting to improve the distributed approach in Chap. 3 by defining a suitable geometrical set. The expected performance of an improved approach is not worse than the corresponding centralized one.
- The set-based fault diagnosis could be extended to deal with different types of faults, e.g. multiplicative actuator and sensor faults. In terms of multiplicative faults, switching control techniques may be used.
- The closed-loop input design to guarantee the admissibility of discrete-time descriptor systems can be integrated into the proposed active mode detection mechanism. Suitable state/output feedback can be introduced for descriptor systems.
- The proposed EMPC formulation can be extended into nonlinear systems. Investigating the closed-loop properties of nonlinear systems is challenging but of interest.
- The proposed FTC can be extended into a robust case. Making use of the proposed set-based framework could be a good option.

About the Author

Dr. Ye Wang was born in Harbin, P. R. China. He received his B.Eng. in Automation from the Northeast Forestry University, Harbin, P. R. China, in 2012, and M.Sc. degree in Automatic Control and Robotics and Ph.D. degree (Cum Laude) in Automatic Control, Robotics and Computer Vision both from the Universitat Politècnica de Catalunya (UPC), Barcelona, Spain, in 2014 and 2018, respectively. From 2015 to 2018, he was also a Research Fellow of the Spanish National Research Council (CSIC) at the Institut de Robòtica i Informàtica Industrial (IRI), CSIC-UPC. In March 2016, November 2016 and January 2018, he was a visiting researcher at the University of Seville, Spain. From February to May 2017, and from March to April 2018, he was a visiting researcher at the CentraleSupélec, University of Paris-Saclay, France. On February 2019, he joined the College of Automation, Harbin Engineering University, Harbin, P. R. China as an Associate Professor. Since September 2019, he is currently a Research Fellow at the Department of Electrical and Electronic Engineering, The University of Melbourne, Australia. His research interests include model predictive control, diagnosis and fault-tolerant control.

Y. Wang, *Advances in State Estimation, Diagnosis and Control of Complex Systems*, Springer Theses, https://doi.org/10.1007/978-3-030-52440-1